Career Perspectives
in
Electronic Media

Career Perspectives in Electronic Media

Peter B. Orlik

©2004 Blackwell Publishing
All rights reserved

Blackwell Publishing Professional
2121 State Avenue, Ames, Iowa 50014, USA

Orders: 1-800-862-6657
Office: 1-515-292-0140
Fax: 1-515-292-3348
Web site: www.blackwellprofessional.com

Blackwell Publishing Ltd
9600 Garsington Road, Oxford OX4 2DQ, UK
Tel.: +44 (0)1865 776868

Blackwell Publishing Asia
550 Swanston Street, Carlton, Victoria 3053, Australia
Tel.: +61 (0)3 8359 1011

Authorization to photocopy items for internal or personal use, or the internal or personal use of specific clients, is granted by Blackwell Publishing, provided that the base fee of $.10 per copy is paid directly to the Copyright Clearance Center, 222 Rosewood Drive, Danvers, MA 01923. For those organizations that have been granted a photocopy license by CCC, a separate system of payments has been arranged. The fee code for users of the Transactional Reporting Service is 0-8138-2477-X/2004 $.10.

Printed on acid-free paper in the United States of America

First edition, 2004

Library of Congress Cataloging-in-Publication Data
Orlik, Peter B.
   Career perspectives in electronic media / Peter B. Orlik.—1st ed.
      p. cm.
   ISBN 0-8138-2477-X (alk. paper)
   1. Television—Vocational guidance. 2. Mass media—Vocational guideance. I. Title

PN1992.55.O75 2004
791.4502′923—dc22

                                                                          2004007854

The last digit is the print number: 9 8 7 6 5 4 3 2 1

About the Author

Peter B. Orlik is professor and chair of Central Michigan University's Broadcast & Cinematic Arts Department, a unit he founded in 1969. Previously a mass communications instructor at Wayne State University, he also has industry experience as a copywriter, radio announcer, music director, and television creative services administrator. A longtime member of the Broadcast Education Association, Orlik served two terms on its board of directors, two terms as chair of its Courses, Curricula, and Administration Committee, and has directed the organization's scholarship competition since 1991. He was the 2001 winner of BEA's Distinguished Education Service Award. Orlik is also a member of the Association for Education in Journalism and Mass Communication, the National Association of Television Program Executives, and the National Broadcasting Society/Alpha Epsilon Rho. He is the author of over forty books and articles, including multiple editions of his *Broadcast/Cable Copywriting* and *Electronic Media Criticism: Applied Perspectives* texts. In 2003, Orlik was inducted into the Michigan Association of Broadcasters Hall of Fame, the first educator to be so recognized. He earned a B.A. cum laude, M.A., and Ph.D. from Wayne State University.

To my valued media colleagues, my esteemed former students, and my inspiring wife

Contents

Contributors   xiii
Preface   xvii
Acknowledgments   xxi

Chapter 1   **Performance Functions**   3
    Disc and Video Jockeys   3
    Talk Show Hosts   10
    Newspersons   13
    Television Entertainment Personalities   21
    Industrial Performers   25
    A Word about Agents   27
    Chapter Flashback   27
    Review Probes   28
    Suggested Background Explorations   28

Chapter 2   **Conceptual Functions**   31
    Copywriters   31
    Art Directors and Designers   36
    Program Writers   42
    Development Managers   46
    Industrial Scripters   49
    Music Suppliers   52
    Web Writers   55
    A Conceptual Conclusion   58
    Chapter Flashback   58
    Review Probes   58
    Suggested Background Explorations   59

Chapter 3   **Production Functions**   61
    Audio and Video Engineers   61
    Lighting Directors   67
    Camera Operators, Videographers, and Cinematographers   71
    Film and Video Editors   78
    Directors: Technical, Assistant, and Main   81

Production Assistants 87
Chapter Flashback 89
Review Probes 90
Suggested Background Explorations 90

Chapter 4 **Sales Functions** 91
Station/System Salespersons 91
Station Representatives 99
Web Sales Executives 100
Program Salespersons (Syndicators) 102
Promotions People 105
Production House Marketers 107
Chapter Flashback 111
Review Probes 111
Suggested Background Explorations 111

Chapter 5 **Directive Functions** 113
Program Directors 113
Sales Managers 123
General Managers 125
News Producers and Directors 128
Entertainment Producers 133
Creative Directors 137
Owners 141
Chapter Flashback 144
Review Probes 145
Suggested Background Explorations 145

Chapter 6 **Facilitative Functions** 147
Account Executives 147
Media Services Executives 150
Traffic Coordinators and Directors 154
Audience Measurement Executives 156
Brokers 161
Satellite Services Executives 165
Webmasters 169
Business Development Specialists 172
Chapter Flashback 174
Review Probes 174
Suggested Background Explorations 175

Chapter 7 **Evaluative Functions** 177
Consultants and Program Researchers 177
Media Analysts 181
Regulators 183
Communications Attorneys 188

Contents

Standards and Practices Officials 191
Lobbyists and Public Relations Executives 194
Critics and Commentators 199
Librarians and Teachers 201
Chapter Flashback 206
Review Probes 207
Suggested Background Explorations 208

Chapter 8   *Cueing Up Your Career: One Hundred Suggestions for Breaking into the Profession* 209
Ten Career Tips for On-Air Talent 210
Ten Career Tips for Electronic Journalism 211
Ten Career Tips for Advertising 213
Ten Career Tips for Online Media 214
Ten Career Tips for Engineering 215
Ten Career Tips for Production 218
Ten Career Tips for Corporate Media 219
Ten Career Tips for Sales 220
Ten Career Tips for Programming 221
Ten Career Tips for Management 222
A Final Flashback 223
Review Probes 224
Suggested Background Explorations 224

Appendixes 227
Key Electronic Media Professional Associations 227
Major Unions Active in the Electronic Media 230

Notes 231

Contributors

Chapter 1

Cara Stern Carriveau, Air Personality, 97.9 WLUP "The Loop," Chicago   5
Phil Tower, Talk Show Host, WYCE, Grand Rapids, Michigan   11
Carolyn Clifford, News Anchor, WXYZ-TV, Detroit, Michigan   15
Matt Kirkwood, Meteorologist, WOOD-TV8, Grand Rapids, Michigan   18
Kurt Wilson, Sportscaster and Producer, CMU (Central Michigan University) Sports Network   20
Peter Michael Goetz, Television, Film, and Stage Actor   22

Chapter 2

Dick Orkin and Christine Coyle, Copywriters and Partners, Famous Radio Ranch   32
Marian Lipow, President and Creative Director, Lipow/Stoner Design, Inc.   41
Ronni Kern, Long-form Television Writer   44
David Salzman, Development Executive and Executive Producer, MAD TV, Los Angeles   47
Kevin Campbell, Senior Information Specialist, Dow Corning Corporation   49
Elizabeth Myers, Partner and Composer, Trivers/Myers Music   53
Jeff Dengate, Web Content Developer, NBA Communications   56

Chapter 3

Jay Rouman, Director, Broadcast Technology, Mt. Pleasant Area Technical Center   62
Charles Nairn, Audio Engineer and President, Com Tec, Inc.   65
Matt Ilas, Director of Lighting and Photography, Summer Nights Film and Video   68
Pattie Wayne-Brinkman, Freelance Cinematographer and Videographer   72
Robin Lin Duvall, Freelance Video Editor, *Jerry Springer Show*, Chicago   79
Brett Holey, Director, *NBC Nightly News with Tom Brokaw*   83
Tohry V. Petty, Producer, BASE Productions, Inc.   88

Chapter 4

Paul Boscarino, Station Manager and Director of Sales, Clear Channel Radio of Grand Rapids, Michigan   92

Dixie Gostola, Sales Executive, Charter Media   97
Mike Feltz, Director, National Account Development, Yahoo!   101
Gary Lico, President and Chief Executive Officer, CABLEready Corporation   104
Judy Paluso, Director of Creative Services, CBS/UPN, Detroit, Michigan   108
Ron Herman, President, The Production Café   109

Chapter 5

Jon Bengtson, Operations Director, WEYI-TV, Flint/Saginaw/Bay City, Michigan   119
Eduardo Fernandez, Vice President and General Manager, Telemundo TV-44, Chicago   126
Mimi Levich, Editor and Executive Producer, VOA English Programs Division   130
Nancy Meyer, Television Program Producer   135
Gerald Downey, Vice President and Creative Director, Alan Frank & Associates   139

Chapter 6

Lisa Albyn Drummond, Senior Account Executive and Media Supervisor, Caponigro Marketing Group   148
Pat Wallwork, Media Director and Partner, McKee Wallwork Henderson   152
Gary Blackwell, Broadcast Traffic Coordinator, Doner Advertising   155
Ed Cohen, Director, Domestic Radio Research, Arbitron   157
W. Lawrence Patrick, President, Patrick Communications, LLC   161
Tim Jackson, Senior Director, North American Video Services, PanAmSat   166
Brian Demay, Webmaster, WBQB/WFVA, Fredericksburg, Virginia   171

Chapter 7

Tim Moore, Managing Partner, Audience Development Group   178
Francine R. Purcell, Vice President, Q Score Services, Marketing Evaluations, Inc.   179
James Boyle, Media Analyst/Broadcasting, Wachovia Capital Markets, LLC   182
Kathleen Q. Abernathy, FCC Commissioner   186
Alan C. Campbell, Founding Partner, Irwin, Campbell & Tannenwald, PC   189
Matthew Margo, Vice President, Program Practices, CBS   192
Stephen Serkaian, President, Kolt & Serkaian Communications   197
David Bianculli, TV Critic, *New York Daily News*   200
Glenda C. Williams, Media Professor, University of Alabama   205

Chapter 8

Jennifer Cotter, Senior Vice President of Development, Oxygen Media   210
Lauren Stanton, News Anchor, WZZM 13, Grand Rapids, Michigan   211
Karl Bastian, President, Greenlight Marketing   213
David Antil, Director, Postmerger Integration, ETAS, Inc.   214
Jeff Kimble, Principal Sales Engineer–North America, NEWSkies Satellites, Inc.   215
Michael Franks, Director of Photography, Michael Franks Enterprises, Los Angeles   218
Scott Wallace, Production Manager, Summit Training Source, Inc.   219

Contributors

Tim Hygh, Director of Regional Advertising, WJR/WPLT/WDRQ, Detroit, Michigan   220

Mike Donovan, Director of Marketing and Educational Services, National Association of Television Program Executives   221

Bill Parris, General Manager, Multicultural Media Broadcasting, Inc.   222

Preface

Welcome to the electronic media world! In formally introducing you to a communications environment in which you were immersed soon after your birth (if not before), this book proceeds on the basis of four fundamental tenets.

Tenet 1 is that the electronic media are serviced and directed by true professionals. A *profession* is usually thought of as an educated calling to which someone permanently devotes his or her working life. And *professionalism*, according to media educators George Pollard and Peter Johansen, "is a multi-dimensional concept focusing on the societal, not self-serving, consequences of work. It is an indicator of individual emphasis on social responsibility and ethical performance, the welding of thought to action through the application of the highest standards or ideals in the performance of an occupation for the primary benefit of society."[1]

As you will discover in the following chapters and in the occupational profiles they contain, such professionalism is a prerequisite to electronic media success. Inevitably, our industry is characterized by intensive labor, continual training, substantial stress, and periodic personal risk. What Price Hicks of the Academy of Television Arts and Sciences wrote about television is equally true of the other electronic media sectors: "At best, it's more interesting and fun than most jobs. The professional, financial and personal rewards are usually good to awesome. However, the hours are long, the work is demanding and difficult, there are no 'overnight successes,' the rejection factor is significant, and it's not easy to have a 'normal' family life when you are in production. And sooner or later, everyone gets fired."[2]

In other words, if you want a comparatively stable, predictable vocation—look elsewhere. But if you seek rapid change, unanticipated opportunities, and a zigzag occupational path that might lead from stations to advertising agencies to government service to marketing management and back again, a career in the electronic media may fill the bill. Contrary to popular belief, this is not a field for which you train in six months or can learn in six years. Instead, to devote yourself to a media-focused endeavor is to devote yourself to a thirty- or forty-year internship—the completion of which is always just beyond the next big assignment or job change.

Tenet 2 is that electronic media careers are personally gratifying. Sometimes you can make a lot of money. But more often, you earn a comparatively modest living where limited financial returns are partially compensated for by the rich diversity of people and problems you encounter. Some of the brightest, most industrious, most passionate, and most compassionate people on earth are involved in media-related enterprises. Unfortunately, the industry also harbors some of the biggest sleazeballs it will ever be your misfortune to meet. Part of the gratification you derive from being in this profession is discovering how to tell the difference—and learning to react accordingly. Because of their scope and intrusive place in our society, electronic media operations make a fundamental impact on the way audiences see their world and themselves. If you can feel good about how you contributed to this vision, there are few more intense satisfactions.

Tenet 3 is that the electronic media now encompass much more than over-the-air broadcasting. The field has expanded to include cable, satellite, and Internet components as well as a vast array of associated endeavors such as advertising, public relations, music marketing, and corporate communications, to mention just a few. Given the range of technologies now available, the delivery system (the hardware) is becoming less and less crucial. Much more important is the content we are able to disseminate via that system—and whether that content possesses the capability to inspire, entertain, comfort, educate, and sell (not necessarily in that order). In the United States, at least, our profession did begin with local broadcast stations. But where it is going is much less clear. Technologically, local stations are no longer the preeminent avenue by which we electronically communicate with our audiences. Some authorities would argue that stations are no longer even necessary in the accomplishment of that linking task. What is unequivocally required, however, is the ability to plan, design, target, execute, and evaluate audiovisual messages so that they reach the right people efficiently, effectively, and in the most timely and socially responsible manner possible. Authentic members of our profession are well attuned to the fact that they are engaged in the practice of mediated communication and not merely the broadcasting, cable, satellite, or online biz.

Tenet 4 is that you are reading this book to ascertain whether you wish to launch an electronic media career of your own. Alternatively, you are attempting to better understand how these media operate so as to put them to appropriate use in another profession as well as in your personal life. By the time you have completed the eight chapters that follow, you will have been exposed to information, issues, and people that should serve either of these purposes.

Plan of the Book

The first seven chapters that follow detail the specific operations and responsibilities carried out by electronic media personnel. Collectively, chapters 1 through 7 expose you to the tasks and issues facing professionals in every phase of our business. The book progresses from a look at media performers, conceptualizers, and technicians to those who market and direct electronic media enterprises. Then you are introduced to the men and women who provide facilitative services for and evaluations of our field.

As a summation and launching pad for your own career, chapter 8 then offers specific suggestions for beginning and sustaining your vocation in each of ten broad areas of the profession. This chapter serves to reilluminate insights gained from the book as a whole and helps focus these insights toward realistic development of your own occupational action plan.

Special Features of *Career Perspectives in Electronic Media*

No single author can provide you with the perspectives you need to begin to grasp the complexity of today's media scene. Therefore, interspersed throughout chapters 1–7 are occupational profiles composed by key individuals who hold vital positions throughout our industry. These media practitioners come from large, middle-sized, and small enterprises. Some are relatively new to the field; others are midcareer professionals. Still other profile authors are senior leaders who have devoted three decades or more to their calling. The twin commonalities binding all of these people are a love of this profession and each person's demonstrated ability to succeed in it. Through their written profiles, these dedicated communicators have fashioned brief conversations with you, conversations intended to give an accurate and unvarnished view of what is required to function effectively in our industry today. Ten more experts join us in chapter 8 to collectively offer one hundred specific career tips in order to focus you even more concretely on the pathway ahead. By the time you have finished this book, no fewer than fifty-eight respected young, middle-aged, and senior "media pros" will have shared with you their wisdom and experience.

Each of our eight chapters concludes with three features designed to improve your understanding of electronic media. *Chapter Flashback* provides a brief review of the major subjects presented in that section of the book. *Review Probes* then presents several questions designed to help you rethink chapter material as a self-test of your comprehension. Finally, *Suggested Background Explorations* provides options for further reading about the topics and issues the chapter has raised.

This Book's Ultimate Goal

As production company head Dale Myers counseled at a recent industry conference, "A *career* is doing what you love to do. A *job* is doing what you have to do."[3] Every electronic media position can become a career or a job depending on your ability, your preference, and your particular passion. This text will have succeeded if it helps you to separate the careers from the jobs and guides you to proceed accordingly.

Acknowledgments

First, and most obviously, thanks are extended to the dozens of media professionals who contributed profiles and career tips to this book. Each of these people is a top practitioner in his or her own branch of our industry, and each took time from the intense scheduling pressures that are the hallmarks of our business to share personal insights with you. Any profession remains strong only when its veterans willingly give of their expertise in tutoring their successors. Our contributors have amply demonstrated such a willingness.

Additional appreciation is expressed to Central Michigan University's Broadcast & Cinematic Arts Department for support of this project; to its office manager, Joan McDonald; and to editorial assistant Becky Jordan, who provided a wealth of manuscript services including the reformatting of all of the book's professional profiles.

I am likewise indebted to Mark Barrett at Blackwell Publishing for his friendship and enthusiastic backing of this project and to Lynne Bishop and Robin DuBlanc for shepherding it through production. Writing a text is never an easy endeavor, but it is a manageable one when an author is encouraged by such experienced and congenial editors.

I also wish to express thanks to—and pride in—several of my students of yesterday who, as today's media professionals, have contributed their insightful commentary to this book. These include David Antil, Karl Bastian, Jon Bengtson, Gary Blackwell, Paul Boscarino, Kevin Campbell, Cara Stern Carriveau, Brian Demay, Jeff Dengate, Gerald Downey, Lisa Albyn Drummond, Robin Lin Duvall, Mike Feltz, Eduardo Fernandez, Ron Herman, Brett Holey, Tim Hygh, Tim Jackson, Jeff Kimble, Mimi Levich, Gary Lico, Tohry V. Petty, Jay Rouman, Stephen Serkaian, Lauren Stanton, Phil Tower, Scott Wallace, and Kurt Wilson.

Finally, I extend special acknowledgment and love to my wife, Chris, and our two children, Darcy and Blaine. This text was composed atop my normal teaching and chair responsibilities, which meant that family life had to suffer so that "author life" could proceed. Their support and encouragement never faltered, and for this I am deeply grateful.

Career Perspectives
in
Electronic Media

CHAPTER 1

Performance Functions

To most people, the electronic media denote glamorous enterprises in which the most important personnel are those whose voices stream (or scream) through our radio or computer speakers and whose faces glow on our video receivers and monitors. As you progress through this book, however, you will discover that on-air or "on-stream" jobs are just the tip of the electronic media iceberg. Nonetheless, because media performers do comprise the first and most prominent impression of what our business is about, it is appropriate to begin our occupational analyses with a look at the responsibilities they exercise and the challenges they face. This exploration will be led not just by your text author but by dozens of media "pros" who share their insights with you in specially prepared profiles interspersed throughout the remainder of this book.

Disc and Video Jockeys

The rise of the disc jockey paralleled the 1950s conversion of U.S. radio from network-produced feature-length programs to locally generated music formats. Disc jockeys established a specific identity for their stations and helped string together isolated music selections and announcement segments into a seamless whole. In the 1950s and 1960s, powerhouse AM outlets trumpeted the personas of their music spinners—like those figure 1-1 features—in their quest for large local audiences.

Today's disc jockeys are overwhelmingly on the FM band, of course, to which most music formats migrated when FM came into its own in the late sixties. Depending on the format sculpted by the *program director* (see chapter 5), the contemporary disc jockey can be anything from a high-profile personality allowed extensive talk time to a virtually anonymous presence mouthing little more than song titles, time, and weather.

Either way, the deejays at most local stations are expected to *run their own board*—in other words, to manipulate the controls on an audio console so that the recorded and live on-mic elements of the program are linked one after the other. It is especially important that the *segues* (transitions) between the elements be so tight that *dead air* (silence) is avoided. Contemporary radio listeners expect continuous and instant gratification and have developed lightning trigger fingers that punch up another station at the mere suggestion of a pause in the action.

Figure 1-1. One station's 1960s stable of "top jocks." *Courtesy of "Wyxie Wonderland" by Dick Osgood.*

Performance Functions

Cara Stern Carriveau
Air Personality, 97.9 WLUP
"The Loop," Chicago

Like most radio personalities, my career began in a small market. My first job was located in a trailer in Clare, Michigan. The plumbing often malfunctioned, making the water undrinkable and forcing us to use a porta-potty outside.

Fast-forward to my current job, hosting middays at a top-rated rock station in the nation's third-largest market. The top–of-the-line studio is located in the world-famous John Hancock Building in downtown Chicago. No more porta-potties!

The duties of an on-air personality change significantly as market size increases. At my first job, for instance, when I had a remote broadcast, I drove the station van to the location, set up banners, hooked up the broadcast equipment, met with the client to obtain my talent fee, performed the broadcast, took down the banners, unhooked the equipment, and loaded up the van. Now when I have a remote broadcast, the promotions department arrives early to set up banners and meet with the client. A salesperson obtains payment. An engineer sets up phone lines days in advance, then arrives early the day of the broadcast to make sure everything is working properly. During the broadcast several interns pass out stickers and help in any way needed. I simply arrive for the broadcast and leave when it's done. And no more boring broadcasts from furniture stores, banks, and car dealerships. I'm at concerts and other fun events. In smaller markets remote broadcasts are mainly *sales driven*. In major markets they tend to be *listener driven*.

In a smaller market you likely will be forced to wear several hats. At one job I was morning newswoman/sidekick, promotions director, plus voice tracked [recorded breaks to be played back via an automated system] the midday shift. At another job I was evening host, production director, and helped in the traffic department. I now have the luxury of being able to focus solely on my show without any distractions.

The target demographic at WLUP is men twenty-five to fifty-four. Though our listeners are with us mainly for the music, they have a strong interest in sports. I must relate to them. So daily reading of the sports section and men's magazines is part of my show prep. Most of my audience is on the job so I tailor my breaks to identify with them.

I always skim (record) my shows. Most program directors will review your work and give you solid advice on how to improve. Your job is to implement their vision, so they need to communicate to you exactly what they want. If my boss doesn't review my show I will critique it myself. The day that I think I had a perfect show I'll quit this business. There is always room for improvement. Plenty of DJs would love to have my job, so I cannot take it for granted. And it's smart to always have a current tape in case you need it. A few years ago I worked at a station where all the jocks were suddenly terminated because of a format change. Thankfully, I had plenty of tapes ready to send out and landed another job fairly quickly.

Persistence has carried me throughout my career. When I was turned down for my first job I asked the program director where I needed improvement. I took the advice to heart, worked on my flaws, and sent in another tape. I was hired. In fact, I've initially been turned down at every job I've had! The word "no" to me means try harder.

Work on your craft and stay focused. Best of luck to you!

From the previous paragraph, it is apparent that air personalities (and our industry as a whole) have a great affinity for jargon. Such in-group terminology is usually an appropriate means of streamlining communication with our colleagues. When it spills into our messages to the public, however, jargon can obscure our meanings and force the public to witness the manufacturing process instead of simply enjoying the communicative product that process creates. Air personalities especially must never forget that the audience is there for its own gratification and is not particularly interested in the procedures (and labels) used to package that gratification for them.

To be effective on-air professionals, disc jockeys and other program hosts must accomplish two objectives: (1) establish a sense of personal communication, even personal intimacy, with the listener; and (2) blend in with, and use to the fullest, all the elements provided by the format in which they work. Some personalities, on the one hand, know a great deal about the music they are playing but are unable to parlay this knowledge into a sense of sharing conversation with listeners. At the opposite extreme, other air talents possess a superb ability to stimulate a dialogue with the audience but are so ignorant of the music or other formatic features that the interchange soon proves to be boring or purposeless.

Like it or not, in the carefully integrated package that is today's program design, the communicator and the format content must be not just compatible but also perfectly synchronized. Nationally respected radio consultant Tim Moore has long seen this synchronization as vital to the successful positioning of a station in the public's perception. From the standpoint of the air talent themselves, this means, according to Moore, using the precise amount of talk required to "lubricate" this particular format and the sense of *control* that the on-air host seems to bring to and through the microphone.[1]

It is not simply a matter of how much or how little talk is engaged in but how appropriate the talk seems within the dynamics of the chosen format. As radio creative and programming consultant Tyree Ford reports:

> In those same focus groups [small numbers of consumers assembled to probe their perceptions] where overly talky announcers received negative ratings, the same respondents gave high ratings to the station with a high-profile personality air staff. What first appeared as a contradiction was later realized to be a qualitative statement. If the content and style were not appreciated by the audience, the perception was negative—overly talky. If those elements were appreciated, the perception was positive entertainment. . . . Put simply, it's not just what you say, it's what you say and the way you say it.[2]

This intent cannot be effectively projected by merely copying the style of another personality or caricatured personality type. In fact, successful station executive Bruce Holberg believes that copying is merely the first stage in the professional maturation process. "On air people start by imitating everybody," says Holberg. "Then they try to be like nobody, then they evolve back to the center where they are able to create a workable but still distinct image."[3] Tyree Ford further counsels managers:

> Don't expect the application of stylistic delivery crutches such as: The Puker, The Big Smile, and The Big O to improve your air person's delivery. (If you're not familiar with these terms, The Puker is the hyperextension of CHR [contemporary hit radio] delivery. The Big Smile, prevalent on soft-rock and AC [adult contemporary] formats, sounds as if a pre-formed plastic smile insert has been placed in the announcer's mouth. The Big O is the female talent who sounds as though she is on the brink of orgasm each time she opens the mike.)

Applying these surface fixes is easy and may give a superficial consistency to the air sound. It is also the mark of a mediocre radio station.[4]

There are, of course, a significant number of mediocre radio stations, but ambitious media performers try to move beyond them as soon as possible. The ultimate goal of most radio personalities who want to remain on the performance side of the station business is to appear on a distinctive outlet that dominates its market and to appear during morning drive. This time period, normally defined as from 6 to 10 a.m., usually has far more listeners (and therefore more advertising revenue potential) than any other *daypart* (the term programmers use to refer to the blocks of time into which each twenty-four-hour period is divided).

To succeed in the morning, to tap into people's consciousness at a time when they are once again getting ready to cope with their immediate environment, most radio programmers agree that air personalities must be locally oriented. "I always figure," maintains program consultant Ed Shane, "that if I go into a market and I listen to the station's morning show and I don't understand what they're talking about, they're probably doing it right. Morning shows can't be pulled from one place to another."[5] The same localism principle usually applies to shows in other dayparts as well, of course, and these are where air talent most often train to make the move to the morning, either at their current outlet or a new station.

Midday slots require a less intrusive style that can effortlessly accompany listener tasks at home or in the workplace. With more and more adults engaged in out-of-home employment (and audience-measurement firms adjusting procedures to measure these listeners), the importance of midday listening as a way to make the job more bearable is being amplified. In larger markets, where afternoon commutes can be lengthy, afternoon drive is second only to morning as a revenue producer for the station. Like the morning deejay, afternoon personalities are generally more in the foreground than are those at midday. In the afternoon, however, the emphasis is placed on looking ahead to plans and relaxation for the evening. Meanwhile, for those air talents drawing evening shifts, the style reflects a more laid-back enhancement of listeners' free-time activities rather than that of a center-stage performer who distracts us from the boredom of getting to and from the job.

Most announcer career paths begin at the small-market level and in non-drive-time periods. The more successful practitioners parlay insights from this minimal salary apprenticeship to move up to better time slots and larger cities. A few may ultimately ascend to syndicated program hosting, where their voices may be heard on dozens or even hundreds of stations. Yet, even though they are no longer "local," they are still expected to structure their delivery to blend appealingly and seamlessly with the hopefully distinct image of each station carrying their programs. "Listener loyalty is focused more on people than stations," comments radio consultant Fred Jacobs. "So a good air personality is a vital draw." The deliberate use by Jacobs of the term *air personality* rather than *deejay* reflects his belief that while deejays are satisfied with cueing up records, true air personalities are moving far beyond this to skillfully function as "the ultimate salespersons" for their stations.[6]

Becoming this "ultimate salesperson" and progressing to a major market or national air slot is no simple task. Isolated in a cozy control room and preoccupied with manipulating the equipment, it is easy to overestimate your skill and your impact. As professor Steven Shields discovered in his research into radio on-air staff, "the announcers in this study were found to be rather low in creativity, yet they regarded themselves as possessing a high degree of creative control over their work."[7] To lessen such self-deception, Shields advocates that instructors tutoring aspiring on-air professionals move "beyond preparing students to be merely adequate radio stagehands to instead teaching them more about being on-air magicians: Excellent communicators who have something to say

TOP TEN U.S. RADIO FORMATS
2003

	Number of Stations
1. Religion/Gospel/Christian	2,512
2. Country	2,318
3. Adult Contemporary	1,863
4. Oldies	1,208
5. News/Talk	1,199
6. Sports	837
7. News	769
8. Talk	726
9. Contemporary Adult/Top 40	646
10. Classic Rock	606

Figure 1-2. Dominant formats on U.S. radio stations.

worth hearing, who know how to use the power of the radio medium to evoke clear mental imagery to be shared and enjoyed by their listeners."[8]

The best of these on-air magicians can earn six-figure salaries—even in middle markets. Certainly, such paychecks are far from the norm for radio on-air people as a whole. But they do demonstrate the premium placed on professionals who can truly communicate with their listeners while advancing the identity and format characteristics of the station for which they work. (The most popular formats on U.S. radio stations are listed in figure 1-2.)

Though an on-air local image is important, this image may not always be rooted in reality. The 1996 Telecommunications Act set off a massive consolidation of the radio industry that created huge station groups made up of hundreds, or in the case of Clear Channel, well over a thousand radio outlets. Most of these groups are *publicly held* companies, in which constant pressure to generate dividends forces station management to concentrate more on selling than on programming efforts. As a result, *voice tracking* is often done to increase on-air staffing "efficiency." "Voice tracking is an ideal productivity tool," says radio consultant Ed Shane, "allowing air talents to pre-record their air shifts so as to use their work time producing commercials, appearing live at sponsor locations, or doing a variety of jobs other than waiting for songs to end to deliver a 10-second talk over." But as Shane also admits, "[I]nstead of enhancing productivity, voice tracking is more often used as a cost-cutting measure to reduce the workforce. Just one example is KXFM in Santa Maria, California. The morning show comes from Spokane, the midday voice is from Denver, the night show is from San Jose, and a weekend show comes from Las Vegas."[9]

The greatest abuse of this practice is when out-of-market air personalities have replaced local voices—and then make every effort to convince the listener that the show is still "local." Tom

Performance Functions

Figure 1-3. Pat, Johnny, and animal sidekick Rag Mop on WXYZ-TV's *Pat 'n' Johnny Show. Courtesy of the* Detroit Free Press.

Carpenter, a union executive with the American Federation of Television and Radio Artists (AFTRA), charges that in at least one large station group, "announcers are often encouraged to manufacture public appearances that did not occur.... disc jockeys fabricate live calls to the station. The company coaches employees by giving them 'cheat' sheets about local people, places and events. Is the public served by that?"[10] Clear Channel's president, Mark Mays, counters: "In many markets, voice tracking allows us to boost the quality of on-air talent. This may not be true in New York, but it may very well be in Winchester, Va.... I understand why the talent doesn't like it. When we pay a guy to do two shows, that means somebody doesn't have a job. But, if listeners don't like it, they can punch a button."[11]

Unlike their radio counterparts, *video* disc jockeys are seldom concerned with tying into a given locality. Instead, they must project an aura of national or even international music hipness that assures viewers that the music videos they feature are "where it's happening." Like the audio deejay, the video host must still be the communicative lubricant who smoothly intermeshes all the isolated musical selections. But the veejay carries the added burden of having to resemble physically what he or she sounds like. For a lot of electronic personalities, the beauty of radio is that you don't have to "look like your voice" to succeed.

In the early days of television, this difference was not well understood. With lots of local airtime to fill outside of the network-programmed evening hours, many stations grabbed deejays from their sister radio outlets and threw them on camera. A typical example was the *Pat 'n' Johnny Show* on Detroit's WXYZ-TV (see figure 1-3). Every weekday afternoon, beginning in 1949, a rotund deejay from the co-owned radio station (Johnny Slagle) was paired with a statuesque young blonde (Pat Tobin) to fill an hour or more with record spinning and chitchat. If nothing else was working, the audio engineer would cue up the tune "Rag Mop," and the camera would focus on the show's same-named guinea pig in hopes that the furry little creature would be doing something of interest.

More often than not, he was snoozing under the hot studio lights—but did deflect attention from tubby deejay/veejay Johnny's TV-unfriendly appearance.

Talk Show Hosts

On-air personalities who work in talk formats must cope with a much more unpredictable environment than their colleague music spinners. In talk, the assurance of the prerecorded tune is replaced by the usually live contributions made by people in the studio or at the other end of a telephone line. Whether on radio or on television, talk show hosts must lucidly and revealingly communicate with these contributors while keeping in mind that it is the audience, rather than the caller or guest, who must ultimately benefit from the exchange.

Most television talk shows are recorded for later airing. Thus, inappropriate comments by guests or audience members and inane exchanges that go nowhere can usually be edited out in post-production to remove verbal garbage and quicken the pace. Sometimes, of course, offensiveness can be useful in revealing the topic and enhancing audience interest. But when it is not, judicious editing constitutes a welcome safety net and the assurance that rough spots can be smoothed over before the show is ultimately distributed.

For the most part, this safety net is not available to the radio talk show talent because on radio, live spontaneity is the key to listener attention and caller participation. "In fact," observes talk format innovator Bruce Marr, "talk radio is radio waiting for something to go wrong."[12] To lessen the chances of things "going wrong" while taking telephone calls, talk show hosts usually make use of a *squelch button*. The call is either recorded or sent through an electronic delay loop before reaching the transmitter. If an obscenity or other unwanted phrase is uttered, the host has from four to seven seconds to hit the squelch button and delete the offense before it reaches the airwaves. This delay mechanism is one reason callers must turn down their radios to avoid their own words echoing back to them.

The squelch option is just one small aspect of the rapid-fire and multilevel decision making that a radio talk host taking phone calls must master. As Eric Bogosian, the creator of the play *Talk Radio,* paints the picture:

> The caller is in his sights. The talk show host drops on him like some kind of diving human smart-bomb. He computes a response, an angle to play, a button to push. The talk jock scans his computer terminal for the next call. He watches his engineer (who's watching the clock). The guy behind the microphone is evaluating, calculating and shaping. He is carving fresh chunks of air time out of nothing but human voices and radio waves. . . . Every time the jock sits down at the mike, he starts with nothing, expecting anything. In the next two or three hours, using whatever comes flying at him, he must construct a show. And he must do it "in character." He also has a boss to please. Like any entertainer who punches a clock, he must stifle most of his emotions, whether they be anger, boredom or pity. The show must go on.[13]

As Bogosian's quote indicates, talk radio relies on computer technology to enable the host, and the *call-screener* who usually works with that host, to orchestrate calls more effectively. When calls come in, the software program indicates which line has been ringing the longest and which caller has been on hold for the greatest amount of time. "Callers can be prioritized and updated quickly," reveals *BE Radio*'s editor Chriss Scherer. "The software will also automatically refresh and indicate when a call is active, then reset when the caller hangs up. Many screener programs allow some basic text messaging as well. The call-screener program can display information about the caller to aid

Performance Functions

the host. Some programs also offer database functions so you can track frequent callers or avoid problem callers. Some read caller ID data to further streamline the process."[14]

Clearly, the air personality on a radio or television talk program must possess computer as well as communicative skills. In addition to articulate speech, the talk host must exhibit an interest in and knowledge of a wide range of subjects that may become the topic of conversation. It is essential that these personalities be well read and continually in search of items that might prove compelling to their audiences. Beyond this, it is necessary to cultivate a quick wit and the ability to use that wit in keeping the program interesting and well paced. Long-winded participants must be gracefully cut off and timid participants drawn out, without the hosts seeming to dominate the dialogue. Air personalities thus require a good deal of seasoning before they are ready to attempt a talk assignment, and such seasoned professionals can usually demand significantly higher salaries than those accorded most deejays.

Phil Tower
*Talk Show Host, WYCE,
Grand Rapids, Michigan*

As a radio professional, I frequently have been asked by people why I chose this field. The best answer I can give them is that the field basically chose me. What I mean by being "chosen" is that I have always been a curious person; someone who loves to soak up information. For a long time, I have been driven by the desire to be informed and to stay on top of the issues and events that fill a daily newspaper or radio talk show. I feel this curiosity makes someone a good candidate for hosting his or her own talk program.

There are also three other aspects that every good host brings to the studio every time he or she does a show. First is the ability to pick a good topic for discussion. Second, the skill to present that topic in an appealing way. And third, every good host learns to adapt his or her style to fit the topic and/or guest. Being adaptable also means having the ability to shape a topic to your audience and community. More on this later.

Picking a topic to talk about on a show is not as easy as it may seem at first. As a host, you've got to have your finger on the pulse of what people are discussing at work, at lunch, and even at social gatherings. What issues are pushing people to speak out? Read your daily papers, watch a ton of TV news shows, and even listen to other radio shows. Especially, check out your competition. Also scan your local paper for letters to the editor. Better yet, go online. Visit the chat rooms and forums for an inside peek at what folks are *really* thinking about. This is a great way to stay on top of the topics that are driving conversations all over and will stimulate people to call and respond to your on-air presentation. Presentation of a topic can make or break a show. Every station somewhere has had a host do a show on abortion, gun control, Congress, or any of the myriad issues that talk radio has beaten to death over the years. So how do you make a topic stand out? Step back and examine what aspects of the issue people are *not* talking about. Ask what about the topic is most relatable for the average listener of your show. By relatable, I mean the ability for listeners to say that they can imagine the situation happening to them, or at the very least they have a basic understanding of what you're talking about.

For example, instead of the topic "Should more U.S. troops go to Iraq?" how about asking your listeners to respond to this one: "Would you move to Iraq for one year for a million dollars?" It certainly is a different way to look at an issue that has been talked about for a long time and unfortunately has bored too many listeners with discussions that are too cerebral. You could interview a network journalist/reporter in the city and ask about life in a never-ending war zone. My guess is that your listeners would stay with you to hear others respond to that intriguing question. Remember, it's not the topic necessarily but the angle you take to present it. That really can make all the difference in the world when it comes to lighting up those phones.

Finally, I think the ability to adapt a topic or issue to fit your style, your strengths, and your comfort zone as a host is very important. The main point here is that you should feel that the topic is a good fit for you as a host and the mood of your listeners at the time. Don't just schedule a topic because it fits the guest's schedule or a hole you need to fill. Too many times, hosts talk about things that they are not prepared to cover or basically don't care enough about. Preparation is important but not nearly as important as having the *desire* to really cover an issue. Choose topics that you *really* care about, and your show will sound like it! A good example of adapting a topic to fit your comfort zone would be the Iraq example mentioned earlier. If you don't know all the specifics of an issue, pick a part of it that you do know and make it into a topic that can be easily handled by you and your audience. The "would you do this for a million dollars" idea customizes the topic and makes it much more relatable and involving for your average listener.

Adaptability also means being able to switch topics at the drop of a hat and to talk about the breaking news story or the quickly building controversy at city hall. It's a horrible thing to hear a host covering a boring, overdone topic when there are always other exciting or more current things to cover. You may be able to talk about Congress like Rush Limbaugh, but Rush can't talk about the scandal that involves members of your city council. Even if it's a hot national topic, you should be ready to dump it when a more local or regional issue develops. This way, you're almost guaranteed to get more calls.

There are always other factors to consider as a talk show host. But most importantly, you have to consider that there are tons of shows out there today, and it's getting tougher than ever to make yours stand out. Great talk hosts don't happen overnight. They spend years developing and honing their style. Don't despair, though. Because if you really love being curious, asking different questions, and putting a unique spin on the issues, you're almost guaranteed to be a success wherever your show plays!

Radio or television talk hosts are capable of consolidating extremely loyal followings, although it takes time to build them. Such followings provide the hosts' stations and networks with stable audiences that can be promised with assurance to advertisers. There are concerns within and outside our industry, however, that the power this loyalty confers on the host/personality may be abused—particularly as it relates to social and political controversies. In the 1960s, talk radio's formative years, "[y]ou were really there to answer the phone and engage the listener in what they wanted to talk about," recalls Michael Jackson, Emmy Award–winning talk host whom many consider the "dean of talk radio." "There were restrictions on topics; you couldn't talk about race, religion or sex, and the public didn't want to talk about politics. Gradually, all of these topics crept into talk radio. And as the format began to succeed, management released the reins."[15] With this release, the pronouncements by talk show hosts and their guests began to make a larger impact, sometimes igniting charges that these personalities were functioning as social and political agitators who manipulated the airwaves to advance their own agendas.

What, if any, are the limits to non-elected electronic communicators' use of their media to oppose the actions of elected officeholders? On the other hand, should the electronic media be no more than an uncritical pipeline for the conveyance of information from the governors to the governed? This debate predates talk programming, of course. It has gone on since the beginning of our democracy. Still, talk show hosts are likely to remain at the forefront of the dispute because of the nature of their vehicle. As Wayne Munson points out, talk shows are complex and contradictory because they make "public spectacle of private passions even as the caller remains to a degree private. Yet, they also make *private* spectacle of *public* passions" and thereby penetrate listeners' and viewers' lives in often unsettling ways.[16]

Newspersons

Despite their potential impact, talk show hosts are still primarily *entertainers* who *draw on* public opinion. Electronic media *journalists,* on the other hand, see their job as the pursuit and investigation of news items in order to furnish that opinion with fresh information. Newspersons pride themselves on providing revelations that the public does not have the time or the capability to uncover for itself. Even though this information may not always be world shaking, a premium is placed on its ability to increase audience understanding of events or processes and to suggest key implications for the viewers' or listeners' own lives. Talk show hosts are primarily fielding reflections of existing public knowledge and attitudes, whereas journalists seek to expand that knowledge based on new or recently unearthed factors. Distinguished political commentator Walter Lippmann put it this way:

> The news is not a mirror of social conditions, but the report of an aspect that has obtruded itself. The news does not tell you how the seed is germinating in the ground, but it may tell you when the first sprout breaks through the surface. . . . It may even tell you what somebody says is happening to the seed underground. It may tell you that the sprout did not come up at the time it was expected. The more points, then, at which any happening can be fixed, objectified, measured, named, the more points there are at which news can occur.[17]

In capturing and packaging these "obtruded aspects" of which Lippmann speaks, electronic journalism, unlike its print counterpart, is a *performance* as well as a "report." A reporter or correspondent engages in both activities. Whether covering the cop house (police department), state capitol, or State Department, the reporter does the legwork, cultivates sources, writes the story, and is then responsible for delivering this story over the air. Some authorities view this method as the most credible journalism because the same professional has followed the story from inception to delivery. In short, the person who's been there is the one actually chronicling the event or discovery for us.

Many times, however, the performance (on-air) and reporting (investigating/writing) functions of electronic journalism are parceled out to separate people. A radio announcer reading copy pulled from a wire service machine and a TV *anchor* conveying stories that have been prepared by other staffers are two examples of the performance function in isolation. When John Cameron Swayze on NBC (see figure 1-4) and Douglas Edwards on CBS began hosting the first TV network nightly newscasts in 1948, they elevated the radio news *reader* into the TV news *anchor*, thereby establishing a convention whereby high-profile newspeople could be totally divorced from the news-gathering process. This separation of story deliverer from story preparer is not necessarily good or bad. It just means that an intermediary (the on-air personality) has been injected between story coverage and audience reception.

For persons employed exclusively as air talent, a priority is clearly placed on delivery skills. This means a pleasing, authoritative voice and crisp, clear diction that follows the conventions of midwestern pronunciation the networks adopted in radio's heyday as the most easily assimilated national standard. In television, these aural attributes must be combined with a comely (telegenic) appearance that wears well on viewers. Taken as a group, the station's on-air news team should also

Figure 1-4. John Cameron Swayze hosts NBC-TV's *Camel News Caravan*, a fifteen-minute nightly newscast wholly sponsored by a cigarette company. *Courtesy of* Broadcasting & Cable.

reflect a balance between male and female deliverers plus include representatives of ethnic groups that constitute a prominent part of the market. In short, persons who seek to break into electronic journalism solely as anchor/presenters, without firm credentials in the harsh street world of news gathering, must recognize that their careers are dependent almost entirely on their looks and their sound. "We could argue that a small part of their salary is to be a journalist and a large part is to be a performer," asserts the manager of one major-market station. "And the qualifications there include looks and gender—just as if you're putting on a play."[18] "On the outside, you need an attractive communicator, an effective communicator," says television news executive Mike Cavender. "But the quality that [stars] share is when the camera goes on and they're telling you a story, that's exactly what they're doing. They're not reading script, they're not reciting from a TelePrompTer. They're telling you a story. That's what people like to hear."[19]

Some anchors, of course, are highly regarded reporters who came to their jobs as the culmination of years of hard work in actual news gathering. Even though physical attractiveness is also a mandate for them, their credibility gained on the "news beat" side helps compensate for scars, wrinkles, and gray hair. Many of these anchors, especially at the networks, try to keep their hand in reporting by working on special assignments or flying to the scene of breaking stories, such as an earthquake, insurrection, or signing of a major peace accord. The danger in such expeditions, of course, is that their presence is seen as mere window dressing that detracts from the actual work on the story being compiled by lesser-ranking journalists. On the other hand, a seasoned local or network anchor is indispensable in interweaving the strands of a chaotically unfolding event. "The true test of an anchor," asserts St. Louis TV critic Eric Mink, "is how he behaves and conducts himself when everything around him is going wrong. The anchor becomes the linchpin of the coverage."[20] Research conducted by professor Mark Harmon supports Mink's belief. Harmon's study of anchor longevity found that "regular local television news viewers valued friendly anchors, anchors who were calm in a crisis, and anchors who are involved in the community."[21]

Carolyn Clifford
*News Anchor, WXYZ-TV,
Detroit, Michigan*

Being a local television anchor and reporter may bring you a measure of fame in the town where you work. But beyond the glamour is a whole lot of hard work. It is also an investment of your time: years of learning and living up to the challenge of informing the public accurately. It's using a powerful medium—television—to influence people, perhaps even positively change lives in a responsible manner. Not all careers allow you to lend a hand and change things for the better, but you really can do it in this business.

That's a tall order and a whole lot to ponder as you set out in television news. But the fact is that while much has changed since I started fifteen years ago, something remains the same. Television is still far and away the most popular way for Americans to get their news. Just think about September 11th coverage or the war in Iraq and the live television coverage of those stories. Another thing that remains constant is change: whether expanding the hours of local newscasts or enhancing the technology behind the scenes.

There is no substitute for learning the craft. I choose my words carefully here, for practicing journalism is a *craft*, and there are sets of skills peculiar to television. At the core of it all is my belief that you must hone your writing skills. You can only do that under the pressure of a real deadline for air. But once you have that skill, you will always be employable, especially in an industry that never seems to have enough good storytellers who can write well.

I started at a television station in Lansing, Michigan. I had a news director, Ross Woodstock, who I still call to this day for advice. He demanded that everyone in the newsroom learn every job in the newsroom. For me, that meant writing, editing, producing a newscast, working an assignment desk, reporting stories in the field—and only after all those things, sitting in the anchor chair. Some people think that being an anchor is easy—and overpaid. Well, it's not as easy as it looks. And if you do your job seriously and do it well, you can never be paid enough. Starting in this business as a reporter is often no different in pay than working at a fast-food counter. The pay comes later, when you are considered a valuable—a *franchise*—player.

To be sure, some people do get to sit in the anchor chair without having the fundamentals to keep them grounded. If their management doesn't see their shortcomings, the viewer usually does. They don't last long. I'm grateful I paid my dues and worked my way up. It built a strong foundation, but it also gave me great self-confidence. Ten years after getting started, I achieved one of my professional goals: to work in my hometown of Detroit, a Top 10 market.

Realizing your dreams in this business is never guaranteed. Sometimes, beyond your skills and talents, it still comes down to good luck and good timing. But here is what you need to know about luck. You can make your own luck by working hard, and you can excel at your work if you have learned all the basics and have practiced your craft. Once you do that, beyond simply "holding a job," you can actually practice the craft of journalism, giving voice to the voiceless, uncovering wrongdoings, educating the public, acting as an advocate for change. And here is the best part: you can do it every day. It is just up to you.

There is nothing inherently deceptive or deficient in employing skilled presenters to only *deliver* the news. In some environments, this practice is mandated by policy or labor considerations. Irish broadcast journalists, for example, are strictly divided into two union groups: the on-air newsreaders belong to Equity (the national actors' union), and the reporters and writers belong to the National Union of Journalists (along with print newsgatherers and writers).[22] In Ireland, the United States, or any other country, the separation of performance and news-gathering personnel can still make for accurate and efficient journalism. This is provided, of course, that the content standards for news remain uncompromised and that the presenters do not try to masquerade as authorities who have gathered the stories themselves. When the surface look or sound of a news program is allowed to override researched substance, you create a nearly worthless veneer.

An inherent characteristic of electronic news is that it is delivered to the audience via dynamic human speech rather than by static symbols on a printed page. For radio and television journalism, this situation can make for either greater clarity and audience involvement—or worthless babble and audience confusion. No matter how the on-air and creative aspects of electronic news are divided or consolidated among the professionals involved, all of them must realize, as veteran television reporter Byron Harris puts it, that "they have been entrusted with the communication of information. And society is only going to be able to continue to have the pillars that it has if that information has real content."[23]

Every step of the way, electronic reporters/writers must compensate for the fact that they have a much narrower framework in which to cast their piece than do their print media counterparts. A half-hour news show often includes fewer words than even a single newspaper page. Brevity, clarity, and conciseness thus become crucial. The great majority of potential subjects cannot be accommodated at all, and the subjects that are covered must be expressed in the fewest words or clearest pictures possible. Fortunately, through the judicious use of sound bites and video clips, the electronic media allow the subjects and aura of a story to flow directly to the audience without the need first to transcribe everything into print description. The writing that is done can then be devoted to putting this actuality into a context that deepens listener or viewer comprehension.

For the radio or television journalist, the trick in all of this is to be able to grasp the essence of the story quickly so that only the most meaningful sounds, sights, and words are selected to put the piece together. The result will almost always have to be brief, but it need not be fragmentary. Instead, the astute electronic newsperson will have set reasonable parameters for what can be covered and then will stay within those boundaries to deliver a narrow-focus insight that makes relevant sense to the newscast consumer. As CBS News correspondent Susan Spencer explains: "You have to adjust to the fact that television pieces are not documentaries. You may know an awful lot, and everything that you know probably helps what you end up being able to put on the air, makes it more clear. But you can't say everything you know or you'll consume the whole broadcast.... Half the time when I get ready to write a piece, I think 'This is the one I can't do. This is hopeless. No way can I cram all this stuff into two minutes.' But it's always gratifying when I do it."[24]

Spencer's point has even greater ramifications for radio, of course, where there are no pictures to help cover the topic in the time allotted. This makes radio news a very difficult enterprise—and, unfortunately, a shrinking one. Consolidation of the radio industry has meant less and less attention to local news and consequently fewer and fewer jobs for radio journalists. "Radio news," writes trade reporter Marc Fisher, "which has played third chair to print and TV's concertmasters ever since John Cameron Swayze first read the headlines on the home screen, is tumbling into a time of swift and merciless change. More and more stations are going news-free. Many, perhaps even most, stations that feel compelled to offer some news are sacking their reporters and anchors, and out-

Performance Functions   17

sourcing the news without ever letting the audience in on the trick."[25] Stations accomplish this outsourcing by hiring state or regional services that provide voiced, customized "headline packages" that can be easily dropped into the regular music format. This practice might be referred to as the news version of voice tracking.

It should not be surprising, therefore, that what remains of the radio news enterprise is not very attractive to potential practitioners. "The narrowing of the talent pipeline stems from the polarization of the business," Fisher adds, "with those young journalists seeking celebrity drawn to television and those wanting to do serious, gritty reporting lured to print. Radio's unique quality—the chance to tell stories without the multi-level editing of print or the equipment and crew restrictions of TV—is mostly ignored by young reporters."[26] Seeing an opening by which to attract unserved listeners, public radio stations have moved into this void with a combination of national news from NPR (National Public Radio) and local coverage generated by small news departments of their own. Some have further broadened their coverage by rebroadcasting international news programming such as that provided by the BBC (British Broadcasting Corporation) World Service.

The Web constitutes an additional opportunity for electronic journalists, usually as an extension of existing print, broadcast, and cable news operations. *Convergence* (the combining and disseminating of content across several media platforms) means new opportunities for news professionals who have learned to compose stories for multiple vehicles. Therefore, the broader the electronic journalist's platform competence, the greater his or her employment options. Similarly, the more widely educated this journalist becomes, the wider will be the range of subjects on which she or he can quickly hone in. A firm grounding in history, political science, and economics is almost always helpful, with a background in business, science, or the arts a required plus in developing special-beat expertise. Genuine electronic news professionals pride themselves on being more than just conduits for raw facts. Despite the time/length limitations inherent in their formats, most also strive to develop a context and perspective for those facts. These qualities can come only from an authentic knowledge base that such professionals bring to each story assignment. Therefore, Radio Television News Directors Foundation head Deborah Potter advises that "today's students need training to deal with the pressures they will face on the job. They need to learn where to find information in a hurry, and how to produce more than one version of a story on a tight, daily deadline."[27]

Weather and sports are more specialized beats but require the same adaptability to multiple media channels that is now demanded of general news journalists. Increasingly, weathercasters not only perform on-camera duties for their television station but also maintain a weather page on the station's website, voice audio forecasts on affiliated radio outlets, and perhaps even prepare the weather column for the local newspaper. While there are still some weather *reporters* on the air, more and more stations and networks now demand that the person delivering the weather be a certified meteorologist as well as an effective performer.

As Vincent Condella, chief meteorologist for WITI-TV in Milwaukee, observes: "I think that a lot of people have said, 'I want to get into television. I just want to get my face on the tube. How can I do it? . . . I could wave my arms in front of a weather map, that kind of thing. I've got a shtick.' A lot of times weather was the comic relief between news and sports. Now, in the first two years of an undergrad program, you usually weed out the people who are really serious and those who aren't."[28] Weathercasting is no longer something that can be taken casually or delivered off-the-cuff. Most meteorologists agree that every minute of on-air weather requires one hour of off-air preparation. The availability of sophisticated graphic/forecast systems (boxes) like AccuWeather only means that viewers have come to expect that more visually appealing and detailed illustrative information be wedged into each of those minutes. "It's a strange dynamic—the boxes can do more, so we can

Matt Kirkwood
*Meteorologist, WOOD-TV8,
Grand Rapids, Michigan*

Would you like to be the part of a broadcast that impacts *everyone* who is watching or listening? If you answered "yes" to that question, consider stepping into the world of weathercasting. Whether a storm is brewing or a picnic is being planned, weather is often the universal draw that encourages people to watch the news.

The main role of a weathercaster is to provide an accurate and up-to-date forecast. This combines the art of communication with the science of forecasting. If you want an advantage climbing the career ladder in this particular business, I would advise you to earn a degree in meteorology with a communications/broadcast minor. This degree will not only give the broadcaster insight and understanding of weather processes but will also provide the expertise necessary for explaining complicated meteorological factors in an easy-to-understand way. Even with a degree in meteorology, be prepared for the occasional forecast to go awry. At times this will place you in the path of public criticism. So in these instances it is important to realize that weather is not an exact science and not to wear your emotions on your sleeve.

The one constant in this industry is computers. From analyzing forecast data to creating your weather graphics, these machines are indispensable. Because we live in the "computer age," you are expected to have good computer skills. TV stations often spend a good chunk of their budget on weather software, and the burden is on you to explore the software's graphical capabilities. Once the forecast is finished and the graphics are made, the moment of truth arrives: the presentation of your forecast. There are great scientists who cannot communicate, and there are great communicators who could care less about thermodynamics and the dew point. Success comes from consistently and accurately explaining the science of weather in a way that everyone can understand it. If you frequently achieve this balance you will capture an audience.

The physical presentation might be the oddest part of the whole job. You are expected to deliver the forecast while standing in front of a blank wall, looking at your *backward* image on a few small, strategically placed monitors while you ad lib to the camera. The manner of your delivery should be confident, straightforward, and personable. What makes the process even more challenging is that you will always have a finite amount of time to deliver the forecast. The choreography may seem difficult at first and even look humorous at times, but with practice it will become natural.

I'm often asked the question, "Why did you choose this career?" Most of all I enjoy that weather is always changing and people rely on me to prepare them for that change. There are days that I simply advise my viewers to put on sunscreen. Then there are *severe* weather times where I am in a critical position to help protect life and property. During severe weather, TV is one of the few live venues to relay information to a wide public. If you are timely and accurate with this information, you can ultimately save lives. When all goes well during these circumstances, I often get an adrenaline rush followed by a sense of accomplishment.

If you want to impact lives by meeting the twin challenges of forecasting and live delivery, weathercasting might be the career for you!

Performance Functions

do more," says WDBJ-TV Roanoke's chief meteorologist Robin Reed. "There's no net loss in the time I have to spend on graphics; it's the same as before—we're just making more dynamic and more accurate renditions of what we're trying to say."[29]

For weathercasters, social attributes often are more valued than physical ones. As critic Elayne Rapping has commented:

> Weather people are usually the homiest, wittiest . . . and most informal members of the news team. They are often loved by community members, who will choose a particular news station as much for the weatherperson as anything else. If there is no one at home to commiserate with about the eight days of rain we've just had; no one to complain to about having to cut the grass, miss the softball game, or whatever may happen, we've still got good old Bill What's His Name to share it with. He understands just how we feel. He too forgot his umbrella, or had to shovel his walk four times in one week.[30]

These social attributes may be even more important than are the radar and graphics systems that outlets continually promote. "You can have a weather department with every whiz-bang gizmo out there," says Bryan Busby, chairperson of the American Meteorological Society's Board of Broadcast Meteorology, "but if these guys and gals are perceived as stuffed shirts or not involved in the community, they're probably not going to be as successful as the weather department that came to my kid's school, or that I saw at the Rotary Club, or that hosted a telethon."[31]

This is not to say that accurate forecasts are less important than on- or off-camera personality. For weathercasters are allowed almost no margin of error, and their mistakes are immediately apparent to every member of their audience. "If I give 19 forecasts—perfectly accurate, right on the money—and I miss one, that's the only one I get calls and letters about," laments Vince Condella. "That's the nature of the business. If we nail it—I mean, nail it right on the nose—it goes totally unnoticed."[32]

Sportscasters run the same risk when they make game predictions, of course. But unlike performers on the weather side of the desk, they can concentrate more on in-game or after-the-fact reporting than pre-event prognostication. Sports reporters are expected to cover much more than game stats. They must help us to savor victory and overcome defeat. To be successful, says professor Michael Seidel, on-air sports talent "must go beyond analysis and amusement; they must, after a fashion, console. . . . You long for the familiar voice of the broadcaster to hint, even if by the soothing assurance of the terms of his employment, that tomorrow is another day."[33]

The range of stories covered by the sports reporter tends to vary by market size. Researchers Ray Carroll, Laurie Lattimore, and Bill Erwin found that "whereas professional sports dominated in the largest television markets, small-market stations gave greater emphasis to collegiate and high school athletics." Therefore, large-city sports coverage is actually less diverse than that in smaller markets.[34] This is readily explainable. Large markets are home to major league teams that command the interest and loyalty of most of the sports fans throughout the metropolitan area. Top-tier collegiate teams also tend to be located within the coverage pattern. Meanwhile, the many high school leagues found in a large market make it difficult to feature any one of these without boring the vast majority of viewers. So large-market stations restrict most of their attention to "big league" coverage. Less populated areas, conversely, usually lack major pro and collegiate teams—so coverage of big league events by stations in these smaller markets is inevitably perceived as "secondhand." However, interest in local high school and small-college sports is greater and more universal because there is no "big league" activity in the locality to siphon off fan interest. These local events

Kurt Wilson
Sportscaster and Producer, CMU
(Central Michigan University)
Sports Network

It's all Ernie's fault! Like most kids in Michigan growing up in the late sixties and early seventies, I snuck my radio under the covers at night to listen to Ernie Harwell "paint the picture" of the Detroit Tigers game in some faraway place. The magical image that flowed through the airwaves would eventually lead me into this crazy business that we call sportscasting.

To the outsider looking in, the sportscaster has the most glamorous job in the world! The best seats in the house, traveling to distant places, showing up right before game time to go on the air, actually being paid to talk sports! Mr. or Ms. Outsider, you are wrong on all counts. The seats are cramped and uncomfortable. A seventeen-hour bus ride with a minor league baseball team is less than glamorous. If you show up right before game time you won't be in the business very long. And the pay? Oh yes, the pay. For a majority of sportscasters, the guy asking you if you want fries with that value meal probably has a healthier financial portfolio than you do.

All right, with that disclaimer of doom out of the way, welcome to the greatest job in broadcasting. All of the above is true, but hard work, preparation, a little talent, and a supportive family will set you above the rest.

Be prepared to do anything, move anywhere, and get no money from your first job in broadcasting. I got my foot in the door as a minor league baseball play-by-play announcer because I agreed to wash the team's laundry and clean the team clubhouse as a side job. I know of one guy who drove the team bus to road games, called the game on the radio, and drove the team home again. It's not glamorous, but it's called "paying your dues." At this level of sportscasting, you live with the players, ride the bus with the players, and learn to sleep in strange positions in the luggage compartment of the bus. At an away game you set up your own equipment, fix various problems that occur (sometimes in the middle of the game), dodge rain and lightning from your broadcast position on a cardboard table sitting amongst the crowd, and keep your eye on rowdy fans who would just as soon dump their favorite beverage on you as listen to you call a game-winning grand-slam home run for your team.

If you're good enough, lucky enough, and people like to listen to you, you'll have a chance to move up the ladder. There are very few sportscasting jobs in the college and pro ranks, and the key to getting there and staying there, besides the talent to "paint the picture," is being the most prepared. I devote at least eight hours per game to preparation. Two or more of those hours will be spent compiling depth charts, stats, and everything that I'll need to have right in front of me for that particular game. I'll spend a couple of more hours reading media guides, researching on the Internet, and finding out everything I can about the teams. Hours more will be spent memorizing names, numbers, and faces so I know both teams without having to look at cheat sheets all game long. Going to practices (when allowed), talking to coaches, players, and sports information directors are all good places to get information. On game day, depending on the complexity of the equipment setup, you'll want to get to the stadium or arena between two and five hours before game time. Setting up and testing equipment is your first priority, and knowing how to fix an impending problem with the setup is the difference between getting on the air or not.

Once you're on the air, be prepared for anything. I've been knocked off the air by power surges, wires have been kicked out by people walking by, and diving basketball players have smashed equipment (and me!). During the game you should

Performance Functions   21

be able to keep score if it's a basketball or baseball game, keep notes on drives in football, and remember to read the live commercials that pay your salary. Keeping score and a running tally on what has been going on during the game is an art form that truly separates a good play-by-play announcer from an average one. With people tuning in and out all game long, updating the earlier action is a must.

If you're dedicated, prepared, willing to wear six different hats to pay the bills, have a family that knows (and accepts) that you'll be on the road a lot (sometimes on birthdays and holidays), and people like to listen to you, maybe someday you'll be the person that some youngster will be listening to—painting the beautiful picture of sports.

---

can also be covered on site rather than relying on warmed-over reports from outside the market. Smaller-market sports reporters and their crews therefore broaden their coverage over many more teams and sports than is typical in the large-market setting.

When it comes to actual broadcasting of games, the best sportscasters in both large and small venues realize that an effective sportscast never simply reproduces the contest. It adds, deletes, and rearranges elements to create a more continuous and involving scenario. As Sarah Ruth Kozloff points out, the television viewer, for instance, "sees the events filtered through the control room, which switches from crowd shots, to the cheerleaders, to the coaches, to the action; which flashes back to pregame interviews; which forsakes real time for slow motion and freeze frames; and which repeats the same play over and over. The viewer is no longer simply watching the game, but rather a narration of the game in which various choices have been made concerning temporal order, duration, and frequency."[35] As the game proceeds, the skilled sports commentator begins to see a storyline emerge and works with the producer to arrange between-plays elements to enhance that storyline's evolution. In this way, even a lopsided contest can retain audience interest because of the additional narrative that the production is weaving.

Whereas they used to keep such storylines to themselves as underlying structure for their commentary, sportscasters now explicitly reference this structure and even use graphic inserts to spotlight it. This helps to make the game more comprehensible to more fans and makes it easier to quickly orient latecomers or channel grazers as to how the contest has developed to this point. Hopefully, such signposting encourages the just-arrived audience member to stick around. In describing the blueprint by which he built ABC Sports, pioneering ABC sports producer Roone Arledge put it this way: "What we set out to do was to get the audience involved emotionally. If they didn't give a damn about the game, they still might enjoy the program."[36] This is an appropriate guidepost for every electronic sports reporter. But it does not come easily. Emmy Award–winning NBC sportscaster Dick Enberg confided to the author that his rule of thumb is one full day's advance preparation for every hour he will be on the air.

Television Entertainment Personalities

If the road to on-air success in electronic news, weather, or sports seems tortuous, the path to fame in entertainment television can be equally if not more difficult. In each of these performance contexts, personal talent, training, luck, and unquenchable determination all seem to play a part. But in addition to these attributes, television entertainment personalities must understand the uniqueness of this expressive medium if they hope to prosper.

Television series acting, for instance, is not like stage acting or even like appearing in the movies. The late actor-director John Houseman, who worked in both the film and electronic industries, isolated the differences between them this way: "Television puts more emphasis on the spoken word; it lessens the importance and effectiveness of the reaction shot, which is the most basic element of most film performances; it encourages a more naturalistic mode of acting; its emotional curves tend to be shorter, intended for a more direct and immediate effect on the viewer."[37]

In other words, involving viewers in the events on the small screen, and involving them in spite of family-room distractions and the brevity of most television programs, presents actors with a tremendous focusing challenge. As media scholar Horace Newcomb observes, "Even when landscape and chase become part of the plot, our attention is drawn to the intensely individual problems encountered, and the central issue becomes relationships among individuals."[38]

Under such conditions, an actor's believability becomes a function of concentration and controlled movement as well as physical appearance. Certainly, concentration is an essential part of any actor's art, but it must be especially intense in television. Director George Schaefer, for example, recounts that actors in a theater try to develop a bond that reaches beyond the footlights in order to gauge and promote a unanimity of audience reaction. In creating television's tightly framed world, however, Schaefer finds that he must convince his actors to ignore what's out there in the darkness if they are to focus their performances properly.[39]

Peter Michael Goetz
Television, Film, and Stage Actor

Even though I have a demo tape with excerpts from more than forty television and film productions, which helps to remind casting personnel of my potential for any given project, many times I must visit an office peopled by anywhere from three to twenty casting directors, producers, network executives, directors, and others. They conduct a job screening complete with prepared or unprepared readings and/or interviews. Many of my days consist of traveling from one studio to another for various chats about the latest projects.

Once winning the option of participating as an actor in a film or TV piece, my responsibility is to do an enormous amount of preproduction preparation. Unlike theater, very little time is allotted in television for acting rehearsals. Nor is there time for examination of historical significance, experimental rehearsals, character development, or evaluation. Television actors are thrust immediately into the scene. The director and technicians determine the camera angles and blocking marks based on our first rehearsal on the set. Actors can only hope that in the final cut, the editors retain at least a semblance of what the actor put into the scene.

It's very difficult for the video/film actor to totally control the outcome of his performance. For example, I once played a sinister character who possessed some eccentric redeeming qualities. These qualities were created to keep the character three-dimensional, more human, and just plain more interesting. I chose to play him rather inscrutably and mysteriously until one moment when he smiled and reacted to a very tender moment. That "moment" lasted only a few seconds (hopefully in a poignant close-up). However, the editor at that juncture selected a quick-cut reaction shot of my fellow actor instead. My character revelation was lost. This gave my finally-aired portrayal a very bitter and unredeeming quality. The editor, in short, had removed the key to my character which I had worked so hard to insert into the production.

For the most part, acting in film and television can be a consistently rewarding experience. If nothing else, there is a certain satisfaction in hitting the exact marks, remembering the precise moment we put the cigarette to our lips for continuity, maintaining the 8-foot distance between the car we are driving and the camera truck, being aware of holding our posture so we don't shadow our fellow actors, not being distracted by the maneuvering crew, and somehow even managing to remember our lines.

Note: Mr. Goetz's electronic media credits include continuing roles in such series as *The Cavanaughs, The Faculty,* and *The Buccaneers (Masterpiece Theatre)* and guest-star appearances on *St. Elsewhere, L.A. Law, Midnight Caller, Golden Girls, Twin Peaks, Matlock*, and several others. He also has a wealth of movie credits and major roles in Broadway productions.

Even in the case of situation comedies with live studio audiences, performers cannot play to that house. Instead, they can use audience reactions only as a partial guide in styling their delivery to the intimate camera and the viewer perspective it represents. Add to this the breakneck production schedule of television as opposed to film or stage work, and it becomes clear how rigorously distilled the process of video acting must become. "You don't have time to do your scenes," objects veteran series actor Ken Howard. "You have to be ready in two takes. That can be an incredible burden. I think many good feature-film actors would be crushed on TV."[40] Usually, a one-hour drama has to be shot in eight days and a situation comedy episode in four or five.

One measure of a television performer's success or failure in connecting with viewers—despite these productional tribulations—are the Performer Familiarity and Appeal Ratings, or *Q Ratings* for short. First used in 1963 by New York–based Marketing Evaluations, the Q Ratings annually survey some ten thousand households as to their attitudes toward approximately fifteen hundred performers and celebrities in twenty-one categories. From these surveys are derived the Performer Q scores. The same firm uses similar methodologies to determine the appeal of famous athletes (Sports Q), animated characters (Cartoon Q), company and brand names (Product Q), and—as figure 1-5 evidences—television programs themselves (TV Q).

Subjects' appeal and familiarity are broken down by how each fares with consumers in forty demographic categories. In other words, the Q Ratings tell how many respondents actually recognize the names of performers/celebrities and then how much these respondents like them. The demographic categories are divided by age, sex, income, Nielsen county size, geographic region, occupation, and race and ethnicity. The Q Rating is derived from the percentage of all respondents who indicate the personality is "one of my favorites" divided by the percentage of all respondents who are familiar with the personality. The decimal point is dropped. For example, if 60% of respondents recognized the person, and 21% listed that individual as "one of my favorites," the Q Rating would be 35 (21 divided by 60).

One result of the formula is that a relatively little-known performer can achieve a high Q Rating, pointing to a potentially effective product spokesperson outside the ranks of the superstars. Thus, if only 25% of respondents were familiar with the personality, but 18% listed that person as a favorite, the Q Rating would be a formidable 72 (18 divided by 25). Q scores should not be the sole determiner of casting decisions, however. As Marketing Evaluations president Steven Levitt points out, "We caution clients to do more research beyond Q Ratings. They only tell a part of the story. You cannot infer 'credibility,' 'appropriateness,' 'qualifications,' or 'fit with the product' from likability ratings."[41]

The Q Ratings do constitute an important tool by which advertising agencies can zero in on spokespersons who appear to test well on name recognition and familiarity with the demographic

PROGRAMS AIRED: 12/30/02 - 2/9/03
UPN TUESDAY 8:00 PM - 9:00 PM

BUFFY THE VAMPIRE SLAYER
GENERAL DRAMA

	FAMILIARITY			POSITIVE TVQ									NEGATIVE Q		
		NORMS			NORMS		SCALE POINTS							NORMS	
	TOTAL FAM-ILIAR	PGM TYPE	ALL PGMS	TVQ	PGM TYPE	ALL PGMS	ONE OF MY FAVS.	VERY GOOD	GOOD	FAIR	POOR	NOT SEEN	NEG Q	PGM TYPE	ALL PGMS
TOTAL SAMPLE	41	29	34	17	31	26	7	5	12	11	6	59	42	21	23
6 - 11	35	16	21	34	36	39	12	8	5	7	3	65	29	28	22
12 - 17	55	25	30	15	27	28	8	11	16	11	9	45	37	28	26
18 - 34	51	32	39	18	28	26	9	5	13	15	9	49	46	25	24
35 - 49	34	31	37	12	29	22	4	1	12	12	5	66	48	21	25
50 AND OVER	33	35	34	15	35	28	5	5	11	7	5	67	38	16	20
18 - 49	43	32	37	15	28	24	7	3	13	13	7	57	47	23	24
18 AND OVER	39	32	36	15	31	25	6	4	12	11	6	61	44	20	23
25 - 54	38	32	38	14	30	24	5	3	12	13	5	62	47	21	24
TOTAL MALES															
6 AND OVER	42	28	33	17	27	24	7	5	13	11	6	58	41	24	25
18 - 34	53	31	38	22	25	24	12	7	14	11	9	47	38	26	25
35 - 49	39	31	36	18	27	21	7	2	12	12	6	61	46	22	26
50 AND OVER	36	31	31	11	29	23	4	4	13	9	6	64	43	20	24
18 - 49	46	30	38	20	26	23	9	5	13	12	7	54	41	24	26
18 AND OVER	42	31	34	18	27	23	7	4	13	11	7	58	42	22	25
25 - 54	44	32	37	18	28	22	8	4	12	13	7	56	46	23	26
TOTAL FEMALES															
6 AND OVER	39	30	34	17	34	28	6	5	11	11	6	61	43	19	22
18 - 34	51	33	39	14	30	27	7	4	12	19	9	49	55	24	23
35 - 49	30	31	35	4	30	22	1	1	13	11	4	70	50	19	24
50 AND OVER	30	34	35	18	39	32	5	6	9	5	5	70	33	13	18
18 - 49	39	31	37	10	30	25	4	2	12	15	6	61	53	22	23
18 AND OVER	38	33	37	12	34	27	5	3	11	11	6	64	47	18	21
25 - 54	35	32	37	9	32	25	3	3	12	13	4	65	48	19	23
INCOME															
UNDER $20,000	48	35	39	24	32	30	12	5	13	9	9	52	38	24	24
$20,000 - $39,999	42	30	35	15	30	25	6	6	13	12	5	58	41	22	24
$40,000 - $59,999	34	25	31	11	25	23	4	2	13	9	6	66	45	21	23
$60,000 AND OVER	39	28	30	15	32	26	6	7	9	12	5	61	44	19	23
$75,000 AND OVER	32	27	28	13	35	28	4	8	10	6	4	68	33	16	20
EDUCATION (ADULT)															
HIGH SCHOOL OR LESS	41	32	36	18	32	28	7	3	12	13	6	59	45	21	22
SOME COLLEGE/DEGREE	37	31	35	12	29	23	4	5	12	9	7	63	43	19	24
OCCUPATION (ADULT)															
WHITE COLLAR	37	31	35	13	30	23	5	4	11	12	5	63	46	20	24
BLUE COLLAR	43	31	37	12	29	25	5	4	14	13	7	57	47	23	24
RACE															
NON-BLACK	38	28	33	17	30	25	6	5	11	10	6	62	42	22	24
BLACK	58	41	42	15	33	31	9	7	19	16	7	42	40	17	19
ETHNICITY															
HISPANIC	47	34	40	21	32	30	10	3	9	16	9	53	52	20	20
COUNTY SIZE															
A	40	30	35	17	32	26	7	5	12	10	6	60	39	21	23
B	46	31	35	17	31	27	8	5	12	14	7	54	46	22	24
C & D	35	27	32	15	28	25	5	4	11	9	6	65	42	21	23
REGION															
NORTHEAST	31	29	32	14	31	25	4	3	12	8	4	69	38	20	22
MIDWEST	40	28	34	17	30	24	7	5	13	10	5	60	39	24	26
SOUTH	44	31	35	16	30	28	7	7	12	12	6	56	42	20	22
FAR WEST	43	31	33	20	32	27	9	3	10	12	9	57	48	23	24
VIEW PRIME TIME (ADULT)															
HEAVY	45	43	45	19	36	31	9	6	10	13	7	55	44	16	20
MODERATE	43	30	35	11	29	22	5	3	15	14	6	57	47	20	24
LIGHT	24	16	21	8	19	17	2	1	9	5	7	76	49	29	29

Figure 1-5. A program TV Q Report for *Buffy the Vampire Slayer. Courtesy of Francine Purcell, Marketing Evaluations, Inc.*

group(s) that clients are attempting to reach. In addition, casting directors and network and studio executives also use such numbers in selecting actors most likely to deliver (or enlarge) a program's target audience. If this all seems callous and calculating, it must be remembered that nationally marketed products and programs represent investments of many millions of dollars on the part of the companies who produced them. Personalities famous enough to be "Q-ed" probably owe much of their notoriety to such investments and stand to benefit additionally if their profiles suggest appropriateness for future commercials, series, or movies.

The same talent is seldom effective in all of these placements, of course. So when casting for television projects, legendary network and studio executive Brandon Tartikoff used what he called the *Airport Factor*. In Tartikoff's words, "The big difference between movie stars and TV stars boils down to this: 'If I saw this star in an airport, would I go up and say hello?' TV stars must be likable, approachable; movie stars can be remote, dangerous, bigger than life. Even TV stars who play unlikable characters . . . have a certain twinkle."[42]

Another group of television's on-air professionals who are regularly Q-Rated (and increasingly important parts of syndicated programming's profit picture) are game show hosts. The challenge for these performers, however, is that they must not only register well with viewers at home but also with onstage contestants and a studio audience. In addition, game show hosts must mesh so tightly with the program's fabric and pacing that they control tension build without seeming to dominate it. Above all, game show personalities must bring the same sense of immediacy to their televised tournaments as anchors and reporters seek to inject into their newscasts—and for the same viewer-snaring reasons. In a review of the immensely popular *Wheel of Fortune, TV Guide*'s Merrill Panitt isolated the key attributes for a game show ringmaster when he observed: "The host is Pat Sajak, a former television weatherman and public-affairs show moderator. A smooth operator, he is pleasant enough, asks the contestants where they're from and what they do without getting in the way of the game. . . . As game-show hosts go, he's as good as they come."[43]

Sajak later unsuccessfully tried his hand as a CBS talk show host—a seemingly similar performance situation but one, as he found, in which the central performer must be always prepared to take center stage. When a talk show lags, there are no spinning wheels or lit game boards to pick up the slack.

Actors, like game and talk show hosts, can make comfortable livings in *first-run syndication* (properties that have never appeared on a network and are sold directly to stations). Though syndicated projects lack the same-time nationwide visibility of a network series, they are also free from sometimes stifling network oversight and breakneck production schedules. As trade reporter Elaine Warren discovered, "Most actors working in syndication contend that the advantages far outweigh the disadvantages. In general, the actors say, on the higher-quality shows, the salaries are comparable to what the networks pay, the pace is more relaxed and there is no such thing as a standards and practices department monitoring every line."[44]

Industrial Performers

Though their work is less visible, we should focus for a few moments on the performers who work in corporate video. This burgeoning field of organizational training, communication, and sales presentations is, in many ways, the fastest-growing segment within the electronic media. In some cases, the on-camera professionals involved in corporate video spend all of their time serving this enterprise. In many instances, however, industrial media actors are also engaged in separate *mass* media projects, too.

Corporate video pays reasonably well, so many an actor uses its roles to keep bread on the table between public performance opportunities. Because the earnings are good, often at established union scale, industrial television directors expect that actors in their productions will be as committed to these projects as they are to consumer television and film roles. As corporate television production company head Kevin Padden bluntly puts it, "Don't do me any favors. If you think corporate work is slumming, don't slum. Either you want to do corporate work or not. There's no need to give an excuse for doing corporate work, and I certainly don't want to hear it, especially for what I pay talent."[45]

"Who cares whether it's corporate?" asks New York stage actress Chris Casady, "I want to work!"[46] Though few actors earn a living through industrial video exclusively, corporate work provides dramatic opportunities upon which they can build their audition reels and creative range. Performing in an industrial production is no easier, and requires no less talent, than performing in commercials or broadcast series. Inevitably, many corporate presentations are highly technical in nature. Thus, actors in them must not only be convincing as characters but must also assimilate a great deal of unfamiliar if not downright complex information. To convey this information in a believable and interesting manner takes intellectual curiosity as well as dramatic ability. This is especially true if the role is that of actual spokesperson for the company. In such an instance, the image of the entire firm is on the line, and the actor is expected to project that image to carefully manicured perfection.

This concern for dramatic quality is a relatively new trend in the industrial media. As corporate video expert Fred Cohn recalls, "It wasn't long ago that the cast list of a typical business program may have been headed by the human resources exec with the gimcrack Kirk Douglas imitation or the gregarious woman in accounts payable who played the lead in her high school production of *Oklahoma*. Such amateurs have sunk more than a few productions—inspiring derisive laughter rather than retention with their inept or grandstanding performances. But now, corporate video's actors are likely to be professionals; people who make their living on stage, screen and broadcast TV."[47]

From a corporate producer's point of view, using trained actors is always preferable to being forced to rely on workers and executives from within the organization. As trade writer Carl Levine points out, for professional talent, "appearing in front of a camera is their chosen career. They have studied how to portray characters and emotions. . . . It's a lot easier to direct and set schedules for professional talent than for employees. Generally, corporate executives don't like to be told what to do and when to do it, whether it's in a TV studio or their office."[48]

In industrial no less than in mass electronic presentations, then, a professional result requires professional performers. The comparatively obscure but lucrative world of corporate video will likely continue to prosper. The market for such production services "is not saturated," reports *Video Systems* editor Ned Soseman. "Many companies haven't thought of video as a solution. Others have the need but haven't the time or resources for critical evaluation. Most of this untapped market simply needs to be sold."[49]

The prosperity of corporate video enterprises will continue to depend on how well their presentations conform to the media expectations of their audiences. Industrial video may not be quite as slick—or anywhere near as expensive—as a typical broadcast network presentation. But it cannot ignore the fundamental performance requisites that make such mass audience programs intriguing to watch and understand. "We're all up against broadcast television and its network-quality look," cautions corporate video sales executive Robin Pride. "Because your audience is accustomed to it, your creativity is challenged. . . . You must remember that the quality of your video will be forever judged and evaluated at the level of network prime-time programs."[50]

A Word about Agents

Most actors and many major-market news and other on-air personalities utilize the services of agents. In exchange for a percentage of salary earned, an agent's responsibility is to secure work for, and advance the careers of, the performers she or he represents. Agents can't usually get you a job, but the good ones can get you *considered* for jobs. For actors, this means getting you into auditions. "Actors live in a lottery system. We signed up for it years ago," says Jim Bracchitta. "Each time I go in and put myself up against anybody out there."[51] A well-positioned agent will get you into more of these lotteries and with better odds. The difficulty of getting in can't be overstated. Approximately 135,000 performers belong to the two main talent unions: the Screen Actors Guild (SAG) and the American Federation of Television and Radio Artists (AFTRA). But among SAG's 98,000 members, 80-90% are unemployed at any given time, and less than 10% make more than $15,000 a year at their craft. Fewer than 10,000 of the two unions' members make a full-time living as performers.[52] In such a brutally competitive environment, a talent agent can make a critical difference. Not surprisingly, however, the chances of getting a good agent to represent you are often inversely related to your need for such an agent. The more in demand an actor's services are, the more agents will want to represent that performer due to the larger salary such a person can command.

Agents are less essential for on-air news talent—at least until that talent gets established. Former news director Scott Jones believes that newspersons working in markets below the Top 60 seldom should be paying an agent. If an agent wants to represent them, they typically should not have to pay that agent until he or she helps them land a Top 60 market job. In addition, Jones advises news talent:

> Don't listen to a potential agent's success stories. Who cares whom your agent represents; just make sure he does well for you.
>
> Remember you are not your agent's only client. But if you call your agent at 10 a.m. and leave a message, you should expect a phone call back that same day. . . . If it is taking a day or two to call you back, dump him.
>
> Don't hire a suck-up! You do not want an agent that news directors love. . . . If news directors love an agent, there is a reason for that. Many times it is because he is a pushover and won't get you the best deal.[53]

Whether he or she is representing entertainment or news talent, in the final analysis, an effective agent is someone who knows the business, is known in the business, and can develop a deep and direct relationship with the talent he or she represents. "A good agent," writes William Morris Agency executive vice president Sam Haskell, "is a father or mother, a brother or sister, a son or daughter, a best friend, a confidant, a psychiatrist, a tennis or golf partner, and a business associate. If you meet an agent who can't be most of the above, keep looking."[54]

Chapter Flashback

Even though many people believe that on-air media performers are the most important part of our profession, such personalities are just the most noticeable players in a multifaceted enterprise. Disc jockeys exhibit a variety of styles in a variety of formats but share the universal responsibility of providing an appealing linkage between the station and the listener. Usually, this linkage must resonate with local community needs and character. Talk show hosts, in addition, must be consummate conversationalists with broad interests and a constant desire to learn more about any subject that might prove of importance to their listeners.

Newspersons discover and present information the public does not have the time or expertise to gather for itself. Sometimes news presentation and news gathering are handled by different individuals. There is nothing wrong with such a practice as long as the content standards for news remain uncompromised and the people who function solely as presenters do not try to masquerade as authorities. Unlike print reporters, electronic journalists usually must reduce stories to their briefest essence, but without distorting the subject in the process. Despite this inevitable condensation, top electronic news professionals still strive to develop a context within which their reports can be better understood.

Electronic media actors often function under severe time constraints that put a premium on concentration, fast work, and cultivation of a sense of approachability that projects through the television screen. Q Ratings are often used to measure the appeal of individual performers and programs among their target audience members. In addition to network projects, first-run syndication shows offer qualified actors and game show and talk show hosts further opportunities for on-camera exposure. The industrial/corporate sector is growing and depends on performers and creators who can take often dry and highly technical information and sculpt it into clear and interesting program modules that compare favorably with network television. Agents can assist actors and on-camera news talent to secure employment by getting their clients considered for parts and postions. However, a close working relationship between agent and client must be developed if the process is to function effectively.

Review Probes

1. What are some of the differences between the job of disc jockey and that of talk show host? Some of the similarities?
2. List advantages and disadvantages of dividing news gathering and news presentation functions so that these are handled by different individuals.
3. What are some reasons that electronic news stories are almost always much shorter than their print counterparts?
4. To be successful, in what ways must the broadcast of a game be different from the actual event as experienced at the stadium? What are the reasons motivating this difference?
5. What are some of the ways in which the task of video actors differs from that of their film and stage counterparts?
6. Describe advantages and disadvantages to seeking a performing career in the industrial/corporate sector.

Suggested Background Explorations

Arya, Bob. *Thirty Seconds to Air: A Field Reporter's Guide to Live Television Reporting.* Ames: Iowa State University Press, 1999.

Bernard, Ian. *Film and Television Acting.* Woburn, MA: Focal Press, 1993.

Blum, Richard. *Working Actors: The Craft of Television, Film, and Stage Performance.* Woburn, MA: Focal Press, 1989.

Davies, D. Garfield, and Anthony Jahn. *Care of the Professional Voice: A Management Guide for Singers, Actors, and Professional Voice Users.* Woburn, MA: Focal Press, 1998.

Hausman, Carl, Philip Benoit, Frank Messere, and Lewis O'Donnell. *Announcing: Broadcast Communicating Today.* 5th ed. Belmont, CA: Wadsworth, 2004.

Hawes, William. *Television Performing: News and Information.* Woburn, MA: Focal Press, 1991.

Henson, Robert. *Television Weathercasting: A History.* Jefferson, NC: McFarland, 1990.

Hyde, Stuart. *Television and Radio Announcing.* 10th ed. Boston: Houghton-Mifflin, 2004.

Reese, David, Mary Beadle, and Alan Stephenson. *Broadcast Announcing Worktext.* Woburn, MA: Focal Press, 2000.

Rich, Carole. *Writing and Reporting News: A Coaching Method.* 4th ed. Belmont, CA: Wadsworth, 2003.

CHAPTER 2

Conceptual Functions

In the case of most entertainment and corporate informational electronic media content, the people who appear on camera or on mic are different individuals from those who have conceived the material that these performers are delivering. This "conceiving" or *conceptual* function is handled by writers, visual artists, and musical collaborators whose labors create the framework within which performers can work. This chapter examines the role played by these conceptualizers.

Copywriters

In advertising, the professionals who write commercials, public service announcements, station and program promotional messages (*promos*), and similar types of continuity are known as *copywriters*. Unlike newspersons (see chapter 1), copywriters seldom if ever voice their own work. Instead, they script it out for performance by on-air station talent or the professional actors and singers who make at least part of their living by bringing these messages to life.

Copywriters are employed by stations, cable systems, or advertising agencies. They also work *in house* for corporate, institutional, or governmental units who prefer that their audio/visual communications be prepared by their own staffs rather than contracted out to an advertising or public relations firm. Some copywriters even work for themselves as freelancers. Whatever the setting, it is the copywriter's task to fashion those short persuasive messages that are the lifeblood of a free-enterprise electronic media system. Unlike entertainment program scripts, which are the province of a comparatively few East or West Coast specialists, pieces of copy must be fashioned by writers at almost every station and cable system in the country as well as by their colleagues in the advertising and public relations firms who use the electronic media on behalf of their clients.

The typical piece of copy is ten, fifteen, thirty, or sixty seconds long. Nevertheless, like the full-length script, it must tell a tale or reveal a truth within the time allotted. That is why many writers discover that the experience of copy creation is invaluable preparation for a career in long-form entertainment writing. Despite their brevity, commercial and other continuity pieces exemplify all the requirements of media form and content demanded in radio/television news as well as in series-length script formulation. In fact, electronic media copywriting has evolved beyond the audio/visual hawking of goods. Audiences now accept the best commercials as entertainment in their own right, with entire television specials built around collections of the copywriter's art. Because the production costs of nationally distributed TV commercials may regularly be ten to fifteen times that of comparable-length portions of the shows they sponsor, it is not difficult to see why top copywriters are highly prized professionals.

This is not to diminish the importance of *radio* copy and the people who write it. For years, radio was often thought of as a second-class medium, a kind of "television with the picture tube burned out," as writer/performer Stan Freberg once protested. Electronic media copywriters saw the building of their TV reel as the way to fame and glory and looked on radio assignments as hardship duty. As Jerry Siano, then-chairman of N. W. Ayer Advertising, recalls, "When television came along people started watching that. Radio didn't change. But our perception of it did. We forgot a quarter of a century of people watching their radio sets. And this might be what led to the development of the 'secret' to bad radio advertising. . . . Just stuffing the commercial full of twaddle, wall-to-wall words, endlines and some cheap music is the surefire formula for ineffective radio advertising."[1]

Dick Orkin and Christine Coyle
Copywriters and Partners, Famous Radio Ranch

Each year, the nation remembers and laments the date when an assassin put a bullet into the head of John F. Kennedy, a beloved president. We humans have a need to ritually memorialize major events in our personal as well as national lives. We remember even if the events were the most minor of the most minor and involved a look, an utterance, a posture, an aroma, an odor—anything at all. The two of us suspect that all humans have this inclination, storing away a special recollection securely locked in our memory bank. Finding the stimulus to retrieve it can trigger our storytelling process.

The deeply puzzling question for us is: Do we both recall simply because we have a strong recall gene or do we recall because we need the memory for our work in telling stories and dramatizing a product or service benefit?

"Where do you get your crazy ideas?" is the question repeatedly asked of us, and it reminds us of our particular writing and performing method at the Famous Radio Ranch. We won't raise your hackles of jealousy by telling you that the memories and stories just come to us. Because they don't. It's more like we need to put into play something that is part conscious and part unconscious, and that a kind of silent incantation summons the memory. For example, if we are at work developing radio spots for a home security system, we'll wait a few minutes, not pushing for the idea and certainly not contriving one, but just waiting, not for hours but minutes.

Suddenly, much to our surprise, a specific memory forces itself into our consciousness. We firmly believe that this particular memory appears at that moment because it will serve the making of the commercial at hand.

For example, recalls Dick, "An image of my father suddenly comes to me. Then I ask certain specific types of questions of that memory. Or, if you will, I engage in a dialogue with the memory. The memory is this. In the middle of the night, my father gets up to use the bathroom and suddenly notices a man about eight feet away. He is actually seeing his own dim reflection in a bureau mirror. He begins shouting at his own reflection and frightens my mother awake. *Who are you? I said, who are you? I'm warning you—I have a pistol and I won't hesitate to use it!* My mother, now fully awake, realizes what's happening and tries to tell my father, but his angry words drown her out. And, needless to say, he never owned a gun in his life, except a water pistol when he was seven or eight. My dialogue with this memory begins by probing how I can use this memory to talk about the home security system. And ideas tumble forth one after another."

Still working on the home security system, Chris volunteers this counterpart memory about her mother: "My mother calls a male neighbor in the apartment next door, begging him to get out of bed at 3:00 a.m. and come to her and her husband's rescue because someone is in the living room. And my father argues with her that in making the call, she's casting aspersions on his manliness and he finds it embarrassing. Again, as Dick did, I enter into even more dialogue with this memory: how will this true experience help in creating a spot for the home security system? More ideas for story-spots tumble forth."

We see this method as the way to engage in effective *advertising storytelling* rather than mere *announcing*.

For us, an *announcement* is a linear package that states explicit fact after fact in often cliché language and in a dull rational recitation. In contrast, a true *advertisement* is most often an implicit presentation. It is packaged through a story that allows the single USP (unique selling proposition) to be taken in by the listener through conflict and contrast, either comedic or melodramatic.

By the end of our memory quest, we have more than enough material to develop radio spots that reach out first to the heart and then to the head of the listener. Virtually everyone will identify with the experience of the two parents. And, after producing the spot and hearing it aired, we have learned through the years that these stories of real people in real situations—perhaps slightly exaggerated—demonstrate that emotion and humor can always remember what the head may forget.

Today, radio is being looked at differently, in part, it is true, as a result of the rising cost of television time that has priced many smaller advertisers out of that medium. But the rehabilitation of radio is also due, argues former Radio Advertising Bureau president Bill Stakelin, to the renewed realization that "radio is visual. It is one of the most visual if not the most visual medium in existence today. Where else can you see 5,000 albino rhinos stampeding down Pennsylvania Avenue? Only on the radio. Radio is visual, and as you plan for it, as you create for it, as you write for it, think of it visually."[2]

This visual imagery through solo sound is not, however, easy for a copywriter to capture. Everything the conceptualizer wants to say must be wholly transportable via words, sound effects, and music. There is no camera lens to convey automatically what the product looks like and how it works; the radio copywriter must explain these things through audio-only sensations. Because of this challenge, the creation of radio messages "offers something unique in this day and age," points out advertising executive Ed McCabe, "the opportunity to stand up and be counted, which is why radio can be a scary medium for creative people who aren't so sure of themselves. . . . Radio separates the doers from the talkers."[3]

These doers, the true professional radio copywriters, realize that the audio medium requires the same respect for language as does print, but a respect that must be rationed into segments of a minute or less. Also unlike print, the radio message cannot be reread by the audience. Nor can listeners slow down the speed at which they consume it. A radio spot exists in real time at one predetermined velocity and then disappears until the next occasion it is scheduled for airing. If the listener doesn't comprehend it the first time, there may not be another chance.

Yet, despite all these drawbacks, a copywriter can make the radio message more integral to a consumer's life than expressions via any other mass medium. This is because appropriate sound cues from the creator can trigger target listeners to recall specifics from their own past experiences or future dreams in order to complete the picture. The well-fashioned radio spot thus becomes a part of the listener, and the listener becomes a part of the spot. Note how the commercial in figure 2-1, written by Don Poole, brings skiers' excitement and fears to vibrant, pictorial life.

> **(MUSIC: 'BLUE DANUBE WALTZ' UNDER)**
>
> **ANNCR:** Imagine. It's a clear, cool January morning and the fresh snow spreads out in front of you sparkling in the bright sun. You tighten your boots and pull your goggles down over your eyes. You turn your skis downhill. And you go.
>
> You're skiing, and you look great.
>
> Then one ski crosses over the other. And gracefully you plant your head in the deep powder and all you can see is white. Then you see blue sky. Then white snow again. Then sky, snow. You're tumbling.
>
> Your skis are gone now. So are your poles. Your goggles are down around your neck and your neck is down around your ankles. And it's getting worse. You're picking up speed. Under the lift now, you hear the laughter and you wish that you had come up to Loveland early this season and tuned up your technique.
>
> You kick yourself because you could have skied Loveland for only $15 before December 24th. But you didn't come up to Loveland early, and just look at you now.

Figure 2-1. Exploiting radio's visual potential. *Courtesy of Don Poole, Karsh and Hagan Advertising, Inc.*

When a copywriter is given a television assignment, of course, a somewhat different set of opportunities and challenges presents itself. On the plus side, in proportion to the number of words used, television copywriters probably earn more money per word than do writers for any other medium, even when the vital but unspoken words expended on a script's video explanations are figured into the computation. More importantly, from a psychological standpoint, copywriters in TV advertising experience the fun of dealing with a communications vehicle that is more flexible than any other channel human beings have contrived. And a vehicle that is becoming *more* flexible all the time. The danger, however, is that the writer is so enthralled by television's technical possibilities that the resulting ads are too diffuse. The creator tries to show/present too much in a single spot.

Keeping the message clear and objective-oriented is always a prime communicator concern. This concern is further amplified in video given the multitude of available tools but the limited amount of airtime that any video spot has in which to use them. Successful television copywriters always remind themselves that, no matter how state-of-the-art the technique, it is irrelevant if it does not promote the one central point of the message. As veteran TV conceptualizer Jim Dale cautions,

Conceptual Functions

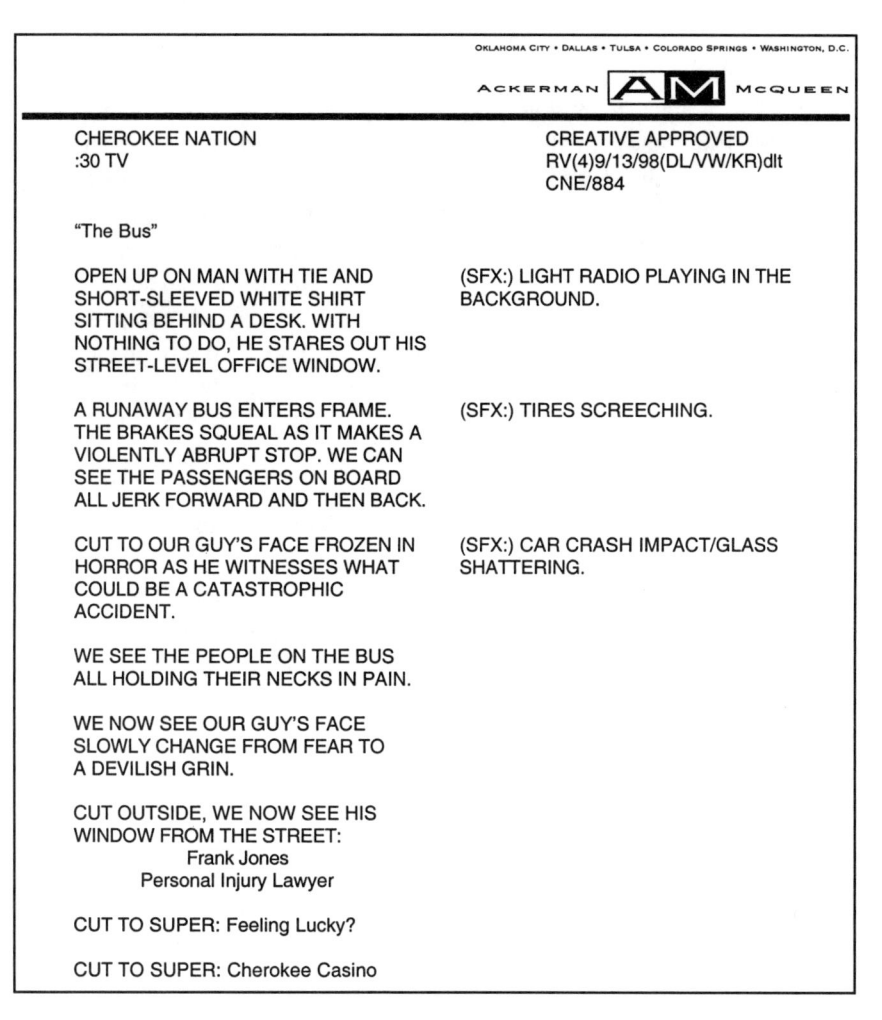

```
                              OKLAHOMA CITY • DALLAS • TULSA • COLORADO SPRINGS • WASHINGTON, D.C.
                              ACKERMAN  [AM]  McQUEEN

CHEROKEE NATION              CREATIVE APPROVED
:30 TV                       RV(4)9/13/98(DL/VW/KR)dlt
                             CNE/884

"The Bus"

OPEN UP ON MAN WITH TIE AND         (SFX:) LIGHT RADIO PLAYING IN THE
SHORT-SLEEVED WHITE SHIRT           BACKGROUND.
SITTING BEHIND A DESK. WITH
NOTHING TO DO, HE STARES OUT HIS
STREET-LEVEL OFFICE WINDOW.

A RUNAWAY BUS ENTERS FRAME.         (SFX:) TIRES SCREECHING.
THE BRAKES SQUEAL AS IT MAKES A
VIOLENTLY ABRUPT STOP. WE CAN
SEE THE PASSENGERS ON BOARD
ALL JERK FORWARD AND THEN BACK.

CUT TO OUR GUY'S FACE FROZEN IN     (SFX:) CAR CRASH IMPACT/GLASS
HORROR AS HE WITNESSES WHAT         SHATTERING.
COULD BE A CATASTROPHIC
ACCIDENT.

WE SEE THE PEOPLE ON THE BUS
ALL HOLDING THEIR NECKS IN PAIN.

WE NOW SEE OUR GUY'S FACE
SLOWLY CHANGE FROM FEAR TO
A DEVILISH GRIN.

CUT OUTSIDE, WE NOW SEE HIS
WINDOW FROM THE STREET:
              Frank Jones
          Personal Injury Lawyer

CUT TO SUPER: Feeling Lucky?

CUT TO SUPER: Cherokee Casino
```

Figure 2-2. The importance of verbal description in a video script. *Courtesy of David Lipson, Ackerman McQueen.*

It all keeps coming back to the idea. Being forced to have one. And having everything riding on the idea being great. When you have it, there's always a way to bring it off. When you don't have it, all the techniques and tricks and hot songs and New Wave and blah blah blah aren't enough. And the funny thing is, when you do have a strong idea, finding a way to pull it off isn't even as hard as you thought anyway. Because people *want* to work on a great idea. They can smell it on a storyboard or a script and they start pulling rabbits out of hats from the beginning.[4]

Simple or sophisticated, a television announcement generally must *be* visual in order to attract and hold viewer attention. Therefore, copywriters usually find themselves writing the pictorial descriptions first. They let the sense of sight carry as much of the revelation task as possible and then supplement as needed with music, sound effects, and spoken words. In conceiving television impressions, the best and most extensive writing is not heard by the audience but instead will be seen through the produced result of the verbal video design. In the ultimate extension of this principle, notice in figure 2-2 how many vibrant words have been used to set down the concept for a thirty-second commercial promoting Cherokee Casino. Yet none of these words are actually spoken.

This necessity of conceptualizing pictures before words is a task shared by television copywriters and newswriters alike. This is because both journalists and commercial creators are inherently storytellers. And "in doing a television story," declared longtime news correspondent Charles Kuralt, "[y]ou must always know what picture it is you're writing to. That is, you never write a sentence without knowing exactly what the picture is going to be. I will say that again. My philosophy is that you must always know what picture accompanies the words you write. You cannot write for television without knowing what the picture is."[5]

The next time you sit down to watch television, try turning off the sound of the commercials and news items to which you are exposed. How many of the messages compellingly convey the main point with the visual alone? And of these, how many tempt you to turn up the audio to acquire supporting information? The results of this little experiment should help you see just how important pictorial factors have become to today's messages and the writers who help sculpt them.

Copywriting today can be a high-stress occupation—particularly in large agencies where the flat 15% commission structure clients formerly paid has been reduced or replaced altogether by more stingy fee-for-service arrangements. "My sense is that one thing that happened as the 15% commission started to go away is that agencies began to cut back on the number of people in the creative department," observes veteran advertising executive Robert Levinson. "And this meant that there wasn't as much time to work on a campaign. . . . That brings along with it a lot of changes. And everyone as far as I can tell is kind of stressed out in the creative departments at big agencies."[6]

Still, copywriters working for large shops, small shops, and as independent freelancers can derive great satisfaction in seeing the messages they fashion receive wide exposure. Sometimes, the lines they pen become part of the national vernacular, like "Where's the beef?" "Can you hear me now?" and "When you care enough to send the very best." Even local spots can assume great prominence in elevating local businesses, events, and personalities. In short, the small number of words the copywriter is allowed can become inordinately important. "But here's the thing," warns Los Angeles creative director Sally Hogshead. "The process of creating brilliant ads isn't glamorous. It's sitting down and working. Hard. For a long time. Long after the night crew has stopped vacuuming the hallways. . . . Talent can get you a great ad. Work ethic can get you a great career."[7]

Art Directors and Designers

Electronic media writers compose verbal descriptions of needed visuals, but the actual realization of these visuals depends on the expertise of other professionals. In the case of commercials and the other short messages we call *continuity,* this pictorial expertise is provided by *art directors*. The art director possesses the graphic and layout skill necessary to translate the copywriter's words and concepts into finished products suitable for showing to a client or to the production people responsible for capturing the message on video or film.

In many instances, the copywriter comes to the project first and carves out the general idea, which is then refined in collaboration with the art director partner. At other times, however, the pair starts the assignment together, contributing more or less equally to the commercial's formulation. In either case, the art director almost always then follows the project into production to make certain that the specified look and design of the spot are adhered to in realizing the finished message.

Most often, copywriter/art director television teamwork is directed toward the creation of a *storyboard,* a sequence of selected sketched stills that helps keep the message visually oriented. An example board appears in figure 2-3. With paired visual and audio blocs, each storyboard frame or

OPEN ON THE CAMERA FLYING THROUGH THE CLOUDS.

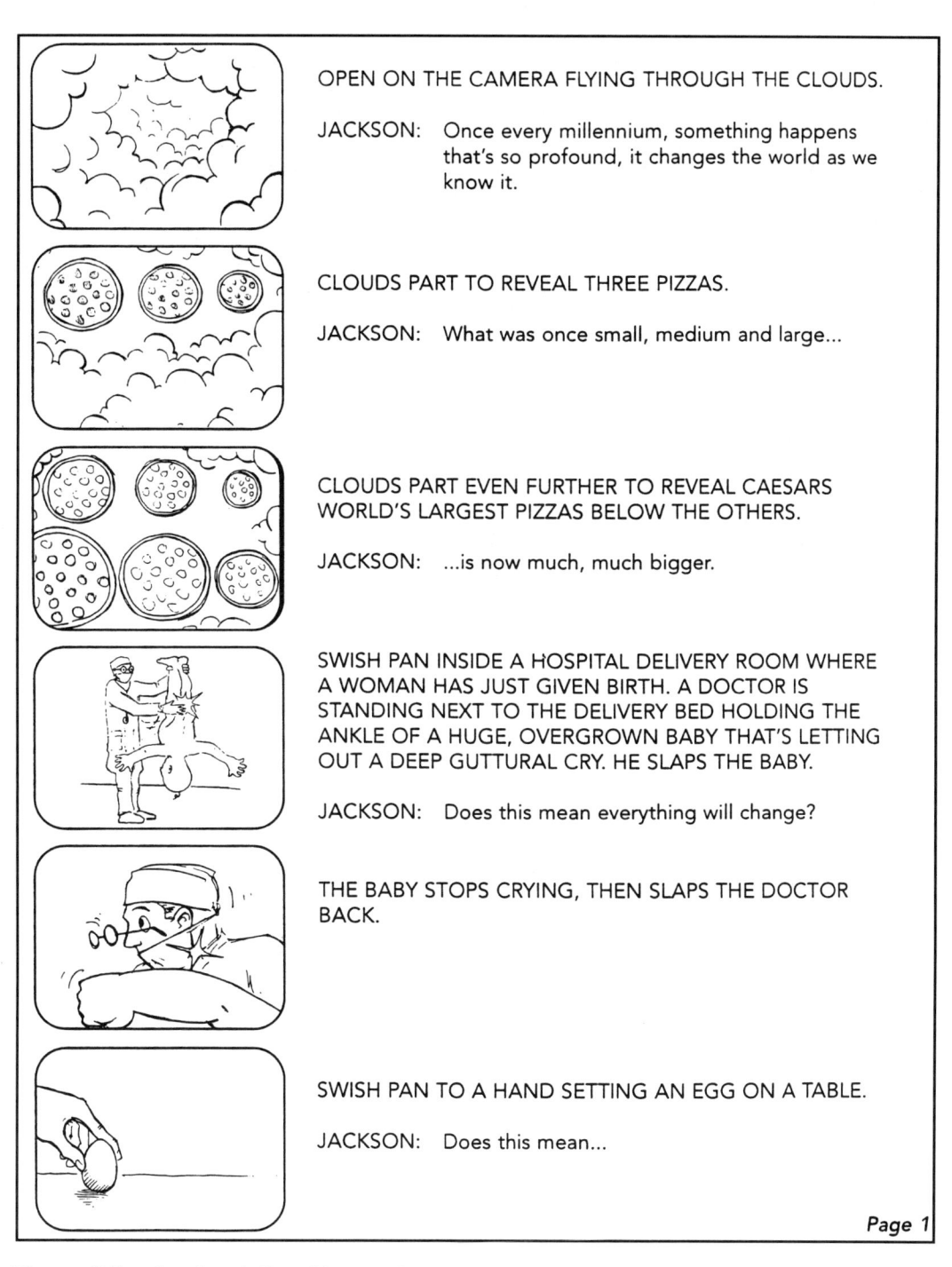

JACKSON: Once every millennium, something happens that's so profound, it changes the world as we know it.

CLOUDS PART TO REVEAL THREE PIZZAS.

JACKSON: What was once small, medium and large...

CLOUDS PART EVEN FURTHER TO REVEAL CAESARS WORLD'S LARGEST PIZZAS BELOW THE OTHERS.

JACKSON: ...is now much, much bigger.

SWISH PAN INSIDE A HOSPITAL DELIVERY ROOM WHERE A WOMAN HAS JUST GIVEN BIRTH. A DOCTOR IS STANDING NEXT TO THE DELIVERY BED HOLDING THE ANKLE OF A HUGE, OVERGROWN BABY THAT'S LETTING OUT A DEEP GUTTURAL CRY. HE SLAPS THE BABY.

JACKSON: Does this mean everything will change?

THE BABY STOPS CRYING, THEN SLAPS THE DOCTOR BACK.

SWISH PAN TO A HAND SETTING AN EGG ON A TABLE.

JACKSON: Does this mean...

Page 1

Figure 2-3. Storyboard (first of 3 pages) for a 30-second spot. *Courtesy of Catherine Abate, Cliff Freeman and Partners.*

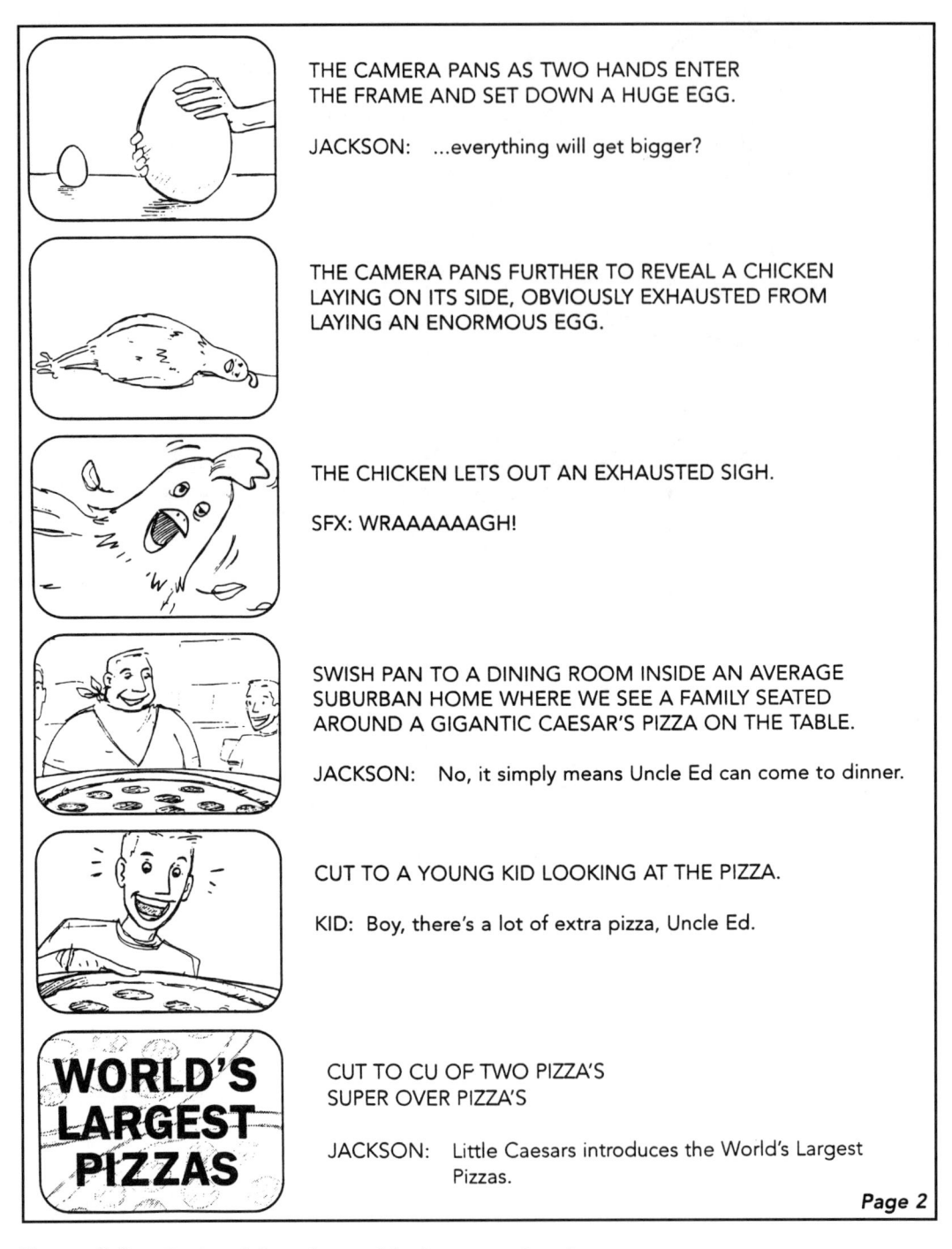

THE CAMERA PANS AS TWO HANDS ENTER THE FRAME AND SET DOWN A HUGE EGG.

JACKSON: ...everything will get bigger?

THE CAMERA PANS FURTHER TO REVEAL A CHICKEN LAYING ON ITS SIDE, OBVIOUSLY EXHAUSTED FROM LAYING AN ENORMOUS EGG.

THE CHICKEN LETS OUT AN EXHAUSTED SIGH.

SFX: WRAAAAAAGH!

SWISH PAN TO A DINING ROOM INSIDE AN AVERAGE SUBURBAN HOME WHERE WE SEE A FAMILY SEATED AROUND A GIGANTIC CAESAR'S PIZZA ON THE TABLE.

JACKSON: No, it simply means Uncle Ed can come to dinner.

CUT TO A YOUNG KID LOOKING AT THE PIZZA.

KID: Boy, there's a lot of extra pizza, Uncle Ed.

CUT TO CU OF TWO PIZZA'S
SUPER OVER PIZZA'S

JACKSON: Little Caesars introduces the World's Largest Pizzas.

Page 2

Figure 2-3. *Continued.* Second page of the 3-page storyboard.

38

Conceptual Functions

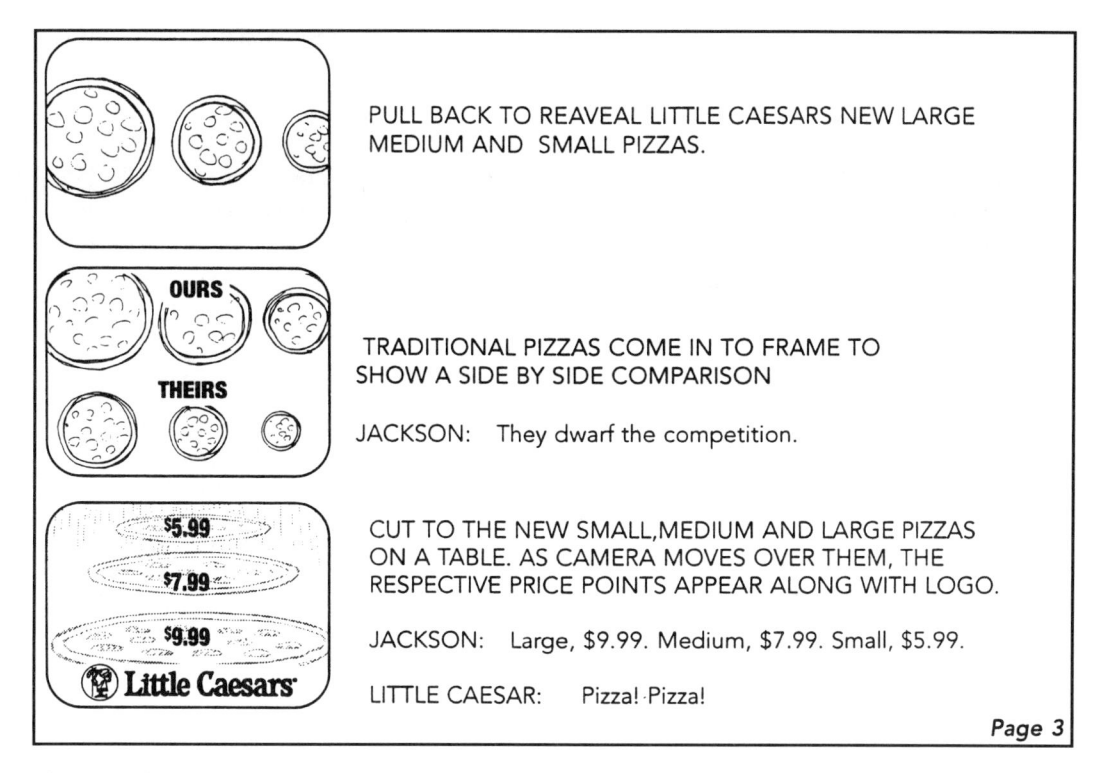

[Storyboard panel 1]
PULL BACK TO REAVEAL LITTLE CAESARS NEW LARGE MEDIUM AND SMALL PIZZAS.

[Storyboard panel 2 — OURS / THEIRS]
TRADITIONAL PIZZAS COME IN TO FRAME TO SHOW A SIDE BY SIDE COMPARISON

JACKSON: They dwarf the competition.

[Storyboard panel 3 — $5.99 / $7.99 / $9.99 / Little Caesars]
CUT TO THE NEW SMALL, MEDIUM AND LARGE PIZZAS ON A TABLE. AS CAMERA MOVES OVER THEM, THE RESPECTIVE PRICE POINTS APPEAR ALONG WITH LOGO.

JACKSON: Large, $9.99. Medium, $7.99. Small, $5.99.

LITTLE CAESAR: Pizza! Pizza!

Page 3

Figure 2-3. *Continued.* Last page of the 3-page storyboard.

panel strives to illustrate the sequence of pictorial action, the camera settings, angles, and optical effects desired, and the dialogue, music, and sound effects being considered for use.

As it is being prepared, the storyboard provides the art director with the opportunity for effective interaction with both the visual concept and the writer who is helping to define it. This interaction can be most beneficial when the art director is not afraid to make suggestions about the spot's soundtrack and reciprocally encourages visual recommendations from the copywriter. Art directors have been known occasionally to come up with a better word or phrase than the writer had originally captured. And copywriters, in their own scattered moments of pictorial insight, do stumble on more compelling visual ideas than the artist at first had in mind. Because the professional well-being of both depends on their *collective* ability to derive a successful audio/video communication, the art director and the copywriter each has something to gain from pooling rather than departmentalizing their brilliance.

Today, most successful art directors understand that for a concept to stay on track, continual dialogue with their wordsmith partner must be maintained. "Traditionally," laments artist Alexander Molinello, "art directors or illustrators take copy and run away with it and then come back with something that the writer never saw that way."[8] When the artist physically works with the copywriter on the board, however, the result is much more likely to be a unified idea rather than two mutually exclusive ones.

The television art director's first quick sketches, suitable only for showing to his or her creative teammate, constitute what is referred to as a *rough* or *loose* storyboard. Later versions that are polished enough to exhibit to the client are called *refined, tight,* or *comprehensive* boards. In short-deadline situations or low-budget projects, however, it is not uncommon to proceed directly from a *rough* into actual production, provided that the rough is definitive enough to secure executive approval.

With modern computer technology, the distinction between rough and refined boards continues to narrow. Personal computers and inexpensive software programs now allow art directors to expand, contract, shift, and otherwise modify their frames on a video monitor before printing out their best effort. Typeface and size can be chosen and experimented with at will. When the client wants changes, the stored image can be retrieved, adjusted, and reshaped accordingly. Such capabilities can cut the computer-literate art director's time expenditure by at least one-third, with an even greater proportional saving in art materials and photoprocessing costs. Perhaps even more importantly, the advent of computer-assisted storyboards permits the artist to sit down at the screen with the copywriter and to manipulate images together—before anything gets locked into uncompromising ink.

The graphic computer does have its downside, however. For its very ease and flexibility can engender visual ideas that are more imitative than communicative. As international award-winning art director Francesc Petit warns,

> The facility and speed provided by the machine lead the professionals in a very dangerous direction—they cease to think. Many junior, and also senior art directors, instead of looking for an image that is really new—an invention—now leaf through magazines, look for an appropriate image for their purpose, scan [it]—and presto, in a few minutes, there it is, apparently a beautiful ad! . . . We shall spend some years seeing these horrors that advertising has been producing lately, copies of copies, until the clients discover that they are being cheated with false creative work.[9]

Overreliance on software wizardry is a danger faced not only by advertising agency art directors but by artists and designers working in other occupational settings throughout our industry.

Among these other settings are very similar jobs available within television stations, production houses, and cable and broadcast networks. But the job title changes slightly to *graphic artist, design artist,* or *scenic designer*. Simply put, the graphics or art department in a media outlet is responsible for coordinating the overall look and visual image of that outlet. "Every job in the electronic media somehow relates to image," wrties Lynne Grasz, former head of the Broadcast Designers' Association. "Whether it is on the management or technical side, your image is your bread and butter. . . . No one in electronic media is more crucial to image than the image makers themselves—the electronic media designers."[10]

Graphics used on air must be tied in unmistakably with those printed on sales brochures and corporate stationery and emblazoned on station buildings, vehicles, and cameras. Most all of this emanates from the station's or network's *logo*, the memorable insignia that should instantaneously characterize the media facility in a distinctive and high-profile manner. Once the logo has been derived, graphic artists carefully oversee its use and reproduction so that the emblem remains consistent and uncluttered no matter what the application. Examples of such radio and television station logos are included as figures 2-4 and 2-5.

For their part, *scenic designers* translate the logo and all other elements of line, shape, and color into the coherent formulation of the studio sets on which the cameras focus. In production companies, this job also extends to the dressing of outdoor sets and the selection of shooting locations. The creation of series programming and movies further demands the talents of wardrobe and makeup artists. Even though such jobs are mainly limited to program production companies, makeup and similar talent-appearance specialists are also employed by major stations and the networks to enhance the visual appeal of news and other on-air personalities. "Nowadays celebrities favor their

Conceptual Functions

Marian Lipow
*President and Creative Director,
Lipow/Stoner Design, Inc.*

At three years old, I was caught drawing a sizable mural on the wall of my bedroom. My parents got the message, purchased me a sketchpad and supplies, and started me on my way. The end result has been that I love my work and I take solace in the fact that I make my living as a creative entity.

After art school, I came back to New York. Like most new graduates, I found work as a print designer (for *Penthouse* magazine of all places—and weren't my parents proud!). Over the course of the years, I transitioned from print to broadcast design and from designer to creative director. Now I'm a principal in my own design company.

I've been fortunate. I've worked with, and for, most of the leading television and multimedia companies in the world. I've worked on some amazing projects and collaborated with genius from time to time. That's inspiring.

I see my job as creating concept-driven design solutions for challenges that are set before me. These range from designing complete show packages, to live-action shoots, to concept development for documentaries and show content, to the on-air promo telling you about what's next at 8:00 p.m.

Design is, and always will be, about telling a good story.

I was extremely fortunate because Tyler School of Art in Philadelphia prepared me exceptionally well. What I wasn't prepared for was the need to understand and speak the language of business in addition to the language of art. So I recommend taking some business classes. I wish I had. If art directors and designers are to take a seat at the table of business, you'll need to learn the language of business along with honing your creative expertise.

There are a few key tenets that I believe art directors should follow:

- Do the hard part first. Understand the core promise of the brand, product, or service you're working on. In the case of television outlets, the personality of the brand begins with a programming promise and an on-air design.
- Don't be the flavor of the month where it is often style over substance and form over function. Following trends is imitation—not innovation.
- Tell a good story. All elements of a design should serve the story you're trying to convey.
- Be flexible. It allows you and your concepts to grow over time.
- Be inspired. Take time to recharge your creativity so that the blank piece of paper you face can produce unexpected and inspired results.
- It may sound a bit harsh, but the reality of being an art director or designer is being able to strike a balance between your creativity and serving your clients' business needs. As Michael Gass, one of my mentors, always told me: "It's wonderful to see it as art, but it's still a business, so get over it, Sadie!"

I have several mantras, but my favorite is: *Substance over style, function over form*. If it doesn't have a concept, then it doesn't have a soul. If you want a career in design, follow your passions, enjoy the ride wherever it leads you, and have an amazing life.

Figure 2-4. A format-signifying radio station logo. *Courtesy of Austin Hale, KYYK/KNET.*

Figure 2-5. Locale-inspired TV station logo. *Courtesy of Julie Miyagi, KITV-4.*

personal stylists," observes Hollywood reporter Larry Gerber, "while a diverse collection of freelancers scramble for jobs. They range from artistic beauty specialists to guts and gore gangs who do prosthetics."[11]

Like their art director counterparts in the advertising agencies, all of these graphics, design, and stylist experts play a major role in determining the visual values and consequent impact that programs, commercials, and other forms of continuity (such as station promos and IDs) have on the viewing audience. Even a seemingly static news or interview set depends on such artists for the appropriate conveyance of mood and character. And, though radio may initially be a sightless medium, it, too, relies on artists to extend and magnify its image via such devices as logo development, print piece design, and remote unit decoration.

In the completion of this range of assignments, today's artistic implements, as previously mentioned, are likely to be computer driven. As inventoried by video designer/developer Damon Rarey, these include tools like

> a frozen frame of video, a magazine picture captured with a video camera, electronically created type from a character generator, shapes created free-hand by the artist with an electronic stylus and tablet, and an array of graphical effects, such as edging, embossing, gradations, tiling, air-brushing, grids, curves, circles. . . . Still, the qualities most sought after by people hiring in the field have not changed: basic design talent and an enjoyment of whatever medium is being used.[12]

Unlike air personalities, designers are not themselves in the electronic media's public eye, but these conceptualizers are key determiners of what engages that public eye's attention.

Program Writers

Having just discussed the role of artists in entertainment program production, we now turn our attention to the professionals who script such programs.

In the United States, most television entertainment programming is created in New York or Los Angeles. A significant number of daytime soap operas are produced on the East Coast, while the rest, together with most game shows, situation comedies, dramas, and action/adventure series emanate from the West. Because the production of television entertainment for national network or first-run syndication distribution is an immensely expensive enterprise, a relatively small number of companies compete in this phase of our industry. This also means there is a relatively small number of jobs for program writers, at least compared with the thousands of copywriters who make a living at or through advertising agencies and local stations.

Conceptual Functions

Scriptwriters (also called screenwriters, staff writers, or just plain writers) are, of course, responsible for coming up with the words and actions that on-screen talent conveys. (We used the term *on-screen* because feature-length *radio* dramatic programming in the United States is, unlike the case in most of the rest of the world, virtually extinct.) Many of these scriptwriters are freelancers, preparing one-shot projects for various studios and producers. These assignments might range from creating an additional episode for a just-extended series to a movie screenplay to an industrial promotion or training film. When these writers receive up-front approval and a guaranteed pay for the project regardless of whether it is eventually produced, they are technically known as *contract* or *commissioned* writers.

Staff writers, on the other hand, are ongoing employees of the studio, production company, or network. Often, their employment is contingent on the continuation of the series on which they are working, but the more well established among them may have negotiated a longer-term arrangement through which they are shifted to other projects within the company if their current show is terminated. For further job security and creative control, the most successful scriptwriters eventually become writer/producers or writer/directors, thereby acquiring a greater role and financial interest in the entertainment properties they are developing.

To help them secure equitable salaries, contractual protections, information on new project opportunities, and insurance and pension coverage, electronic media scriptwriters join the Writers Guild of America (WGA), which maintains major offices in New York and Los Angeles. The WGA also helps its members locate reputable *agents,* those important intermediaries who help match their writer-clients with producers in search of scripts or scriptwriter services. As is the case of the performers' agents discussed in chapter 1, agents servicing writers are particularly important in the earlier stages of their clients' careers. Once a writer has put together a string of creative successes, the agent's main function may turn to analyzing and sweetening project offers the writer receives from competing program developers.

In fact, recent trends in the industry have bolstered the fortunes of a small number of star writers and lessened the prospects for everybody else. Trying to protect themselves against flops, today's studio executives are more comfortable paying a "name" writer more than double what a veteran but less well-known scripter might command. "They think an expensive writer will get it right the first time," explains one writer/producer. "And if he doesn't, the executive has protected himself by using a pre-approved writer."[13] It is not that studio and network executives don't appreciate the importance of writing, but that they appreciate it so much they are unwilling to entrust it to unknown hands. "Stars may bring viewers to the stable," affirmed top CBS executive Les Moonves, "but only quality [and] consistent writing will keep them there."[14]

There has never been a surefire path to a thriving, or even marginal, scriptwriting career. Looking back, Stephen Cannell, who created such shows as *The A-Team, Wiseguy, 21 Jump Street,* and *Hunter,* was an undiagnosed dyslexic who successively flunked out of three private schools. After college, however, Cannell's imaginative talent and urge to write would not let him be satisfied working in his father's interior design business. A trade publication interview reveals:

> Determined to become a writer, Cannell sat himself in front of a typewriter every evening after work, including Saturdays, and pounded out stories. Sometimes he would collaborate with a college fraternity brother while their wives played cards. . . . After four years of writing on the side Cannell sold his first script—for $5,000—for an episode of *It Takes a Thief*. That gave him the nerve to tell his father he wanted out of interior design. . . .
>
> In the next six months Cannell got assignments for two scripts, one for *Mission Impossible* and

the other for *Ironside,* but he was nearly broke. Rescue came from a source that could qualify as a Cannell plot twist. His house was robbed of the Cannell wedding silver. The insurance brought him $8,000, enough to carry him until payment for his scripts came in.

Cannell's first big break came when he sold a script for *Adam 12* and was hired as story editor on the series. That led to a writing contract at Universal. He began furiously churning out scripts—20 in his first year under contract—and it wasn't long before Universal had him developing and creating his own series. *Rockford Files* was created in 1973, followed by *Baretta* in 1975.[15]

In contrast, Hugh Wilson, who gave birth to *WKRP in Cincinnati, Frank's Place,* and *The Famous Teddy Z,* first tried to break into advertising in New York City after earning a journalism degree. When job hunting got him nowhere, he was forced to become a shipping clerk at the Armstrong Cork Company. There he met two other young clerks, Jay Tarses and Tom Patchett, who were trying to build careers as stand-up comedians. Years later, after working his way up from copywriter to president of a small Atlanta advertising agency, Wilson founded his own commercial production company. On a 1975 trip to Los Angeles to shoot a commercial, he visited his old friends Tarses and Patchett, who were now writers for MTM Enterprises. They helped Wilson secure employment as an apprentice director, but he spent whatever time he could composing unsolicited scripts. A year later, he was a regular writer of episodes for the original *Bob Newhart Show,* and a year after that he created *WKRP in Cincinnati,* which ran for four years on CBS.

Unfortunately, CBS moved the series around in the schedule eighteen times. "Strangely enough," recalls Wilson, "the week they announced the cancellation, we were ninth. I used to joke if your mother doesn't know when you're on, nobody does. The happy ending is that it went into syndication and far exceeded many times over anybody's expectations."[16]

Ronni Kern
Long-form Television Writer

To start with the truth: no one sets out to make a career writing TV movies. If you dream in two-hour talkies, you aim at the big screen. If you're wedded to writing for TV, you focus on episode, where all the money and network energy is. Given that regular network movie nights are an endangered species being only partially superseded by a Balkanized landscape of niche networks making two or three movies apiece, it is hard to believe anyone wakes up in the morning thinking, "Gee, I want to write long-form." So why do I wake up every morning (okay, most mornings) deliriously happy that that's exactly what I do?

First of all because it is a constantly changing canvas and I love learning new things. Another way to put that is that I'm a voracious student with a very short attention span. At the moment, my office floor is covered with thousands of pages of legal documents and news articles. My hard drive is filled with thirty hours of interviews from a (very hot) week in west Texas. All for one two-hour movie about Tulia, Texas, and the miscarriage of justice there.

On other projects I have spent months learning about: shortwave radio operations in the thirties, translating Old French to decipher tenth-century documents on the real Queen Guinevere (priestess warrior and childhood friend of Lancelot, not the hotty invented by a twelfth-century Frenchman to titillate the ladies of the court), search and rescue tracking in the California desert, and enough Hebrew to write about the women of the Bible without thousands of years of "interpretation" getting in the way.

And while I may be an extreme case and the despair of agents who would prefer to pigeonhole me, the fact is that the vast majority of TV movies are fact based. Most involve research. Most involve conveying that research in an accurate and (hopefully) engaging way in eighty-seven minutes punctuated (or punctured) by six or seven commercials. And most involve the legal documentation of every line and every action. If you thought footnotes were only for academia, think again.

But in compensation there is the second-best part of writing TV movies: the amazing people you meet. Napoleon and Churchill and Hitler may get their time slots, but by and large TV movies honor the small heroes: the men and women who grapple with life on pretty much the same level as we do and yet somehow do astonishingly more.

I am grateful to have been taught to look at the world through the eyes of the homeless teenager who made her way from the Bronx to Harvard; the Boston housewife who could look beyond her rape to embrace wholeheartedly the child it produced. I have written many fiction scripts and in all honesty prefer that sort of writing. But I will never regret that writing TV movies demands that you get out into the world.

And then, of course, deal with the reality of getting that world on screen. If you're lucky you only have one producer, one development exec, ingenious stars, and a respectful director. If you're not lucky you'll have a phalanx of quarreling producers and stars who scream at you for daring to put words into their mouths. It helps to have the skin of a rhinoceros and the diplomatic skills of Henry Kissinger. It also helps to develop the flexibility to rewrite summer California for winter Halifax without resentment and on the spot. But compared to feature films where you can languish in "development hell" for years and still never get into production, the development/production ratio in television is very high. And writers, at least occasionally, are listened to. Who could ask for a more interesting life than that?

As Wilson's background demonstrates, copywriting can be a valuable stepping-stone to a program-scripting career. Former copywriters and current successful screenwriters Tom Parker and Jim Jennewein testify that

> advertising was a terrific training ground. There are a lot of similarities. As copywriters, we were always working on deadlines and that's helped a lot because on a movie, between rewrites and polishes, your life is nothing but deadlines. . . . In advertising, you learn how to present your ideas to the creative director, then to the company you're creating the advertising for. Today, we'll get a script to rewrite. We have to figure out what we're going to do with it, then go back and pitch it back to them in 15 or 20 minutes. That's all the time we have to convince them.[17]

Although Parker and Jennewein are speaking of movie projects, the process for series pitching and creation is not very much different. "When you present a drama series to a network," reveals veteran writer/producer Josh Brand, "they always want to know what's the franchise. What they mean by that is what is going to generate stories over four or five years. They understand if you say a medical school or a law school or a police station. For them stories aren't generated from character. They come from a profession."[18] That Brand's own cocreation, *Northern Exposure,* was able to place more prominence on character was, as he recognized, a rare exception to the norm.

For those seeking employment in series or feature-length writing despite such creative limitations, there is a self-directed apprenticeship that must be served regardless of previous background. "You must start with the drudgery of episodic writing to break into the business," maintains studio executive Peter Roth. He and many of his colleagues advocate authoring episodes for existing series to develop an understanding of the format. These probably won't be purchased by the shows for which they were written but will demonstrate a mastery of technique and dramatic insight. Good scripts can only come from "the writer's honesty of experience and passion of their belief," Roth adds. Thus, even for many of the scripters he employed while at Cannell Productions, "we go through five-hour therapy sessions, probing their life to see or find something they can use."[19]

Once a writer has attained success, this occupation can still be frustrating. For "the studio people know that the cheapest fixes you can make are at the writing stage—before actors, locations, and sets—so as they dither," complains industry critic Richard Schickel, "they drain the script of its freshness, often enough of the very impulse that attracted them to it in the first place. . . . So no matter how much experience and prestige a writer acquires, he remains essentially in the same position as a first-time writer—infinitely rewritable, instantly dismissable, his only reliable reward (or consolation) being the check."[20] "Writing is absolutely as torturous as ever, word by word, page by page," adds now-successful scriptwriter Ron Bass. "You want the studio, the actress and the director to love what you've written and it kills you when they don't. . . . It's still the same thrill when it works and you watch an audience enjoy something you had a hand in. But it does not get any easier." [21]

Development Managers

Depending on whom one talks to, development managers either assist or inhibit scriptwriters in the accomplishment of their conceptual tasks. Development people are those officials at studios, networks, and independent production companies who are charged with evaluating and sculpting potential projects. They make pitches and recommendations to programming heads as to which of these projects should be "green lighted" for actual production. This decision involves a commitment of multiple millions of dollars, with the executives making this decision looking for the best chance of a return on investment. In-house development staff then work with the actual producers and writers of approved shows in shaping the property to meet what they see as the specific needs of their network or studio.

Networks have specific "holes" to fill in their schedules. They seek shows best calculated to fill these holes by delivering a desired target audience and facilitating flow with programs that will come before and after the new show under consideration. But many projects are pitched without a specific slot in mind. Instead, these proposals have been packaged around particular performers and/or writer/producers, with the appeal of these "name players" constituting the project's driving force. Major talent agencies are usually involved in the creation of such packages—either initiating the project themselves from the clients they represent or putting the vehicle together in response to a request from a network or other media distributor. "A concept alone is rarely enough to make a sale," says the Fox network's former vice president of series development, Danielle Claman, "but a seemingly mundane concept in the hands of the right writer can make a series jump off the screen."[22] Sometimes development staff at the studios assemble proposals for the consideration of their development counterparts at the network. As all the major networks are now part of corporations that also own studios, this process can increasingly become a totally in-house activity.

Because both developers and writers are involved in the conceptual process, there need not inevitably be conflict between them. Skilled development professionals bring essential marketplace perspectives to the creative process, perspectives that increase the chances the show will be a monetary success for all involved. As professor Eileen Meehan once observed, "television is not reducible solely to manufacture nor to artifact. Rather, television is a complex combination of in-

David Salzman
*Development Executive and
Executive Producer,
MAD TV, Los Angeles*

Television program development has undergone significant change in the past few years though the fundamental construct is the same. There are buyers (broadcast networks, cable channels, satcasters [DBS companies transmitting direct to consumers], and other distributors such as syndicators) and sellers (program suppliers such as independent producers, production companies, and major studios).

In the 1980s the sellers would develop program ideas and verbally pitch them to the buyers. That changed in the 1990s as the financial interest and syndication rules (which prohibited networks from owning and selling the shows they carried) lamentably sunsetted. Networks and/or their parent companies now were allowed to compete with other sellers of programming. The result is analogous to the owner of a vital bridge securing the unbridled right to install tollbooths and, instead of charging a fee, taking ownership, partial or full, of each vehicle using that bridge. This reduced independent producers to virtual tenant farmer status. It resulted in decreased creative freedom and program diversity as well as egregiously unfavorable terms for producers.

In the 1980s, if one developed and produced a successful network program, the rewards were commensurate with the show's level of performance. Home-run ratings brought home-run profits to the developer-producer-owner. Today, networks often command full ownership and copyright control, the producer's fees are lower, rights to renegotiate terms are nonexistent no matter how successful or long running the show is, and ultimate creative and financial control rests with the buyer. Worst of all, a studio developer can be released from his/her own series without cause. This means if a network/channel has a creative difference of opinion or simply wants to make a change, the developer/producer is dumped from the program he/she birthed.

The above background information is important if you are to understand the new development process.

The process begins with an idea for a program which is fleshed out by writing a treatment. Most commonly, this takes the form of a 1–3 page narrative which explains the show concept and key elements. The treatment may include a format rundown which chronologically lists, segment by segment, each program element and its estimated duration. It may get into casting, point of view, look, cost, and other relevant variables. Generally it is a good idea to register one's treatment with the Writer's Guild of America (WGA), which can easily be done for a nominal fee.

A pivotal question to ask oneself about the program idea is: Who are the most likely buyers? Since there are so many distribution platforms, many highly targeted to affinity groups (especially thematic cable channels such as Discovery, MTV, Sci-Fi, and Lifetime), one has more prospective buyers with defined niche audiences than ever before. Back in the golden days (pre-1987) when there were only three broadcast networks and very few viable cable networks, young program developers were thrust into a narrow if not impenetrable content universe. Today the opportunity to break in is vastly improved, though the upside rewards are more limited.

Access is a key issue since it is difficult to get a meeting with a buyer unless you and your development work are well known and respected or you have a licensed agent or manager who is able to open doors. However, given available desktop technology, you have a leg up on your program-developing predecessors not only because the galaxy of buyers has expanded but also because of the technical capability to quickly and inexpensively mount PowerPoint presentations or miniproductions, which convey the flavor and thrust of your passion project.

Many developers look at the landscape to see which programs are garnering big audiences and what specific buyers seem to be interested in. Then they add an element or clever twist to what's working and try to sell the show, typically within a 15-minute verbal pitch that starts with the show title and one-line description. For example, let's say your show idea is a *Queer Eye for the Straight Guy* type of makeover or a talent competition like *American Idol,* where the ultimate champion outsurvives fifteen challengers who are eliminated by celebrity judges and Internet voting. While this copycat approach to program development may cause severe nausea, it occasionally works. Because most buyers are both risk averse and creatively unadventurous. Also, program development is a little like the game of telephone in that the executive you pitched relays his/her impression to a superior who then talks with the programming head who in turn confers with the president and others.

The fresher and more complicated your program idea is, the greater the chance for mistranslation and rejection. On the other hand, if you are trying to sell a Hispanic *Friends* set in San Antonio, little will be lost in translation. But the network's response will probably be a "pass" since the notion is fairly obvious. The network reaction may be: Why not go to the number-two writer on *Friends* and ask him/her to do this? Then we will own it.

Program development is easy in a superficial sense. But thinking through a show idea in depth, getting in front of a buyer, making that person "see your show" and want to buy it from you is not easy. Nor is the development process itself a breeze. It frequently involves pitching the buyer, securing a "development deal" (generally the ordering of a script, if it is a drama or sitcom, or a low-budget run-through or presentation if it is a game show, talk, or reality show), delivering the script or presentation to the buyer, getting a pilot order (production of an actual episode, often the "premise plot" episode), convincing the buyer to love it and order a number of initial episodes, then praying that the program scores well with viewers, executives, and advertisers.

Development can take a year or two from idea capture to getting on the air. However, if you believe that your idea is at least as good as *CSI, Everybody Loves Raymond,* or *Fear Factor,* the sirens of program development may possess Odyssean irresistibility.

---

dustry and artistry. . . . While we recognize that economics set the parameters, we must also recognize that television is a very peculiar sort of industry—a culture industry that reprocesses the symbolic 'stuff' from which dreams and ideologies are made."[23] Experienced development people can help writers to identify the promising "stuff" that needs enhancing in order to better attract target audiences and advertisers. Hopefully, these are elements already in the treatment that the writer has underexploited, not outside elements a development person tries to jam down the writer's throat. In other words, a development executive who works as collaborator instead of a dictator can be invaluable in helping scripters to unearth treasures they themselves have buried.

On the other hand, intrusive and insensitive development people have always been a writer's worst nightmare. Upon his 1994 induction into *Broadcasting & Cable's* Hall of Fame, successful situation comedy creator Garry Marshall (*The Odd Couple, Happy Days, Laverne and Shirley*) explicitly described this terror's variations: "People ask me the difference between doing television in the '70s compared to the '90s. I always had a nightmare in the '70s that I was doing The Last Supper as a movie of the week and the network would come to me and say, 'Judas isn't testing well. Could you give him a dog or something?' In the '90s as I direct The Last Supper, for financial reasons two of the disciples have Coca-Cola cans in front of them."[24]

The development despots of which Marshall dreamed are more nitpickers than conceptualizers, reducing the creative process to, in the words of legendary CBS commentator Eric Sevareid, "being nibbled to death by ducks." Similarly, *The West Wing*'s creator Aaron Sorkin recalls when, in a script

Conceptual Functions

from that show, "one of the characters mentioned something about Jewish slaves in Egypt 3,000 years ago. The network asked to see the research."[25]

The best development persons, be they at studios and networks or independent production companies, are neither nitpickers nor nibblers but rather conceptualizers in every sense of the term. They work hard to forge a clear vision for the project and help writers to polish their proposals and scripts in ways that seek to ensure that a show evolves into an economic and artistic success. Because this success is so elusive, development personnel can expect to change jobs and companies frequently.

Industrial Scripters

The importance of good scriptwriters is increasingly appreciated in industrial/corporate productions, which have a much richer tradition than most people realize. "In spite of what they taught you in film school," writer/producer Chris Jones comments, "the first talkie was *not* Al Jolson in *The Jazz Singer*—it was some Western Electric executives giving a plant tour in 1924." Jones further points out that the first industrial film was a black-and-white silent film made a hundred years ago by International Harvester. "The film was extolling the virtues of a horse-drawn thresher."[26] Clearly, electronic wordsmiths working in the industrial sector today are laboring in a mature field with well-established techniques and professional expectations.

The largest companies often employ scripters on their regular staffs on the assumption that full-time employees are in the best position to grasp the complexities of the business. Yet, even in these instances, an outside video writer may be retained to ensure that the content does not clog the program's communicability. "Creativity cannot be parceled out or pushed through amplifiers," warns

Kevin Campbell
*Senior Information Specialist,
Dow Corning Corporation*

Video, audio, computers, networks, streaming, virtual meetings, audio conferencing, video conferencing, large events, editing, budgets, graphics, long hours, constant skill building, problem solving, career changes, counseling, and the list goes on: just a few of the things you need to know as a corporate media writer.

It's a career filled with challenges, excitement, and constant change. You need to have a strong writing background, solid communications skills, creativity, a touch of media technology knowledge, and the ability to work with a variety of people who don't have any idea what you do. It's a very different writing environment than feature film, commercial TV, or even independent production. Because it's about the business. You're not there to write award-winning videos, training programs, or events. You're there to communicate concepts and ideas, act as a trainer, and support the objectives of the business.

This doesn't mean that in the corporate environment you'll never write or create award-winning programs. The corporate media sector is filled with some very creatively written productions. But in the corporate world you always have to remember the reason your skills are needed is to help the business be successful.

So, with that said, what's the work of a corporate media writer like? It's a dynamic world that can change as fast as the stock market. You'll work with many highly skilled individuals from diverse backgrounds, from the top of the company down, and from both inside and outside your organization. It's not just saying, "Let's make a videotape of this really exciting multimedia CD." It's about listening, understanding what your counterpart wants to accomplish, and then making a recommendation as to the best way to approach the subject. Sometimes it's even recommending a simple approach.

You can't have an ego in the world of corporate writing. If an executive is not comfortable with the way you wrote something, you'll most likely be making a change. And did I mention tight deadlines? It's not unusual to receive changes even when you are in the middle of shooting something—new information that must be included in addition to what you already wrote.

The other thing about writing for corporate media is legal review. Many times you will find yourself going through rewrites because the legal department objects to the wording. This is particularly true when writing about products and services.

So what are the skills that a writer would need to be successful in the corporate media environment? First and foremost is the ability to write clearly and succinctly. Everything from e-mail to streamed audio communications requires this skill.

But remember, writers who are really good with the written word are not always good at the spoken word. You have to develop your skills for the medium.

Second, learn the business. It's important for a corporate writer to understand the business and understand the issues. This means you need to learn to ask questions and do your homework. It's part of that constant learning I mentioned in the first paragraph.

Third, know a little bit about the technology that the corporation has available to produce what you write. Remember what I said about writing? You need to understand the medium before you can write for it. A video production is very different from a streamed audio message. The words you choose paint the visual.

Fourth, develop your problem-solving skills. Your job is to help others communicate effectively. They have come to you because they can't do it alone and need a good solid solution to the communications challenge they face.

If you want to enter the corporate world as a writer, expect to wear many hats and expect to constantly learn. The hours can be long. Contrary to popular belief, it's not 8-5, and the deadlines are tight. But it is a lot of fun. There are very few media production jobs that will expose you to as much variety as you gain in a corporate environment.

---

freelance writer John Morley. "No matter how business-like or systematic we make the process, going from facts and objectives to an exciting, emotional experience requires an intuitive leap of imagination."[27] The video scriptwriting (rather than subject-matter) expert is usually the best person to capture this emotional experience for the camera. As wordsmith Jon Sellew of Magicwords puts it, "It's far more important to understand the psyche of the target audience than it is to be an expert on the content. Let the client provide the input, while the writer maintains the perspective."[28]

Unfortunately, because companies communicate much more often in print than through radio or television, many in-house training or public relations executives believe they can automatically compose an effective video script. After all, they've written brochures, training manuals, and press releases for years; what can be so different about TV? Thus, people who would never attempt to create a situation comedy script or a broadcast commercial try to write an industrial video that demands the skills inherent in both. As the audio/visual scriptwriting team of Carl DeSantis and Phyllis Camesano point out, this is a prime example of "the print media trap." They therefore urge people planning a corporate video not to "fall for the myth that 'a writer is a writer.' What you want is someone who writes exclusively for the screen. Someone who does a lot of brochure writing, PR releases or even magazine articles probably is not a scriptwriting pro. When you are deciding on a

Conceptual Functions

Figure 2-6. Amateur scripting, like amateur performance, invites industrial video disaster. *Used by permission. Concept: Larry Hover, Copyright Sony Video Institute.*

scriptwriter for an important project, never use a portfolio of print media articles as a factor in your decision. It is like hiring a great plumber to do some carpentry work."[29]

In industrial no less than in mass electronic presentations, then, a professional result requires not only professional performers but also professional conceptualizers. (Otherwise, a scene like that captured in figure 2-6 might result!) A constant challenge for industrial scripters is to keep their script and their client focused on the ultimate objective of the piece rather than the tools and techniques it utilizes. "With rare exception," corporate producer Bill Miller warns, "clients have limited exposure to media production. The bulk of their experience comes from watching television. When they see something they like, they want you to duplicate it. But what they see might not be the right style for their product, their company, their audience, or the upward mobility of your career."[30]

The comparatively unglamourous but lucrative world of corporate video will likely continue to prosper as more and more companies use video to communicate to employees and customers via telecast, cassette, disk, and Web stream. But much of this new work, and an increasing amount of the old work formerly done within the client companies themselves, will be accomplished by freelancers and small outside production companies. In the "lean and mean" environment of the twenty-first century, enterprises try to outsource as much work as possible to lower costs and payrolls. Consequently, freelance specialists are slated to be the key players in the industrial script arena.

In-house or freelance, the prosperity of corporate video enterprises will continue to depend on how well their presentations conform to the media expectations of the employees or clients who make up their target audiences. Industrial video may not be quite as slick—or anywhere near as expensive—as a typical broadcast network presentation. But it cannot ignore the fundamental production and content concepts that make such mass-audience programs intriguing to watch and understand. "We're all up against broadcast television and its network-quality look," cautions corporate video sales executive Robin Pride. "Because your audience is accustomed to it, your creativity is challenged. . . . You must remember that the quality of your video will be forever judged and evaluated at the level of network prime-time programs."[31]

While they must be just as skillful as their mass media counterparts, corporate scriptwriters are not penning programs that will be judged by box office receipts or rating points. "In the jungle of business television," writer/producer Phillip Stella advises, "you have but one job—to help accomplish your employer's mission by either helping to increase revenue (profit) or to decrease expenses (costs) or both. That's what you do. Producing cost-effective or result-oriented multimedia tools may be how to do it, but the law of the jungle demands that your only reason for existence must be to add value to the bottom line. . . . That no-budget, talking-head piece can still be a creative challenge by striving to make it credible, focused, audience-driven and, above all, results oriented."[32]

Music Suppliers

Cost and ability are also vital but underestimated factors when it comes to the creation of production music. Nevertheless, everything from the most specialized industrial video to the most widely popular network adventure series tries to exploit music to bring a sense of depth, mood, and pacing to the visual story being told.

Like art directors, music specialists come into the electronic media after having been fully trained in their own craft. They then adapt their extensive knowledge of music to the particular requirements of the spot or program soundtrack. As veteran film and television conductor/composer Robert Emmett Dolan once explained: "The musician who works in recording, films, or television is inevitably involved in collaboration with professionals from a broad variety of fields, including sound engineers, tape or film editors, cameramen, etc. Unless he has some knowledge of their possible influence on his music, he will be working with a disadvantage."[33]

By the same token, all of these other electronic media professionals, as well as the writers and performers whom Dolan did not list, must be aware of the tremendous contributions that music can make to the success or failure of a project.

Usually, the commercial or program creators have some idea of the type of music bed they want to accompany the words and actions. This idea is then taken to the music specialists, who flesh it out. The largest studios have such experts on staff, but the usual practice is to contract with an outside musical production firm or individual composer. Some of these firms and individuals specialize in the creation of instrumental and lyric treatments for commercials and station promos. A smaller number concentrate on full-length scores for industrial, broadcast television, and feature film projects. In any event, the music supplier's task is to provide tones and rhythms that advance the purpose of the message in question. Successful producers of music for media clients understand that their creations are there to support, not upstage, the product or plot. Production music that calls at-

Elizabeth Myers
*Partner and Composer,
Trivers/Myers Music*

On Tuesday we get a call from the agency producer on a Pontiac job. "Would you be interested in working on a new campaign for the 350-horsepower GTO?" he asks. "Of course we would, sounds like fun," we think. On Wednesday the advertising team comes to our studio to tell us about the upcoming commercial. We listen carefully to the writer and the art director talking about the concept of the spot. Together we discuss whether or not the film should be tightly scored or whether the music should tell a story that is different from the visual images. We also talk about what feels good and we play some musical ideas. Then we actually record a presentation track even before any film is shot. This process is called a *prescore* and is something unusual in our business because most often the music is composed after the film is shot. But this time the schedule demands that the music come first. So that's what we do. Everybody leaves happy.

On Friday the actual rough cut of the film arrives. The agency feels that the prescore doesn't work anymore because it's too "ethereal." The decision might have been precipitated by the fact that there are also bigger problems than we imagined: the story line of the commercial is not as clear as the agency wants, and the celebrity voice-over has a different sound than expected. Neither of these issues are musical problems, but the agency is looking to us for a quick solution. Since the music is one of the last elements added to the commercial, if there are any perceived weaknesses now is the time to try and fix them.

On Friday morning we grab a cup of coffee and start all over. Five days later I stand in front of ninety musicians recording the new track. This is the music that will hopefully make the world want to drive the new GTO. On that day, I feel like the luckiest person in the world because the sound of the music is more beautiful that I could have ever imagined.

But arriving at that moment didn't just come by luck. Every day I use the skills that I learned from my master's degree in composition. I also credit my musical mentors: Nadia Boulanger from the Paris Conservatoire and Robert Ward, the Pulitzer Prize–winning dean of the North Carolina School of the Performing Arts. They taught me the subtleties of the compositional art. I still practice the piano every day; it tunes up my soul. I consider a solid background in music theory and composition to be a prerequisite for becoming a composer for TV and film.

While you are in school, you might also consider taking a course in Human Psychology 101. Knowing how human emotions work will help you become a better composer. And understanding human nature will definitely make you a better businessperson. Not only will you have to compose and perform music quickly, but you will also have to decide how much to charge and how to conduct your business. Some people hire secretaries, business managers, reps, and even ghostwriters. Other people, like us, prefer to keep a small boutique-type business and deal directly with the client. That way, the quality of the product is never diluted by the quantity of jobs. We also feel that we have a definite style of composing and film scoring to offer our clients.

Many directors and advertising professionals are surprisingly uneasy when it comes to music. For that reason it is imperative for the composer to become verbal about music. The wild-haired reclusive Beethoven would probably not be able to make a living in today's world. It doesn't matter how great your music sounds in your own living room, if you can't get it on the airwaves of our pop culture–driven world, you'll never make rent. Promoting your music is different from selling your music. It has to be done diplomatically, yet honestly and without fear. You must believe in yourself.

tention to itself is production music that has failed. Its job is to acquire meaning not for itself but for the associative role it plays in conjunction with the other presentational elements.

It is easy to discriminate between effective and defective musical applications in program soundtracks. Ideally, the music should be a legitimate heightener of audience involvement; but it is sometimes used as a cover-up for dull dialogue or a ponderous story line. As an example of the latter, *TV Guide* critic Robert MacKenzie once wrote this backhanded compliment about a prime-time series: "This weekly NBC hour . . . is providing a lot of work for violinists. . . . If it makes viewers feel better about life instead of worse, and puts all those string musicians back to work, I suppose it must be OK."[34] Obviously, it is not "OK" when the music bed is so massive that it overshadows the characters and situations on which the viewer is supposed to be focusing.

Skilled music suppliers and their astute electronic media colleagues know that music can tap, simulate, and stimulate a wide variety of feelings in even a very short passage. They are also aware that if the music says one thing and the dialogue or narration another it will probably be music's point of view that predominates. Thus, aside from music videos where the tune *is* the message, true professionals do not start off with a music bed and then try to wrap the message around it. To do so can only result in a nice promotion for the melody at crippling expense to the program's main theme or the commercial's selling point.

This same principle applies to the use of *music* libraries—preproduced collections of instrumental beds that a station or corporate audio/visual department can use to enhance commercials, promos, and industrial videos. Because the library's music beds have all been produced beforehand as generic rather than as custom treatments, special care must be taken that a library cut actually matches the media message now being constructed. Music libraries can greatly reduce the cost of beds for their users. However, if the selected cut seems tacked on rather than built into the presentation, it is better to use no music at all.

Therefore, choosing the right library is key. In a properly mastered and organized library, today's digital technology makes possible a wealth of sampling options that should provide a cut of just the right length and character. But "along with the improvements in recording techniques and the development of the digital format, the creativity in a music library still has to be there," states FirstCom-Music House's Robert Jenkins. "A good production music library will have depth and variety that includes different styles and categories of music. Producers no longer have to be satisfied with mediocre when they can have sensational."[35]

Many potentially useful libraries are available from several reputable music suppliers. One reason for their popularity is that every selection in the library comes with full copyright clearance. The *performance* rights to virtually all copyrighted music are controlled by three major agencies: ASCAP, BMI, or SESAC. Virtually all stations and networks hold licenses with these organizations, which authorize transmitted performances. The licenses, however, cover *only* performance rights— permission to play the tune to the public. They do not entail permission to make dubs of the music (the *mechanical* or *recording right*), to link it with an unrelated piece of audio or video (the *synchronization right*), or to use the performance of the particular musicians on the recording for any purpose other than the featured playing of that recording.

Thus, even though the venues in which the production will by played may possess performance rights licenses from ASCAP, BMI, and SESAC, this does not, *by itself*, allow the use of the compositions they license in commercials, programs, or corporate videos. Electronic media content producers, therefore, have four choices: (1) avoid all music use in the production; (2) try to clear all the rights themselves (a frustrating and time-consuming process); (3) commission a custom or preproduced soundtrack from a music supplier; or (4) lease or buy a music library that comes precleared.

Assuming music can't be abandoned altogether, the supplier-created library or preproduced track is usually the most cost-effective option. The length and stylings of the selections, however, must integrate well with the specific media message being constructed. No professional musical supplier can be held responsible if its product is applied like a Band-Aid.

In many ways, electronic media music suppliers are the ultimate conceptualizers. Their work imbues almost everything all the other conceptualizers produce and heightens the listening/viewing experience for both corporate and mass audiences. Increasingly, advancing technology and enhanced budgets mean that music is playing a more and more sophisticated role in video gaming. Now, trade reporter Matthew Mirapaul explains, "Games on DVDs can hold hours of full-spectrum audio, including prerecorded orchestra performances rather than synthesized versions. Faster microprocessors in computers and game consoles can function as sound studios, remixing and rearranging music on the fly so players are less likely to be annoyed by recurring audio loops. Beefed-up system memory can store multiple versions of a musical transition; a game engine can prompt one if a player eviscerates an alien or another if he slips on an eel. And the game industry's success has made developers willing to pay for live orchestra recordings."[36] All of this gives electronic media composers a fertile new field in which to ply their talents. They now have the opportunity to create building-block scores that can be programmed to instantaneously adapt to game-player decisions in musically natural ways. "A great game score will play the player," applauds composer Chance Thomas. "It will get their adrenaline going at all the right moments, give them this feeling of wonder and awe."[37]

To varying degrees, that is what music suppliers throughout the electronic media strive to accomplish with their tonal conceptions. "Music," as composer Richard Wagner once observed, "begins where speech leaves off." Professional electronic communicators know how to tap this additional dimension to best serve client objectives and listener/viewer preferences.

Web Writers

Like composers, electronic media writers, too, are discovering a new outlet for their work and new occupational roles—on the Web. The advent of online advertising in 1994 laid the beginnings of an economic foundation for a variety of content, ranging from static print-like ads to streamed audio and video original programming. The conceptual rules in this mushrooming environment have been slow to evolve because the environment itself is anything but stabilized. At this point, online writing is a mutating blend of creating for print and creating for broadcast. It is electronically delivered, like radio and television, but many of its verbal patterns are conveyed in the print-like world of consumer-read sentences, blurbs, and paragraphs. Some online copy is cast in concise spoken snippets to accompany still and moving graphics—similar to a television spot. But other Internet-delivered material scrolls down the screen to be digested like a full-page magazine layout.

It has become clear, however, that the Web's greatest opportunities relate to its ability to fashion content in ways that parallel the dynamic audio and video messages to which broadcasting has accustomed consumers and the marketers seeking to reach them. Even in the case of a Web staple, the banner ad, research by professors Hairong Li and Janice Bukovac found: "Animated banner ads result in quicker and better recall than non-animated banner ads."[38] Recognizing this and associated realities, the Internet Advertising Bureau (the online advertising industry's trade association) changed the name of its Streaming Media Committee to Interactive Broadcasting in 2002 as a means of better describing the capabilities of their business. "We had the most dynamic medium in the world and we put up billboards as creative," scolded media buying executive Jon Mandel.[39] So today's Web writers are moving rapidly to explore and exploit the new electronic medium in which they work. Sometimes, this is merely the *repurposing* of broadcast content for online delivery. But increasingly, it is the creation of audio/video advertising and program material that is original to the Web.

Jeff Dengate
Web Content Developer, NBA Communications

Prior to the dot-com fallout of a few years ago, many of the larger radio and television groups outsourced their day-to-day website operations to independent Web development companies. Since then, however, most companies have restructured their website departments to operate as a branch of the parent company in an effort to maximize resources while minimizing costs. If you are a content developer for one of these companies, this change has resulted in you working more closely with numerous departments within the organization, relying on both your website development skills and content production techniques to remain an asset to the online and offline aspects of the company.

As part of this larger media group, you can work in the traditional writer-editor-producer setting or you can focus on writing content for online as well as television or radio audiences.

In the traditional writer-editor-producer context, you'll often function within a typical newsroom environment, coordinating the content while working closely with a writer and an editor to help gather the stories that you will manage on the website. You will then primarily be in charge of determining the priority of the content and using your technical skills to best present it online.

If you focus strictly on the writing end, you may work with both online and television or radio audiences and must be able to write copy in the appropriate style depending on your medium. Writing for an online audience is similar to writing for broadcast; it is important to write a little tighter than you would for print because your readers can click away at any second if you do not hold their attention. However, you are not required to be as economical as broadcast because your readers are always able to go back and reread a point they may have missed. One of the key differences between online and broadcast style is you must present your content in a visually appealing manner that the reader is able to quickly scan in its entirety.

A news organization is not the only setting in which you may be employed as a content producer. You also may serve in a public relations capacity, creating content and producing it for use on the company's website or even its intranet portal for internal promotion to the company's employees.

In my current position at the National Basketball Association, I work within the communications group in a public relations role. Managing the day-to-day operations of the media-only access section on NBA.com, I also serve as the primary contract for all internal communications via the company's intranet site. Although the majority of my responsibility lies in being technical contact for the communication group's various websites, I also serve as a writer in an editorial capacity. Working along with two other writers, I actively pursue the story lines the NBA wants to push to members of the media and write feature stories they will carry in their own articles or use in their broadcast material.

A challenge of being a content developer for a professional sports league is working long hours covering both the league's normal business day as well as the games being played in the evening. During the season, a typical business day can be spent coordinating online press releases, conducting interviews with players for feature stories, and attending meetings to determine the best ways to promote an upcoming event. Then the evening will be spent at a game talking with team beat writers and national columnists, followed by making sure the websites are updated so the media members can find the latest information to assist them in filing their stories.

The content developer position is one you may hold in the early stages of your career and can be obtained with at least one to two years' experience in the industry. The opportunity for advancement depends upon which direction you wish to pursue.

You may opt to stay in the content creation side of the business, becoming a writer, reporter, or editor. If you wish to pursue the management side of the business, you may become director of interactive services or managing editor, similar to a news director in a traditional television newsroom. Finally, you could also choose to pursue a more technical direction, focusing on website development. In any case, you will at least need to have a firm understanding of HTML and other Web development technologies as well as strong writing skills.

If there is one general tenet emerging for online writing, it is that the conceptualizer needs to be prepared to craft messages that can be consumed in consecutive stages and that lead in different directions. Unlike one-way conventional broadcasting, what the IAB calls *interactive broadcasting* gives people the power to digest the message in a variety of ways. "This is not about TV," says media analyst Jim Nail. "This is a consumer-controlled medium."[40] Web journalists, for instance, can and should take advantage of this characteristic as much as do their copywriting and program scripter counterparts. "In the genre of television news now being taught for delivery by the Internet, words, pictures, sound, and graphics are combined to present the viewer with a first-hand experience of what it is like to be in a certain locale or situation," explains professor Antone Silvia. "In essence, the multimedia journalism via the Internet begins to fulfill the promise of electronic news since its infancy: the ability to allow the viewer to interact with a story as if he or she were actually there as it unfolds."[41]

Nevertheless, there are limitations to the online medium that creators for it must respect. Many consumers' Internet connections still lack the bandwidth to translate complex video in a fluid, full-motion manner. In addition, the computer screen is nowhere near as spacious as the typical TV receiver's viewing surface. Therefore, MSNBC managing editor Michael Silberman offers these tips, which help to characterize the limitations with which Web writers are learning to cope:

1. Shoot people tighter. Even on TV, you don't want a distracting background. *60 Minutes* learned that a long time ago. It's even more important on the Web. The more pixels that you have to put down the pipe, the more the image degrades.
2. Minimize motion. Action doesn't work well in streaming media, whether it is in frame or is a pan of some kind. The more motion, the worse the image.
3. Use straight cuts, not dissolves. Streaming video doesn't handle dissolves very well because of all the pixels changing. Like motion, that degrades image quality.
4. Divide graphics by four: Take whatever you learned about presenting information on a TV screen and cut it to one-fourth of that.[42]

As distribution technology improves, of course, these limitations will lessen. But like any other electronic media writers, Web scripters must pay equal attention to form and content. "Fundamental principles of writing and storytelling do not change in the new media," producer Stan Ferguson points out. "If it does not work well off the Internet, it won't work well on."[43]

A Conceptual Conclusion

Despite all the technological tools available to conceptualizers in our profession, these tools are not a substitute for talent. "Yes, talent is still king, and even more so today," observes copywriter and brand communications consultant Chad Rea. "And although talent might make everything look easier, it doesn't work any less than someone without it. . . . Talent does not stop thinking when it leaves the office. It does actually come up with great ideas in the shower. And while lying in bed. And while driving to work. And while playing games. And while scanning through books and magazines. Talent is always working. . . . Talent makes life its work, not the other way around."[44] Talent, in short, is the one great amalgamator of all the conceptualizers in our profession, regardless of medium. It does not substitute for hard work. But lack of talent is an unyielding barrier to eventual conceptualizer success.

Chapter Flashback

Copywriters work in media outlets, advertising, and public relations firms, corporate/institutional settings, and independently as *freelancers*. Their brief messages (commercials, promotions, public service announcements) are all designed to *sell* something in the most interesting and memorable way possible. On radio, well-fashioned copy can trigger a listener's recall of past experiences in order to complete the picture in his or her mind. On television, the visual usually carries the bulk of the communicative task. Some of the best writing is often the visual directions to the production personnel who will convert those words into filmed or taped images.

Art directors work with copywriters to sculpt a message's visual values. Often this collaboration takes place over a *storyboard,* which is the static plan for a fluid television concept. Other artists (graphic, design, scenic) work within media outlets and production houses to coordinate the overall visual image of that outlet and the programming it produces. Program writers are responsible for the scripts that become long-form entertainment projects. These creators' duties may range from per-script freelancers to writer/producers who oversee the conceptualization of multiple series. Program writers must learn to live with the frustration that, because their contribution is the cheapest element of a production to change, their work is usually subject to frequent rewrites.

Development managers are employed by studios and networks to select and shape promising program concepts and then oversee these concepts' production. Although their work is frequently resented by writers and producers, discerning development officials can help a project achieve more consistency and more success. Though much less glamourous than the national television shows on which development officials are focused, the industrial/corporate arena is a significant and growing market that depends on conceptualizers who can take often dry and highly technical information and sculpt it into clear and interesting program modules. In this branch of the electronic media as well as in projects intended for a mass audience, music suppliers play a significant role in providing cost-effective soundtracks that can enhance and pace audio and video productions ranging from commercials to full-length movie scores to video games and Web presentations. Web writers also labor in this online world to craft dynamic messages fashioned to take advantage of the Internet's interactivity.

Review Probes

1. Compare and contrast the tasks of the copywriter and those of the feature (program) writer. Are job opportunities similar in both fields?

Conceptual Functions

2. List five major functions that art directors and designers perform in the electronic media.
3. Why do the same small number of writers seem to be used over and over again in the creation of television programs?
4. What are some of the positive roles that development personnel can play in the conceptual process? What are some of the negative practices with which they are sometimes charged?
5. Is it better for an industrial scripter to be an expert in his client's business or an expert in the craft of writing?
6. What are the major rights associated with any copyrighted piece of music? To which of these does an ASCAP, BMI, or SESAC license pertain?
7. Why is certain Web content now being referred to as "interactive broadcasting?"

Suggested Background Explorations

Bliss, Edward, and James Hoyt. *Writing News for Broadcast.* 3rd ed. New York: Columbia University Press, 1994.

Blum, Richard. *Television and Screen Writing: From Concept to Contract.* 4th ed. Woburn, MA: Focal Press, 2000.

Carlin, Dan. *Music in Film and Video Production.* Woburn, MA: Focal Press, 1991.

DiZazzo, Ray. *Corporate Scriptwriting: A Professional Guide.* Woburn, MA: Focal Press, 1992.

Friedmann, Anthony. *Writing for Visual Media.* Woburn, MA: Focal Press, 2001.

Garrand, Timothy. *Writing for Multimedia and the Web.* 2nd ed. Woburn, MA: Focal Press, 2000.

Hyde, Stuart. *Idea to Script: Storytelling for Today's Media.* Boston: Allyn & Bacon, 2004.

Orlik, Peter. *Broadcast/Cable Copywriting.* 7th ed. Boston: Allyn & Bacon, 2004.

Pavlik, John. *Journalism and New Media.* New York: Columbia University Press, 2001.

Priestman, Chris. *Web Radio.* Woburn, MA: Focal Press, 2001.

Russell, Mark, and James Young. *Film Music.* Woburn, MA: Focal Press, 2000.

Schumann, David, and Esther Thorson. *Advertising and the World Wide Web.* Mahwah, NJ: Lawrence Erlbaum, 1999.

Stempel, Tom. *Storytellers to the Nation: A History of American Television Writing.* New York: Continuum, 1992.

Van Nostran, William. *The Media Writer's Guide: Writing for Business and Educational Programs.* Woburn, MA: Focal Press, 1999.

Wulfemeyer, K. Tim. *Beginning Broadcast Newswriting: A Self-Instructional Learning Experience.* 3rd ed. Ames: Iowa State University Press, 1993.

Zager, Michael. *Writing Music for Television and Radio Commercials.* Lanham, MD: Scarecrow Press, 2003.

CHAPTER 3

Production Functions

We use the term *production functions* to isolate the contributions of the technicians, engineers, and other electronic media pilots who construct, maintain, and manipulate the vehicles to which all of us who work in this profession are tied. It is these specialists who take the product of the conceptualizers' labors, capture the performers' realization of this product, and deliver it to media consumers in as electronically pristine and cogent a form as possible. The work handled by these production professionals is often referred to as "below the line" labor and cost. In contrast, the performers and concepualizers we have examined in the previous two chapters constitute "above the line" elements. Below the line and behind the scenes, the people we examine in this chapter work in comparative anonymity to assemble and distribute the content that the more widely recognized performers and conceptualizers have tried to envision.

Audio and Video Engineers

This broad employment category encompasses a great variety of tasks. Sometimes the term *technician* is used to separate lower-level systems operators from the true *engineers,* who fulfill supervisory, design, and development responsibilities. In most cases, however, the engineering department in a station or production facility is home to both groups, with its entire staff conversationally referred to as *engineers.*

Among the more basic technician-level tasks are the monitoring and adjustment of transmission levels; the operation of studio cameras, audio and video control boards, and tape decks; and the setting up of equipment for regular in-studio and remote productions. With advances in solid-state technology, however, a significant number of these tasks have been automated, and others now require fewer technical personnel. In 1988, for example, NBC began converting its single-set news shows to robotic cameras and tape machines. Other networks and local stations have since continued and accelerated this trend. Certainly, not all technical personnel, even in a one-set show, can be replaced by automation. But as NBC production executive Tom Wolzien has pointed out, "The key is the difference between those adding creative value and those who translate commands from the creative staffers to the equipment."[1] Mere "translator technicians," in other words, are on their way out. "At some point," adds a Top 10 market television station executive, "there will be just a producer and a director and a supercomputer. Automation and digital equipment are going to reduce the work force drastically over time. . . . Look at how fast technology has moved. Look how cheap it is. That's what's going to happen to the TV industry. It's on its way."[2]

On the other hand, personnel skilled in maintaining this more sophisticated digital and computer-based equipment and in applying it ("adding creative value") to a host of new production challenges are in great demand. Operating and servicing a state-of-the-art server/effects bank, digital paintbox, or multitrack audio control console with scores of digital inputs demand both mechanical dexterity and artistic sensitivity. Thus, while job requirements on the technical side have risen dramatically, so have opportunities and rewarded responsibility. Joseph Flaherty, longtime senior vice president of technology for CBS, put this new environment into lucid perspective when he observed that

> engineers, neither artists or writers, are the toolmakers to the arts, craftsmen of fine instruments—instruments that expand artistic horizons, but instruments we cannot play. Yet, this is the engineers' *raison d'etre*. We live, as it were, on the wrong side of the tapestry, amid networks of strings, knots and loose ends. While this infrastructure is the vital support to the artwork itself, ours is nevertheless the skeletal view.
>
> But our fulfillment is this: Without our tools, culture would be diminished, and the arts of modern television and telecommunication would not exist.[3]

Jay Rouman
*Director, Broadcast Technology,
Mt. Pleasant Area
Technical Center*

The job of video engineer encompasses a wide variety of duties and requires a unique set of skills. The job also varies widely, depending on the type of facility at which the person is employed.

Large video operations, such as television stations in major markets and large production houses, will usually have a staff of engineers with responsibilities roughly divided between maintenance and operation. *Maintenance* engineers fix broken equipment, install new equipment, and modify existing facilities to accommodate the needs of the production staff. These needs seem to change constantly. In the largest facilities, the maintenance staff will be subdivided according to the type of equipment with which they deal. RF engineers install, repair, and maintain transmitters, antennas, and microwave links. Studio engineers handle cameras, audio equipment, videotape machines and video servers, audio and video editing systems, and anything else that is needed to make the facility function. *Operations* engineers run equipment such as video switchers and audio consoles during programs. Usually they are not expected to repair equipment or make adjustments beyond routine operations.

Most large facilities will do component-level servicing. This means that when a piece of equipment breaks, they will find and replace whatever internal parts have failed. This often requires specialized test equipment, a stock of spare parts, and personnel who are comfortable with the inner workings of the wide variety of equipment found in a typical video facility. As video equipment gets smaller and more complex, component-level servicing has become more of a challenge. Ironically, the vastly increased reliability of modern video gear makes the job even tougher. Since it is common for equipment to run years without failure, when something does quit there will often be nobody around who has ever worked inside that particular system. In the "old days" most equipment required a lot of attention. It was a common sight to walk into a studio and see cameras with the access doors open and test equipment connected to the electronics inside. Now, cameras will often run their entire lives without a failure and there is no convenient way even to open them. But when a studio camera fails right before a show, the director wants it fixed right now.

Video engineers at small facilities have a much different job. Not only do they need to work on all the equipment instead of only a particular type, but they are often the only engineer. Small facilities will sometimes employ engineers on a contract as-needed basis. While this is a great cost savings, there may be no technical person around when something critical goes wrong. Engineers at small facilities are more likely to do system-level maintenance and troubleshooting. They will determine the cause of a failure down to a single piece of equipment or perhaps a single module within a unit. However, they will not attempt to dig deeper. Instead, they will ship the equipment for servicing. Engineers at small facilities cannot justify the cost of the specialized test instruments which might be required for specific equipment, particularly when there is a good chance that this equipment will never fail during its useful lifetime. They also simply do not have the time necessary for component-level troubleshooting, which often requires a fair amount of detective work and almost never follows a predictable schedule.

In addition to routine maintenance and repairs, video engineers at all operations should be involved in the planning of new facilities. Engineers who have actual production experience are particularly valuable since they can see the relationship of the equipment to the ease of creating the final programming product.

Within a station or production facility, the *chief engineer* is the supervisory head of technical matters. Depending on the size of the operation, the chief may be a department of one (as at many radio stations) or may supervise scores of employees in a number of different job classifications (as in a major-market television station or production house). In the very smallest facilities such as a 500-watt daytime-only AM station or an LPTV (low-power television) outlet, the chief engineer may even be an outside contractor who calls at the station only periodically to check operation of equipment or repair a major component that has broken down. Whatever the case, this professional has usually accumulated considerable experience in the maintenance, installation, and repair of electronic equipment.

The bigger the staff, the greater the likelihood that the chief will delegate hands-on operations to other employees while spending more time in administrative, equipment-ordering, and facility-design activities. "The [chief] engineer has evolved from an appendage to a station management team to being a highly integrated part of the station management team as technology has taken a more prominent role," observes Tribune Broadcasting's vice president of engineering Ira Goldstone. "Also, it has evolved from strictly an audio, video and RF [radio frequency wave] world to one that includes networking technologies and computer systems and the Internet. And the scope of the job, and the reliance on engineering by other departments, become more critical."[4]

The marriage of computer science and radio/television technology also means that today's chief engineer (and many engineers below this level) must also understand computer theory and design as they relate to such systems as audio processors, digital editors, and special-effects banks. Computers are also central components of the automation equipment that runs the transmitter, selects and cues program elements, and keeps track of commercials aired and the billings for them. Even more expansively, computers now function at sophisticated "hubs" out of which an entire group of widely dispersed television stations can be run. Thus, electronic media engineers are now information-processing specialists as well as caretakers of hardware. "Today's broadcast engineers are expected to be versed and capable in a broad spectrum of disciplines," consulting engineer Barry Thomas writes. "They must also act as de facto MIS [management information system] managers."[5]

As in the field of automobile repair, the specter of the self-taught tinkerer armed only with a screwdriver and soldering gun is largely a thing of the past. "Your chief engineer needs to hire people with both computer and verbal communication skills," advise the experts at National TeleConsultants. "Today's typical trouble-shooter has to call a distant 'Help Line' and talk knowledgeably with a manufacturer's service department, or even sometimes, the engineers who built the system. 'It goes all wiggly when I put the whatziwhozit in the thingamabob' won't cut it. . . . The most important tool in today's maintenance department is probably the telephone."[6]

Besides mechanical, computer, and communication skills, engineers working in the recording industry should possess some musical sense and understanding of how tones and harmonics can be alternated and combined for maximum clarity and comprehension. Consulting engineer David Scheirman puts it this way:

> The sound industry is a "specialized" industry. It takes special skills, tools, interests and training to be an effective participant in the industry. The subtle differences between the desired sound for a Broadway show or a corporate theatre presentation—the not-so-subtle differences between the type of gear used and how it is adjusted to present a symphonic performance or a heavy-metal rock concert to a large crowd—each takes a specialist who knows his or her own field of expertise.[7]

In addition, recording engineers who work with musicians must master a special brand of behavioral discipline in order to properly advance the creative process. As industry veteran Julian McBrowne reminds such facilitators:

> Like it or not, recording is a communal art form. Do you wince at bad notes? Yawn between takes? Take phone calls during overdub sessions? Remember, performing artists thrive on reaction, and you are being watched. There's no escaping the power of the person who sits in the engineer's chair. Even when you're "just doing your job," everyone looks to you for a reaction to their performance. . . . Unless you're coproducing the session, try to adopt an attitude of supportive neutrality; it'll make it easier for you to deal with the inevitable "How was that?"[8]

In many instances, however, sound jobs involve not the exciting world of mixing musicians but rather the capturing of prosaic words of some corporate spokesperson or public newsmaker so that volume levels remain stable and audio reproduction is clear.

Music recording may be much more exhilarating, but it is also less lucrative. Recording technology professor William Moylan advises that "the 'glamour' positions, and many of the more creative positions, do not pay well (save for a few 'name' people). Like musicians, music engineers are underpaid for tasks that appear to be *enjoyable*. These positions are very popular, but are demanding. These are the creative positions that many very talented people pursue because they feel they must."[9] This same statement also might be applied to the sound engineers known as *Foley artists*, although their average compensation may be a bit higher. "Foley is the art of creating and augmenting incidental sounds for television and film," explains trade writer Erin Caslavka. "Getting punched in the stomach, crunching through snow, biting into a crisp apple—all of these sounds are enhanced by Foley artists, who use an unusual array of props to make a scene sound more vibrant."[10] Foley work

Production Functions

Charles Nairn
*Audio Engineer and President,
Com Tec, Inc.*

When students come to my office, they are typically full of questions. No matter what other kinds of queries they have, somewhere in the conversation they get around to these three basic and critical issues: What types of jobs are available in audio? What kind of education do I need to get one of these jobs? How do I get my first job in audio?

Jobs

Jobs in audio fall into four main categories. Audio for entertainment—CDs and DVDs—provides a significant number of jobs. Audio in the media—radio, TV, film—is also a large employer. Advertising audio, which might be considered audio *for* the media, offers much of the remaining employment. The fourth, smaller category is made up of specialty areas, which provide many interesting opportunities. A partial list includes performance audio, either at a theater or with a touring group; recordist for a major symphony orchestra; forensic audio for law enforcement, attorneys, or intelligence services; acoustic/electro-acoustic design for performance spaces; and even teaching audio.

Education

Because talent, hard work, and the ability to function well with people are probably more important than any particular formal education, the following are general guidelines.

While you will encounter everyone from high school dropouts to PhDs successfully working in audio, the more academically appropriate preparation is earning a degree as a *Tonmeister*. Roughly translated from the German (and Germany is where the degree originated after World War II), *Tonmeister* means "master of sound." This degree combines studies in electrical engineering, acoustics, and music. There are several universities in Germany and a couple in England and the United States that offer this major. In general, there are many more applicants for admission than there are positions available, so the admission-selection process is quite rigorous. Because of the quality of students admitted to the program and the excellent education they receive, graduates typically have easier access to the better jobs at the bigger companies.

There are a few schools that specialize exclusively in audio education. Some grant degrees and others are creative, high-tech trade schools. If you are quite certain before you start your education that audio will be a consuming, lifelong interest, these schools are a good choice. They are not recommended for the undeclared major.

Education at a less specialized college level might include communications, music or music education, electrical engineering, or physics. No matter which major you choose, be sure to "do audio" at the school's theater, the campus radio station, the TV facility, and the music school. This exposure to many facets of audio will not only provide good experience but will give you and others the opportunity to confirm your talent and enthusiasm for audio.

The First Job

How do you get your first job in audio? Well, the best way is to do some audio work somewhere—at school, at a local theater or cable station, at a large house of worship, or with a local musical group. These jobs often pay little or nothing, but you will gain experience and exposure. Because the audio engineering community is relatively small and fairly close-knit, someone—assuming you have some recognizable talent—will notice that talent and perhaps pass on your name to a prospective employer. Once you are in the business, almost all job changes will come through networking. You can, of course, try cold calling on employers. But this is a very difficult way to break into the business. If you try this method, be prepared for a first job that has little excitement and very little pay. In such cases, the employer is just bringing you on to size you up, and no business is willing to risk very much for that privilege.

Final Thoughts

The vast majority of jobs in audio are in large cities, with Los Angeles, New York, and Nashville being the largest production centers. Each area of audio has a spectrum of jobs that are more or less creative, more or less technical, and more or less physically and emotionally demanding. Generally, the more the job demands of you, the higher the pay and the quicker the burnout.

One additional caution is that many of the jobs in audio require a lot of long, hard days. This can put a strain on normal human relationships and often takes a toll on "media marriages." When my son was in college, he worked for me one summer, which is our busy season. Late one Sunday evening, we were sitting on the studio entrance hall rug eating a dinner catered by McDonald's. My son said he had just worked eighty hours that week and told me he had clocked me at eighty-four hours. He asked me how I, being older than he, did it. "Practice," I remember saying somewhat wistfully. "Practice."

My personal decision to start a small commercial studio has required me to do it all: design, maintenance, mixing, location recording, and, of course, management. Over the years, I have recorded pop music stars in the studio, internationally famous classical musicians on location, and large musical groups in many performance venues, including Europe.

Challenging? You bet. You deal with difficult clients, lumpy cash flow, more than a hundred tax forms every year, equipment that breaks without asking permission. Fun? I wouldn't and couldn't do this if I didn't love it. Is it for you? My advice is choose an audio engineering career only if you're quite sure you'll love it, too.

---

thus demands as deft a creative touch as does music recording. But whereas the music recording engineer is seeking to mix emotive tones and timbres, the Foley artist is attempting to manipulate often mundane sound effects to enhance the impact of depicted visual action. In both instances, their work is successful when their technical contributions are transparent to the audience.

Because of the diversity of opportunity and the range of skills these opportunities collectively demand, technically oriented individuals should set their personal priorities carefully before pursuing an engineering career in the electronic media. Job options exist not only in stations and networks but also in video and film production houses, recording studios, corporate and institutional audio/visual departments, and conventional and wireless cable delivery systems. Some engineers, such as those supervising transmitters, satellite relays, and a cable system's physical plant, spend most of their time in matters of signal sending and reception. Others, like those involved in outlet control rooms and studios, focus on the ongoing process of program production and arrangement. And still others are actively immersed in creating and recording entertainment and informational projects that the studios and companies for which they work will distribute.

The common bond that links all of these situations, of course, is that they are all equipment-intensive, with hardware changes dictating job shifts. William Moylan's admonition to would-be recording engineers is thus appropriate for technical aspirants throughout our industry: "A very real danger exists to those employed in an area that is dependent on technology, and that often finds itself reacting to a new technology. One can quickly find oneself unprepared to handle a job function. . . . audio [and video] professionals must be prepared to re-educate themselves to be on top of the next new technology. In this business, job security is primarily based on the quality and results of the last 'performance,' not on history."[11]

Figure 3-1. A lighting director focuses a Fresnel spotlight for a studio production. A "scoop," used for general illumination, hangs at lower right. *Courtesy of Neil J. Kirby*

Lighting Directors

Smaller video facilities use regular staff engineers or technicians to handle lighting tasks, whereas larger outlets and production houses employ specialists for this function. Television cameras depend on the proper amount, direction, and quality of light in order to reproduce pictures in which the colors are true and the mood is appropriate to the program's content. Modern cameras require much less illumination than the insensitive monsters of the 1940s and 1950s. But the light that is used must still be expertly arranged. Avoiding unwanted glare on the one hand, or dimness on the other, are only the most obvious of a multitude of concerns that impact a lighting design.

At its most basic, illumination can be considered as a purely quantitative commodity that is provided and measured to ensure the proper operation of the cameras. Herbert Zettl refers to this kind of lighting as *Notan,* or "lighting for simple visibility. Flat lighting has no particular aesthetic function; its basic function is that of illumination. Flat lighting is emotionally flat, too."[12] Were this all that video production required, any production assistant could handle the job with a minimum of training.

Lighting directors like the person shown in figure 3-1, however, seek to facilitate pictures that transcend the two-dimensionality of the television screen by also using illumination to accomplish scene sculpting and visual emphasis. Thus, they move beyond mere Notan approaches into *Chiaroscuro* devices—methods by which illumination can be contrasted with shadow for greater pictorial interest. With Chiaroscuro, says Zettl, "the basic aim is to articulate space . . . to clarify and intensify the three-dimensional property of things and the space that surrounds them, to give the scene an expressive quality."[13] In the hands of a competent lighting director, Chiaroscuro makes illumination a key element for pictorial composition and enhancement of whatever mood is most appropriate to the commercial's or program's objectives.

We can illustrate the importance of lighting and shadow in concept communication by examining the television commercial photoboard in figure 3-2. The shadowy, threatening world of the dark sets up the premise in the first four frames, with ultimate blackness punctuating that point in frame 5. Then, in frames 6 to 10, the promising illumination from the refrigerator brightens our prospects as it introduces the product. The beer itself is lit so as to positively glow against the gloominess of the unknown in frames 11 to 13. Finally, to both reiterate and deflect fear, frame 14's crosshatched shadows seem to project from the murky depths of the fridge, etching Martin Mull's face in horror until he turns to the gleaming product in the last shot. The commercial suggests that although this beer may appear "dark," it absolutely glistens because of its smooth, full-bodied taste.

The Michelob Classic Dark spot was shot on film, so certain of the lighting director's specific techniques were different here than if he or she were setting up for videotape (electronic) recording. The basic artistic principles of lighting, however, remain the same in both instances, and it is this sense of polished communication that lighting specialists can bring to a production.

Matt Ilas
Director of Lighting and Photography, Summer Nights Film and Video

In what profession will you be asked to make people look better than they do, turn day into night, set entirely different moods based on a sometimes shaky idea, and do it all up to twenty or more times a day? Welcome to the world of a lighting director. Is it fun? Hmmm. Is it rewarding? Sure, especially if you like doing work that is most successful when it is not noticed. Is it challenging? Bingo! A great challenge that will burn your fingertips (tip number 1—get some good leather gloves and write your name on them). But like any art form, if it were easy, everybody would do it. If that were the case, that cranky supermodel who arrived at 5:00 a.m. with too little sleep and too much last night showing on her face would do the job herself. That would leave nothing for the lighting crew to do but eat the doughnuts.

As lighting director, my day can start off moving fast. Whether the shoot will involve the use of three lights or unloading the entire truck, hopefully I will have some help in the form of grips, so that unloading, setup, and running power can happen quickly. (Grips are people responsible for equipment, set and prop maintenance and adjustment.) One weird thing about this business is that the same object can have many different names. How many names are there for sandbags and clothespins? So working with grips who know the different names for a 2-K light, plus the name I have invented for them, plus how I like to work, is always beneficial.

Next comes understanding the producer's or director's idea of what the piece I am lighting is about. If there is a script, I try to get hold of it at least the day before so some thought can be given to the lighting plan ahead of time. If possible, a trip to the location is in order. A shoot is a lot like a potluck dinner. If you have an idea what the main dish is you can bring something else to the party that maybe wasn't thought of and will enhance the experience. You don't want to stand there wishing you'd brought that one little clamp that would perfectly hold the backlight in just the right spot. Knowing ahead, planning for the unforeseen, and having the right tools to make the impossible possible are what make for a happy ending. The Boy Scout motto of "be prepared" never was more appropriate. Unfortunately, there are times I show up not knowing what the project is about, the mood or look, or if I am trying to match someone else's footage. So thinking fast on one's feet is a must. There will be time later to figure out what you're going to have for lunch.

Once I have established mood, knowing where emotional peaks are, how the project will be used and where (on a huge screen, or after being dubbed down to VHS eight zillion times, for example), I then determine what existing lighting will work, what I need to supplement, and when to start with a clean palette—that is, shut off all lights that preexisted and light the entire scene with my own lights. Basically, I must make the lighting appropriate to the situation. Since the look of video is flat, compared to film, giving the scene as much depth as possible is essential. I try to use existing atmosphere and light that makes sense while making reality just a little better—without my lighting being overtly noticed. And it should have been done five minutes ago. Having fun yet?

Sometimes it is a struggle just getting off the ground. Today's cameras are much more light sensitive—therefore require less light. This fact, complicated by low budgets and whatever is the latest *look du jour*, can make for frustration. Less light should not be thought of as no light. However, time is money, and I have to bear in mind that even though just one more accent light would make a world of difference, the producer wants to roll camera "right now" and "it looks OK as is." Sometimes you must realize that you can do only what you can do and then move on to the next scene. Tweaking is always done, of course, but quietly and between takes. Bottom line: know when enough is enough and too much is too much.

Despite the challenges, there is something exciting about seeing a photographically beautiful shot that cannot be explained as much as felt. This keeps me coming back for more. Every day I am a student of the light, observing what it does, how it affects its subject, and how the result affects the viewer. As the late great still photographer Ansel Adams said, "Photography is painting with light." Without light, and hopefully good light, there is blackness. And then you might as well be doing radio.

For any lighting director, a great deal of work occurs before actual production of the commercial or program. Once the concept and set/location decisions have been made, the lighting director sits down with a set diagram to design a *lighting plot*. The plot specifies the number and type of illumination instruments to be used and where each instrument is to be located in relation to the on-camera action. Rearranging instruments on paper in the planning stage is much easier than moving the actual lights at the time of production, and also much more cost efficient. With a well-conceived plot, the lighting director can devote actual studio time to final focusing, balancing, and special effects enhancements. In this as in virtually every phase of our industry, professionals always preplan; only amateurs think they can wing it. Lighting expert Joseph Tawail puts it this way: "Time *not* spent for pre-production planning will only be spent later while many more people stand around and watch a few guys sweat. The result will cost more, and rarely looks as good as it could. The light plot is basic to a good production for both economic and aesthetic reasons, and one of its byproducts is peace of mind."[14]

The lighting plot is just as important *after* the production has been long completed. San Francisco lighting director Bill Holshevnikoff confesses: "Many times I have lighted a show and have been so exhausted after the job that I just tore down the lights and headed home. It would be months later that I would come upon a similar job and wish that I had made some notes from my last ideas."[15] As in most occupational roles within our industry, organizational skills are a key, if not al-

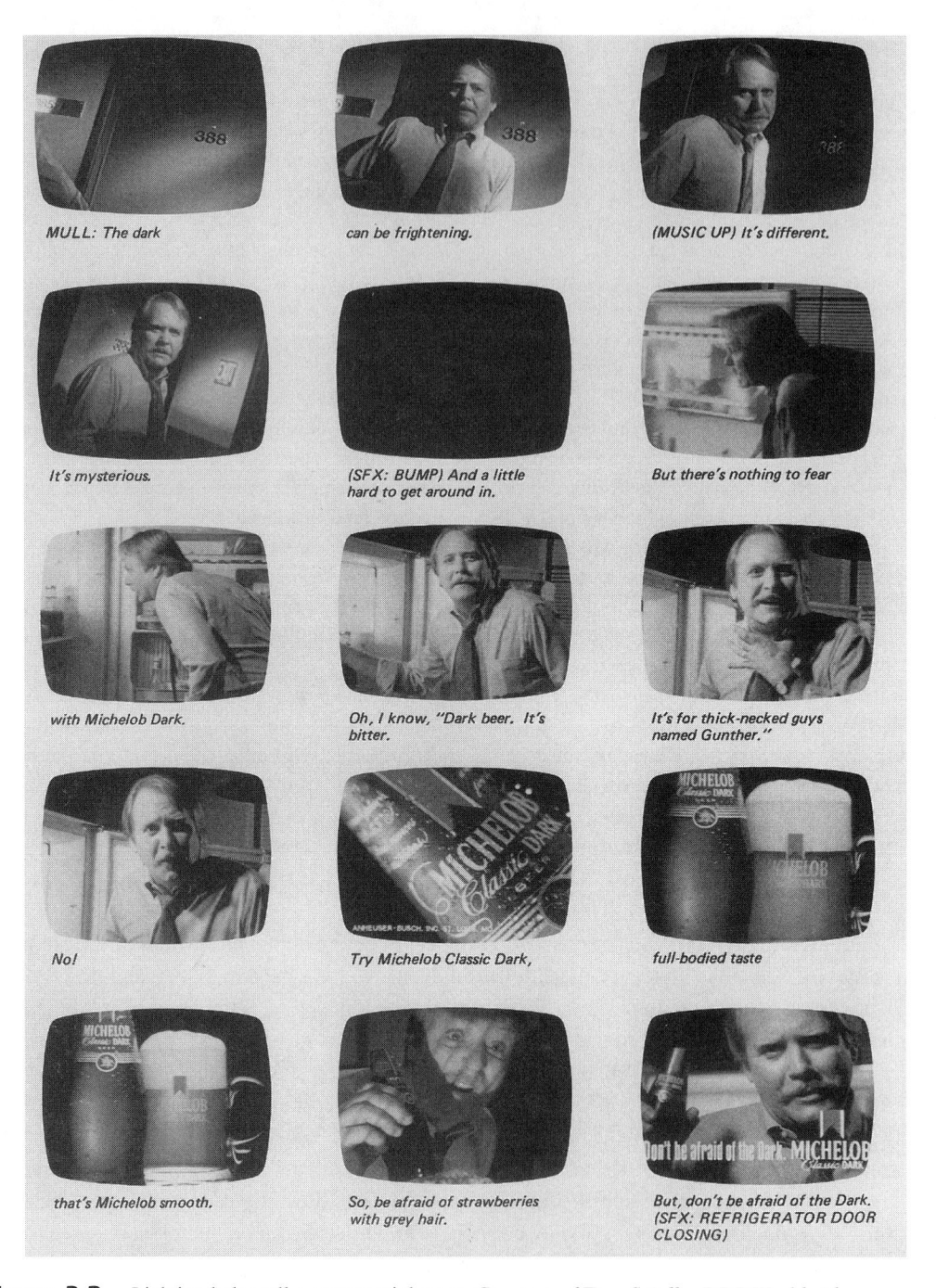

Figure 3-2. Lighting helps tell a commercial story. *Courtesy of Tony Sciolla, DDB Worldwide.*

ways obvious, prerequisite to success. In fact, the need for these skills only becomes more obvious in their absence!

Camera Operators, Videographers, and Cinematographers

The people who manipulate the cameras that pick up the light and the shadows are known by several designations. Generally speaking, *camera operators* function in television studios or at pre-planned events, with their electronic instruments mounted on tripods or pedestals. *Videographers* use highly portable (often handheld) gear to tape location or remote pieces. *Cinematographers* work both in studio and on location but shoot with film rather than videotape or disc. *Directors of photography (DPs)* oversee the total look of long-form film and video productions and coordinate the contributions of both camera personnel and lighting directors. In all of these situations, the fundamental rules for effective shot composition remain constant even though the recording medium and style may vary.

As we mention in our discussion of engineers, prospects for persons seeking to function exclusively as *camera operators* are not particularly bright. At the station as well as the network level, more and more newscasts and interview shows are likely to be serviced by robotic vehicles. And because newscasts and interviews are often the *only* in-studio productions found at many stations and cable-origination facilities, the demand for full-time camera operators is gradually evaporating. Instead, camera manipulation is more and more often handled via remote control by an engineer who can also operate the camera directly on those rare projects where such hands-on work might be necessary. Certainly, production houses specializing in commercial and entertainment program recording will continue to require the services of skilled camera operators, as will the producers of sports events coverage. The number of such jobs, however, is relatively small and is confined to big cities and major media companies.

Because of the flexibility and portability of their gear, *videographers,* on the other hand, are a much more plentiful breed. In today's eyewitness-news environment, videographers capture most of the footage we see in local and national newscasts and often edit it for airing. They are thus essential ingredients of every television outlet's news departments. As much of the success of TV news depends on visual dynamism, adept shooters can distinguish their unit's performance from that of the competition. Everyone may report the same story, and so the best-executed picture is a significant factor in determining what viewers perceive as the best overall coverage. Even a star reporter with keen verbal skills cannot long compensate for dull or mishandled videography.

Visual news gathering is, however, a rough-and-tumble existence in which excitement can gradually give way to aches, pains, and even burnout. Many news videographers, therefore, like the street reporters with whom they labor, eventually seek to move on to other careers. For the videographer, this career move might entail involvement with commercials, corporate videos, documentaries, or even entertainment features that require substantial location recording expertise. Some videographers may also advance to director positions, where they combine their camera abilities with greater overall responsibility for the assemblage of the projects on which they work.

The videographer aspect of our profession has blossomed due to the tremendous technological advances in portable camera/recording equipment. Thirty years ago, remote taping required a tractor-trailer crammed with equipment and cables snaking to the site of whatever camera operators were trying to shoot. Today, a single unit on a videographer's shoulder can move in tight on virtually any locale or subject. In *ENG (electronic news gathering)* situations, the videographer often has the capability of either feeding a live microwave signal back to the station or satellite uplink or recording the event for later transportation or transmission. In *EFP (electronic field production)* applications, the camera/recording technology is similar, but the aim is to capture more polished and preplanned footage from which a commercial or program piece later can be assembled. Whatever the case, the videographer's physical dexterity and sense of pictorial composition are the key framers of what we are able to bring to the viewer.

Pattie Wayne-Brinkman
Freelance Cinematographer and Videographer

I decided, against all odds, that I was going to be a television news photographer. Breaking into what had been a male-dominated industry was a feat unto itself, but WCPO-TV had confidence in my ability and gave me my start. I wanted this profession badly enough to brave midwestern winters, carrying eighty-five pounds of gear through icy winds measuring thirty degrees below zero. I've now traded that bone-chilling cold for blazing hot Texas summers—again, you have to really want it. Enough to withstand the elements, maintain composure under fire, and perform in the midst of extreme circumstances.

There are risks involved as well. At least once in your career, you will find yourself on the scene of an overturned tanker car that is rapidly spewing some kind of toxic substance. I have tempted fate more than once, having prevailed over a midair accident and emergency landing while shooting chopper-to-chopper aerials. It helps to be a little crazy in the news business, but sound judgment is a better bet. The most important requirement, according to my former coworker and thirty-three-year veteran television photographer Jimmy Darnell, is "first and foremost you must be a journalist and then you learn the tools of the trade." Technical knowledge and skill are a must. And because developments in the industry are continuous, it is necessary to stay informed as to what is state-of-the-art.

An accomplished television photographer is aggressive without being pushy, dedicated to the trade, motivated, instinctive, innovative, and artistic. Most of the work involves people, so you must also be adept at public relations. Discipline is imperative when covering a news story that involves tragedy. It is necessary to detach yourself from the situation in order to function, but you must still remain empathetic.

Physical strength, of course, is an asset, but it is more important to be in good shape to protect yourself against injury. Though the cameras are a lot lighter than when I started, the tripods and light kits are still very heavy. A photographer's worst enemy is the cumulative effect of carrying heavy equipment over a number of years. Full days of shooting off the shoulder can be survival of the fittest, and the person who has stamina will prevail. News is a tough business, but not without its rewards.

The news photographer documents events that occur on a daily basis and receives a firsthand lesson in life's realities. There is no typical day. It is that anticipation of the unexpected that draws many people to the excitement of television news. I often miss the surge of adrenaline that comes with covering breaking stories. Yet, for me, there was a greater challenge in *selling* the news product.

My desire for higher production value took me from the newsroom to the marketing department of KXAS-TV, where I was director of photography (DP) for the next eleven years. These years were the most satisfying of my career because I was reunited with my first love, *film,* and produced my best work during this time.

The television marketing department is not unlike a local ad agency (minus the big budgets and large staff). It is responsible for creating the right image for the product. The key to good marketing is knowing your consumer (in this case, our viewers) and then developing a campaign that will attract the marketplace to sample your commodity. The process begins with market research and then interpretation of what you think the viewer wants in a news team. This is followed by a series of creative sessions where the ideas are generated by the executive director, the producer (who also is often the director and writer), the art director, and the DP, who will visually interpret the scripts. After establishing the "marketing plan" and determining the look and feel desired, shooting begins.

Production work is less demanding physically than news only because you usually work with a crew and therefore have help carrying equipment. There is a different type of equipment. There is a different type of stress involved due to the responsibility of having a "budget" and having to produce a film campaign within that budget. Film is not immediate like tape. Therefore, you must wait until the transfer process to see the results of your labor. Fortunately, a wave of energy and excitement flows when you take advertising spots out of the status quo and into a new realm. Combined with the technology now available, the possibilities are as limitless as the imagination. You now have the option to keep video or film in its purest form, or you can magically transform it with shape, size, color, or texture. Because of this trend, I am constantly challenged to produce a look that is cutting edge.

Breathtaking aerials, stylized lighting, and "How'd they do that?" special effects shots are my specialty and garnered me my third Emmy Award the year I left television.

Equipment miniaturization and consolidation have a downside for news photographers, however. To increase the number of news "units" on the street and/or reduce personnel costs, more and more stations and cable news operations are converting to *video journalists*. These "one-person band" professionals shoot, report, and edit their stories all by themselves. The pictorial quality of the piece is in danger of being compromised as a result—because these VJs are hired for their reporting and on-camera skills first, and their shooting skills second. Arguing against the wholesale adoption of the VJ model, Steve Swisher, past president of the National Press Photographers Association, maintains: "Two minds are better than one. If we want to go to idiot cameras (with auto-focus and pre-set) then we can hire idiots to shoot. But news operations that value quality will hire pros to operate the cameras and get the most out of the equipment."[16] On the other hand, WSOC-TV video journalist Michelle Kosinski maintains that working alone actually improves her stories: "It's just me and my camera. People tend to tell me more. They feel closer to me. It's just one on one." Still, Kosinski warns that to succeed as a video journalist, you must be able to "plan and write in your head while you see, hear and haul."[17] The trick is to simultaneously perform all these functions without compromising the technical and pictorial quality of your visual.

This situation is a far cry from the early days of TV news when *cinematographers* gathered the pictures. Before the conversion from film to tape, television news gathering in most cities functionally stopped at about 3:30 p.m. Film crews required time to get back to the station and process and then edit their footage in time for airing on the 6 o'clock local news. (This same time lag occurred for the late news, too, of course; 8:30 was the approximate deadline for what could be shown at 11:00.) Politicians and public relations people were aware of these scheduling constraints and so planned their announcements and news conferences accordingly. Now, however, videographers can capture events right up to the start of the newscast and, if their ENG unit (like the van in figure 3-3) is microwave capable, can even send live reports directly onto the air. These applications are especially exciting, but they also raise serious questions of journalistic responsibility because news directors and editors have no opportunity to screen gruesome or potentially defamatory material before it is sent into viewers' homes.

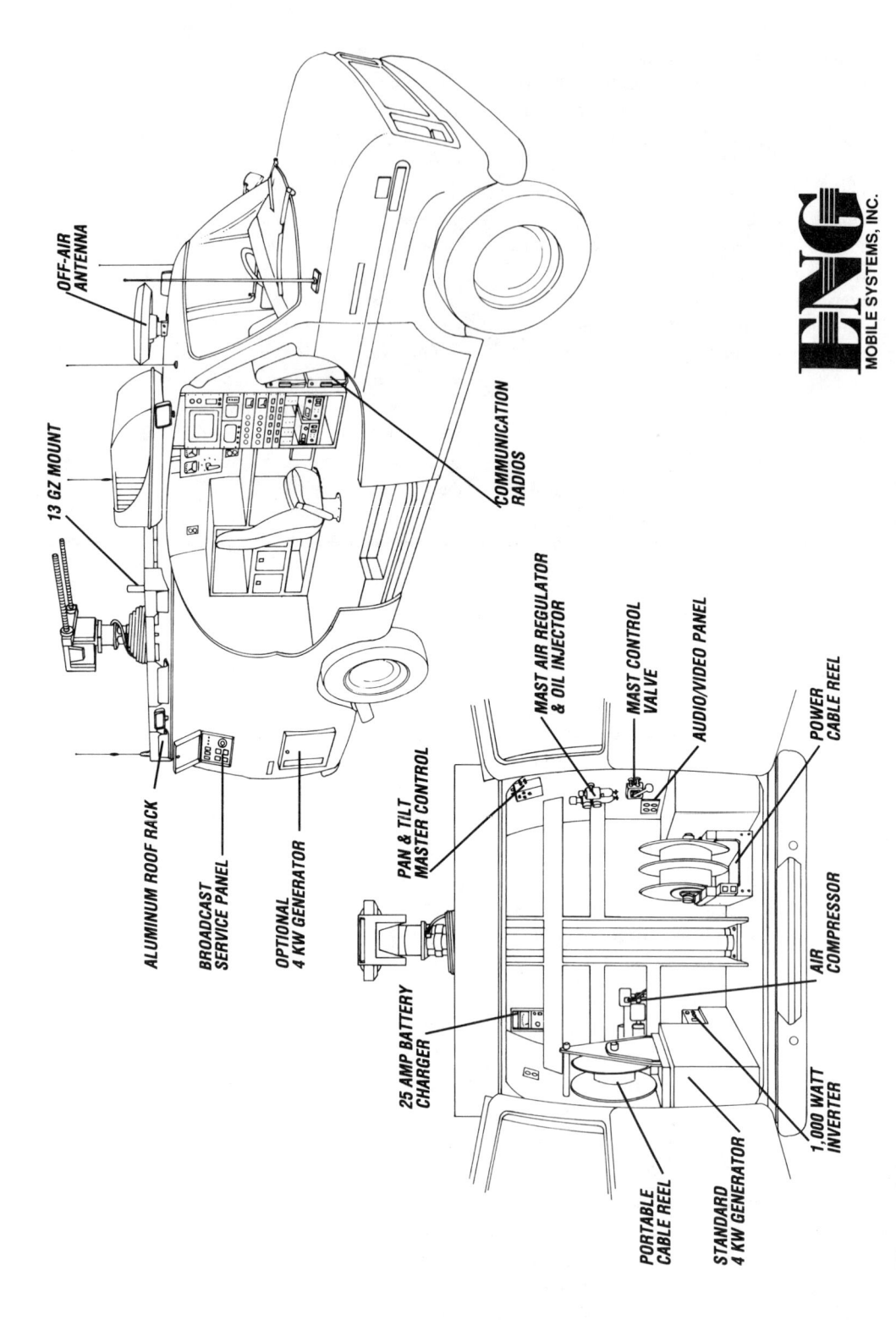

Figure 3-3. Components of a compact ENG vehicle. *Courtesy of Ted Kendrick, ENG Mobile Systems, Inc.*

Production Functions

No longer shooting news footage, today's cinematographers may work in production studios like camera operators or out on location. But because they use film rather than tape, their recording vehicle must be chemically processed before it can be viewed or edited. Unlike their video colleagues, cinematographers cannot simply rewind their recording stock to check if they have the pictures they needed. (A "video tap" can be run off the camera lens to immediately provide an approximate image—but this is of lower resolution than what the film print will capture.)

In commercial, documentary, corporate, and entertainment production, film is still a frequently used vehicle, although much more in mass entertainment than in industrial projects. A number of talented cinematographers thus work in their chemical medium to capture content that is usually then transferred to video for electronic media consumption. In many spots and long-form projects the look of film is still preferred because, as Time/Warner's chief technologist Jac Holzman observes, "Video has a kind of plastic immediacy; film has subtleness and romance. Video brings you up close; film has a softer objectivity."[18] Eastman Kodak staffers further argue that "film origination has intrinsic advantages as a reproduction tool because there is nothing between the camera lens and the recording medium. Even with advancement in solid-state technology making smaller video cameras possible, the fact remains that the image processor, in effect, is carried in the video camera. The result is that film production probably always will provide quality and creative advantages."[19]

However, Kodak's experts made this comment before the first comprehensive demonstrations of *HDTV* (*high-definition television*). Because it doubles the number of lines in a U.S. (NTSC) television transmission from 525 to 1,050 or more (depending on which HDTV system is used), HDTV creates a video image that rivals the clarity of 35-mm film. HDTV thus possesses the capacity to transmit live TV and videotaped images as well as 35-mm film pictures with a uniformly high quality. The film still looks softer, and the video-originated material still looks more like live immediacy, but the depth and coherence of each idiom are now closely comparable.

Film is neither dead nor the adversary of HDTV. Newly developed film stocks provide resolution superior to that of an HDTV video-derived image and can capture acceptable pictures in almost total darkness. Television cameras, HDTV units included, still require significantly higher illumination levels to produce suitably defined images and so cannot attain some of the shadowy nuances that film makes possible. Consequently, many cinematographers and other film advocates see, not the replacement of film with video (as has happened in most news gathering), but rather the production community's combining of film and HDTV technology to optimize signal quality as well as production efficiency. As Jac Holzman instructs:

> A production is not just about capturing images, whether it be on video or film; it is equally about post-production, the editing and manipulation and sometimes digital creation of images to form a synthesized whole. It is here that film, computers and video can serve each other seamlessly and well. Routinely, most dramatic television originated in the United States is shot on film, with the use of video assist for instant replay, and then either edited on film or transferred to video and edited within a video environment.[20]

Computer assistance notwithstanding, picture-framing problems and inconsistencies arise when a project moves back and forth between film and video formats, and these problems occur in both HDTV and non-HDTV environments. Consequently, camera operators must be aware of framing issues as they compose their shots. The shape of a picture is expressed in numerical terms as its *aspect ratio*. The conventional television screen has an aspect ratio of 4:3 (four units wide by three units high). The movie industry refers to this same ratio as 1.33:1 because each film frame is one-third wider than it is tall.

In figure 3-4, art director Barbara Miles cogently illustrates how the evolution of film stock aspect ratio has impacted what we see at the movie house and on our television sets. In each sequence, she shows the relationship between an era's film frame, its realization on a movie screen, and its transfer to the television environment. ("Anamorphosed" pictures are simply those shot for widescreen through a lens that achieves a 2.35:1 ratio, a process known by its Twentieth Century Fox trade name of *CinemaScope*.) In conjunction with Ms. Miles's final (current) sequence, it should be noted that because HDTV is a 16:9 format there would be no need to extensively "letterbox" the video picture. Movie house and television display sizes would be, for all practical purposes, nearly identical. Most entertainment television producers now "future-proof" their programs by composing in 4:3 format but protecting the edges of the frame for eventual 16:9 HDTV distribution.

Despite HDTV's aspect-ratio and picture-quality advantages, it is unlikely that video—even HDTV video—will replace film anytime soon. The "softer" nuance of film is still the medium of choice for the majority of U.S. prime-time television programs and nationally distributed commercials. Cinematographers not only serve this and related enterprises today but also are likely to serve them in an HDTV world where film's mood, and video's editing convenience, can continue in an enhanced collaboration that will result in even higher-quality visual results.

Regardless of the tools employed, cinematographers (as well as videographers who work on long-form projects) are engaged in the process of pictorial storytelling. "I learned that it isn't specific technical things that are most important," says Oscar-winning cinematographer John Toll. "Everyone I worked with had different techniques, just as mine are different from theirs. I think what is important is the idea that great cinematography comes from having a passion for telling a story with images, and having perseverance in making the imagery a vitally important part of the film."[21] "A cinematographer makes decisions on every shot that are like silent words," adds American Society of Cinematographers president Victor Kemper. "The choice of lens, composition, the way the camera is moved, and the use of light, shadows and colors are part of the story."[22]

When the story is a long-form location project, shooters (be they cinematographers or videographers) are usually divided into *first* (principal) and *second* units. Principal photography concentrates on coverage of the main cast members. "Because the cast must be accommodated, often at the expense of camera positioning or the shooting schedule, principal photography calls for the best efforts of the best technicians available to work around or through any and all occurring and recurring challenges," cinematographer Bill Hines points out. "All personnel on the first, or principal, production unit must be flexible, efficient and effective in working with the principal cast members." Second-unit photography, meanwhile, "fills in the pictorial blanks, such as establishing shots, inserts, stunts, special effects, and mob scenes which were purposely skipped or left over by the first unit," Hines adds. "Second unit image capture must be shot to seamlessly merge with footage shot by the first unit."[23] Therefore, despite its designation, second-unit photography is no place for second-class work.

Nor can second-class pictures be excused in corporate/industrial projects. For, as we mentioned in the previous chapter, viewers of these productions have visual expectations honed by thousands of hours of watching Hollywood movies and prime-time network TV. Amateur photography inevitably makes the company commissioning the project seem amateurish and causes quick audience tune-out or derision. A professional shooter, on the other hand, can make even a relatively mundane industrial script watchable and a good industrial script compelling. As veteran corporate producer Bill Miller characterizes it:

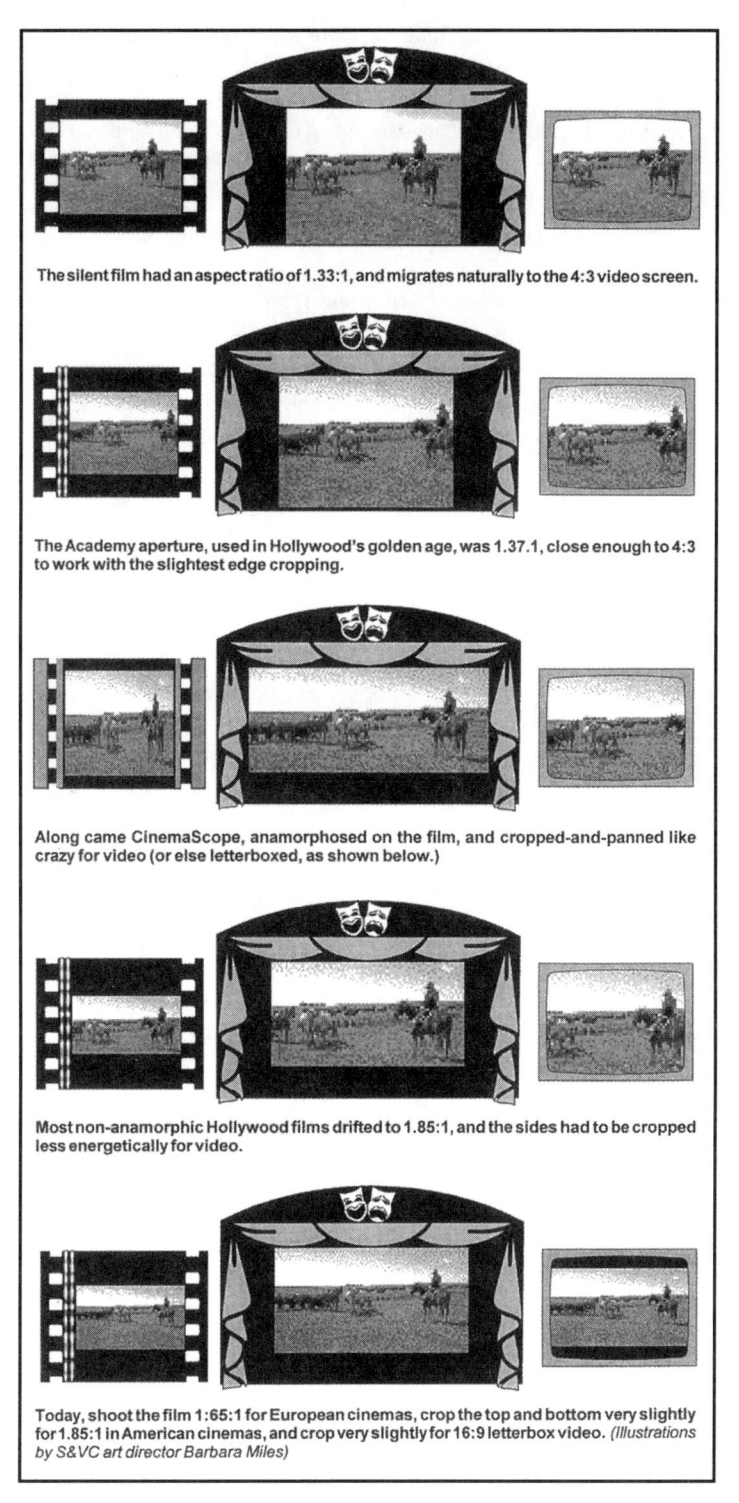

Figure 3-4. Evolution of film and video aspect ratio relationships. *Illustration by Barbara Miles for* S&VC *magazine.*

While a professional can be costly, in the long run he or she may save you thousands of dollars in damage control. If the script is not exactly the way you want it, it's easy and inexpensive to change the words. If the edit doesn't exactly cut it, re-edits can be made. On the other hand, if you are on location and have one chance to get the elephant dancing on one foot, at the right exposure, without the knee-jerk camera moves your mother-in-law brought back from Disney World, then it's time to cough up the coins for a professional. When it comes to shooting video, getting it wrong the first time can be very costly, and even unrecoverable.[24]

Whether working in corporate, long-form entertainment, or news production, and shooting film or video, the professionals most in demand are those who combine a technical understanding of their tools with an artist's sensitivity to the power of a moving image. As award-winning news videographer Bob Tur believes, it all comes down to "really being dedicated and loving cinematography and videography. Having a good foundation in still photography and knowing the mechanics of composing a picture is essential. The best technician might not be able to take a decent picture. The people who know how to compose a picture and know the technical aspects about the camera will be the stars of tomorrow."[25]

Film and Video Editors

It follows from the preceding discussion that the boundaries between film and tape *editing* will continue to blur as editors move back and forth between formats. Editors, of course, are the professionals who manipulate and arrange separate visual segments to further a project's continuity, pacing, and meaning. Editors working entirely in film accomplish this task mechanically by manually splicing together chosen bits (or *rushes*) of film in order to achieve the optimal shot sequence. Video editors, or film editors who work via video, perform the job *electronically* by transferring digitally coded segments onto an electronic master. The centerpiece device in this process, the *electronic editor,* has the capacity to control both the edit/record and playback decks in order to determine precisely where and in what order each edit will occur.

Today, virtually all video-editing professionals work on equipment that uses a standard time code developed in 1970 by the Society for Motion Picture and Television Engineers (SMPTE). One second of a television picture contains thirty separate frames. The SMPTE code gives each frame its own separate numerical designator. The numbers are visible both on special editor picture monitors and on the editing machinery's frame counters. With these numbers as reference points, editing personnel can precisely locate the start and end of the segments that they are consolidating into their master product. In this task, they are frequently assisted by computers that have been adapted to make footage control and inventory tasks easier.

Using the SMPTE code as a benchmark, the computer regulates all the equipment involved in the editing process and also stores all the decisions made by the editor about what to edit. This *decision sequence* or *edit decision list* sets down all the determinations made by the editor for assembling the piece. These include the list of segments to be used and their order of placement, each segment's exact entry and exit points onto the master, and the type and duration of transitions to be used at each point. A *cut,* for example, is the immediate change from one shot/segment to another; a *dissolve* is the gradual replacement of the subject of the first shot with the subject of the second. A *fade-in* or *fade-out* is the progression into or from *black* (a blank screen), and an assortment of *key* and *wipe* transitions provide more distinctive ways of putting one picture within another or pushing one aside in favor of another. With the computer as the control module, editing personnel

Production Functions

can call up all of these decisions in their predetermined order and execute each with to-the-frame precision.

Now that the price of computer packages has declined, even middle- and small-sized production facilities can offer their editors some sort of computer-assisted operation. Usually, this assistance includes the capability to generate graphic and animation effects. Thus, like engineers, editing professionals are being required to become much more computer wise in order to fully exploit the capabilities with which technology has provided them. Even relatively basic projects now regularly use intricate shot sequencing that was formerly all but impossible for even the most skilled and disciplined editing pro. Today's electronic and computer-assisted equipment makes these tasks comparatively easy and trouble free—but only for editors who are in control of the sophisticated systems that await their commands. There is a definite downside for editors to all of this technical flexibility, however. "It's so easy to make changes," says veteran editor John Fuller. "Producers want to see it a million different ways."[26] And creating those million different ways negates the time-saving advantages that computer editing has brought to the process.

Robin Lin Duvall
Freelance Video Editor,
Jerry Springer Show, *Chicago*

While in college, I remember thinking, "I will be glad when I don't need to study anymore!" Well, I'm still waiting for that time to come. With the advent of computer-based editing, software updates (additional features and improvements) seem to come at the speed of light. I have often wished for a "Matrix"-like download device for my brain. I hate reading manuals and would rather just lie back and plug in—but that is just a wish. Beyond the manufacturers' manuals (which can be poorly written and difficult to use) I suggest checking out the Internet for sites devoted to your edit system of choice. User's groups are also very helpful as well as classes, workshops, and seminars. Websites, user's groups, and workshops can also point you to useful how-to manuals and tutorial products. I edit with Avid's Symphony (with Unity), and I find Steve Bayes's Avid Handbook indispensable.

The choice of edit systems has broadened considerably in the last ten years. Before that time, there were just three or four "standard" edit systems used by the entire postproduction world. Currently, you can find an absolute armada of editing products at a variety of price points. Each works wonderfully. Avid, Final Cut, After Effects, and Premiere can all crank out superb projects. I am sure that more manufacturers will be unveiling more products at each and every National Association of Broadcasters Convention.

What I would like to point out here is that the choice of system does not necessarily matter that much. You can do great work on *many* systems. Some of those I mentioned earlier may be superseded by the time you read this. The real trick is to decide what works best for you and your clients. You must evaluate cost, speed, and compatibility when choosing a system. Don't forget to evaluate your own comfort factor as well. If you like working on a Mac, then work on a Mac. If you are more comfortable with a PC-based system, then work on a PC. You cannot do your best work when you are uncomfortable with the environment. Don't be bullied into buying or using a system because it is the "hot" new thing. I've seen wonderful work done even on small home-editing systems. The system doesn't edit. The *editor* edits.

Having said that, I would encourage all editors to continually *experiment* with new systems. If a friend or competitor acquires a new system, ask for a demo. Attend trade shows or user's groups (even for products you may not currently use) and just see what's available. If you have the luxury, attend the NAB Convention. Always test-drive new stuff and keep an open mind.

The postproduction world has moved to a largely freelance base. You may not have a choice as to the type of edit system you must work with. In this case, I would suggest limiting yourself to accepting work which will be completed using systems with which you are familiar. The exception might be a client with whom you have a very solid relationship, and who understands you are using a system new to you.

Hardware and software issues are only one part of an editor's job. A large part of my job falls into the "humanware" category. I must deal effectively with producers, directors, artists, audio engineers, and others. I must be able to communicate, organize, plan, and above all remain calm. It often falls to me to draw up a postproduction schedule and coordinate with other postproduction specialists (like graphics artists). This is like juggling.

I also must interpret what a producer or director is asking for and translate that idea into "the show." Storyboards are often a luxury I must do without. I have to really *know* my clients in order to understand what they are asking for. This skill is difficult to explain. I believe it comes from listening well, then repeating what you hear, verifying you are on the right track. Write everything down!

I must always remain flexible and adaptable. I rarely say no to unrealistic requests, but rather say, "What if we did this . . ." I also must be realistic about time constraints. I will give the producer an assessment of how long an edit will take and offer options for speeding up the process. The final call rests with them, but I never offer an option that I feel is unachievable. Sometimes there are quick and easy options when budget or time constraints are an issue. Don't get stuck in a problem—*there are always options*. If you can't think of an option right away, go to the bathroom. You'd be surprised what occurs to you during a potty break.

Finally, editors often find themselves operating as bartenders. Being stuck in a small, dark room with someone for sixteen to twenty hours at a stretch encourages a certain amount of familiarity. Producers often share personal problems, fears, or concerns with their editor. I sometimes feel I am a psychologist as well as an editor. This can actually lead to loyalty on the part of your clients. When they feel they can trust you, they'll come back. I would say two important things about this part of your relationship: (1) be careful what you share with your clients; and (2) never divulge a confidence. Unless it is illegal or inappropriate, what is said in the edit suite *stays* in the edit suite.

I would also encourage everyone in our industry to join professional organizations. These organizations are a wonderful way to stay connected with others and vent frustrations. It is a great thing to be able to vent about a twenty-four-edit session followed by a six-hour swim meet! Chances are someone else will be able to identify and follow up with a story about a three-day shoot in the rain followed by coaching a weekend soccer tournament. Yea! Someone understands my life! Even in this business, you are not alone.

Unlike the fame often accorded on-camera performers and sometimes bestowed on writers and directors, editors and their craft function most effectively when their contribution goes unnoticed. "I've always seen my job as trying to make the writer's vision come true—as close as I can get to what he had in his mind," states Bob Fisk of Phoenix Editorial. "I'm prepared to support an *idea*. I would describe myself as a pretty much invisible editor. I don't do 'flashy'; I'm looking at the story and I'm making sure that every single shot advances the story."[27] "A good editor is an insurance policy," adds veteran editor Steve Mark. "A claim is made upon him when things go wrong. In any event, he can't allow himself to have much ego. His work is always subject to second-guessing and revision by others, and even if left unchanged, the credit for it will usually go to those who have decision-making power."[28] Still, "it's an incredible craft," editor Jason Rosenfeld affirms, "and I think only other editors can appreciate the joy of it. Because, as they say, it's an invisible art."[29]

Although we have been focusing on visual editing, a similar battery of electronic and computer devices are available to improve the audio editor's performance and product as well. With the multiplicity of sound sources and tracks used in today's music releases and video/film soundtracks, the sound editor or editor/engineer requires more than a good ear and nimble fingers.

As in the case of video, electronic control devices allow the audio editor to keep track of and to manipulate a number of discrete elements in accomplishing the final product, or *mix-down.* Some devices, like the *Lexicon,* further use computer technology to permit the editor actually to compress or expand the soundtrack into other time lengths without altering musical/vocal pitch or comprehension. In other words, a thirty-five-second musical commercial can be squeezed into thirty seconds without listener detection.

All successful editors, those dealing in sight as well as those dealing in sound, develop an innate sense of rhythm and pacing. Some feature editors even use metronomes to help bring a consistent sense of rhythm to a pictorial sequence. "A producer once told me he never hires an editor who can't dance," Jason Rosenfeld reveals.[30] Familiarity with music, theater, and literature serves all electronic media editors well in developing a sensitivity to tonal, verbal, and visual flow, cadence, and build.

Directors: Technical, Assistant, and Main

Film and tape editors work in a *postproduction* environment. They enter the project after the individual rushes, segments, or tracks have been recorded in order to compile them into a final format and arrangement. *Technical directors,* in contrast, have as one of their principal duties the assemblage of multiple video sources in *real time*—that is, during their actual occurrence. Technical directors thus operate the *video switcher* through which all visual sources (cameras, slides, graphics, tape machines, or digital servers) are routed. For this reason, they themselves are often referred to as *switchers*.

By pushing buttons or moving levers and knobs, the technical director (TD) can call up and manipulate any of the available pictures for feeding over-the-air (live show) or to the master tape or disc machine (recorded show). In live programs such as newscasts and sporting events, the source the switcher selects is the visual that the audience sees at virtually the same moment. Sometimes, more than one picture is used in order to produce a split-screen, insert, or *superimposition* (a dissolve that pauses at midpoint so that one picture remains atop the other). Whatever the case, the TD executes the command from the main director, who is choosing from among the various pictorial options displayed on individual monitors.

At many outlets, a duty engineer serves as technical director. When the programming being aired is prerecorded or coming from the network, no main director is necessary. The TD simply follows along in a program log, bringing up either the network feed or the local prerecorded programs, commercials, and other continuity to be played back by video machines in the control room. Only when a local show is being produced is a main director required, and perhaps not even then if the production is very simple, such as a one-camera news-brief feature. At a small-market station, the main director may be only a glorified TD who handles program shot setup and selection responsibilities as a one-person operation. Conversely, at large outlets or production studios, a TD works under a supervising director but exercises a more complicated set of responsibilities. In addition to being the master switcher, the TD may oversee the entire technical crew, with camera operators, lighting and sound personnel, set construction staff, and control room engineers all under his or her control.

In even larger and more complex settings, an *assistant director* (*AD*) position is also provided. The AD's job is twofold. First, he or she helps the director in preproduction planning, making certain that the needed props, personnel, and recorded video footage have been identified and are available at the time the program is to be aired or recorded. Then, during the actual production, the assistant director aids in keeping track of shots, anticipating camera and performer cues, and monitoring the running time of program segments and the show as a whole. If postproduction work is necessary, these same functions also carry over to those editing sessions.

Meanwhile, out in the studio, a *floor manager* facilitates the job of all the control room directing personnel by communicating with camera operators and on-camera talent and by otherwise making certain that things in the studio are running smoothly. Sometimes called a *stage manager* (a throwback to earlier theatrical practice), this person is the ultimate studio traffic cop—receiving headset commands from the director in the control room for relay to others on the set. Camera operators typically wear headsets as well but cannot motion to talent or each other and manipulate their cameras at the same time. Although important, floor manager slots are seldom lucrative, and most people in this position use it as an entry to higher production callings.

Near the top of such callings, of course, is the position of main *director* itself. The director is the key person in the entire production process; this individual shapes the look and feel of the final product that the media audience will experience.

In shooting commercials, the director either serves as the cameraperson or hires another professional to perform that function. Either way, the aim is to accurately translate a script or storyboard into the clearest and most compelling fifteen, thirty, or sixty seconds possible. Commercial directors normally work on a freelance basis or are part of a company that specializes in shooting spots, public relations pieces, and similar assignments. An advertising agency or other creative firm captures the concept on paper and then hires a director to realize that idea on film or video. Much of commercial directors' success or failure depends on how effectively they manage preproduction. Commercial Director of the Year James Gartner maintains that preproduction is "a most important stage. If my art director is making a commercial only 5 percent different from me—and I am making a commercial 5 percent different from the cinematographer, who is maybe 5 percent different from the wardrobe stylist or editor—they add up. Before long you have something that is maybe 30 percent out of sync. So you must define the abstract."[31]

Well-established commercial directors command substantial fees because they can be counted upon to define this "abstract," to bring a conclusive look to their pieces that clients want and to which viewers react positively. In essence, these directors are making micromovies, in which the product must be the star and the plot revolves around how that product dramatically meets a need or solves a problem.

Even though they perform a similar defining and sculpting role for the pieces on which they work, directors of *programs* function in a somewhat different employment context. Within a station, directors are permanent staff members who supervise the visual design and shooting of the news, interview, and talk programming that make up the bulk of most locally produced offerings. If the outlet also shoots commercials for its local advertisers, these same staff directors probably handle those tasks as well. Networks, too, use staff directors for their in-studio news and talk shows. Certainly, the crews and audiences are larger at the network level, but the director's fundamental coordination role is the same as at local facilities.

Production Functions

Brett Holey
Director, NBC Nightly News
with Tom Brokaw

As I write this, I am the director of *NBC Nightly News with Tom Brokaw,* American's most-watched evening newscast for the past seven years. By the time you read this, there will likely have been significant changes in the broadcast, the industry, and probably my career. Tom Brokaw is preparing to leave *Nightly News,* and the impact of that single event will certainly be felt throughout television news. We'll talk about that later. First, a look at my job as it stands today.

To television newcomers, there is often confusion in differentiating the roles of the producer and the director. Basically, the producer (or, in the world of network news, the executive producer) guides the editorial direction and content of the broadcast. The director guides the look, feel, and execution of the newscast.

As director, on a typical night I lead a crew of more than fifty production personnel and technicians who comprise the core group that puts the broadcast together. On a busy night we can marshal as many of the resources of NBC News as we need, and the crew can easily expand to more than a hundred people.

My workday begins with a conference call where *Nightly* producers and representatives of our domestic bureaus discuss the day's events and the topics, stories, and correspondents most likely to be included in that evening's broadcast. Shortly after the morning meeting I have a discussion with the graphics personnel from *Nightly, Today, Dateline,* MSNBC cable, and MSNBC.com about what we are producing and how we will share work and resources through the day. I also meet with the *Nightly* graphics team and key editing people to set a direction for that day's production, monitor ongoing projects, and assign some speculative work for that day's broadcast. The next few hours are usually spent dealing with issues regarding personnel, technology changes (editing, graphics), and advanced planning for future broadcasts.

Midafternoon, the staff reconvenes for our "rundown" meeting. At this time we generate our first realistic lineup for the newscast. The meeting often becomes a lively debate about news events, journalism, and the relative importance of issues to society. Once the rundown is set, the executive producer and I order the elements we think will be needed from our graphics and editing areas. This is another exercise in skilled speculation, akin to reading tea leaves, as the anchor and writers will not forge their copy (which we are trying to illustrate) for several hours.

By 3:30 p.m., three hours before air, the machine is beginning to hit on all cylinders. The next few hours are usually filled with adrenaline-pumped multitasking. It is my job to see that all elements of the program are properly executed on time and to aesthetic standards for which I am the main arbiter. I choose music and graphics that I feel will provide the appropriate context and are visually interesting without being too flashy or detracting. All the while I have discussions with writers and producers to make sure we adapt changes in the rundown, the news, or copy that are radical departures from what we were planning.

There will be several sessions to record promos for the evening's broadcast as well as for tomorrow and the week ahead. Around these recordings I schedule the production of any other elements that will be needed for the broadcast including correspondents' standup shots, graphic illustrations, and "wall beds" (the video montages that play behind Tom in an attempt to help illustrate his copy in a broad manner).

As "airtime" approaches, the list of things to do seems to grow. Satellite feeds pop up into the control room monitor wall with correspondents ready for their live shots. The technical quality and communications for each must be carefully checked. Videotaped stories edited in bureaus around the country and around the world cascade into our record and playback area. Each must be cross-checked for technical standards and editorial accuracy. Graphics that have been tweaked or hastily made to match late-written copy are assembled into playback cues with hopes that I will find a second to review them before they are broadcast. Stories cut in New York are readied for playback, and tons of logistic and editorial information is rushed to the necessary parties, who marry it into an ever-changing road map of how the broadcast will air.

Ready or not, 6:30 p.m. comes and we take to the air with the best broadcast that time, technology, and fate allow us to execute. On a good day, once we're on air my job is like that of an orchestra conductor, cueing every element to play at the precise moment and with the proper nuance to produce a symphony of information, woven into a beautiful broadcast. On a bad day, my job is more like that of a traffic cop, trying to get a lot of noisy objects through a tight space so that we all arrive at the other end without serious injury.

At 7:00 p.m. we refeed the 6:30 program to other time zones and update and repair any parts of the first broadcast that require it. Most days, Tom, the New York staff, and I head for home shortly after 7:30 p.m. There is a small staff in our Los Angeles bureau who can do minor updates to the later broadcast if needed. We provide them with "standby" elements that allow them to get in and out of the pretaped broadcast and add elements for live correspondents.

On an average night, more than 10 million Americans choose to spend a portion of their evening getting information from *Nightly News*, and another 16 million watch our competitors. As a business model, it has never been more profitable. Still we face many challenges aside from Tom's departure and the shifting American lifestyle. (When was the last time you finished dinner before 6:30 and sat down to watch the news?) Many ideas have been discussed as to how to grow the broadcast. Expanding or changing our time slot, finding outlets on the Internet, and exploring time-shifting technologies all seem like obvious options. We are currently available to a potential audience of 50 million in Europe and the Middle East, and that is a market we continue to explore. But each of these options has its own obstacles and limitations. Ultimately, you—as both viewer and future professional—will help determine what becomes of our broadcast and the wider industry.

In-studio news and talk offerings seldom use visual treatments that are daringly experimental, or *avant-garde*. Thus, directors who work in this idiom are not expected to be television artistes but instead must create consistent visual progressions that are easy to follow and that make the on-air talent look as good as possible. The best directors here are those whose work goes totally unnoticed by viewers—whose shot transitions and pictorial selections are so smooth and sensible that they never call attention to themselves but serve only to propel show content into clearer focus.

This same principle holds true for directors who specialize in the coverage of sporting events. Television never merely reproduces a game—though the ardent fan must be allowed to think that it does. In actuality, however, the director is making *video drama* of that game by adding, deleting, and rearranging elements to create a more continuous scenario. Thus, points out professor Sarah Kozloff, the viewer "sees the events filtered through the control room which switches from crowd shots, to the cheerleaders, to the coaches, to the action; which flashes back to pregame interviews; which forsakes real time for slow motion and freeze frames; and which repeats the same play over and over. The viewer is no longer simply watching the game, but rather a narration of the game in which various choices have been made concerning temporal order, duration, and frequency."[32]

As the game proceeds, the experienced director begins to see a single story line emerge and, between plays, arranges elements to enhance that story line's evolution. In this way, even a lopsided

Production Functions
85

contest can retain viewer interest because of the additional narrative that the production crew is weaving. In preparing for game day, a skilled director "has hundreds of story lines rattling around in his head," reveals sports reporter Tim Dwyer. "He feels the pressure of worrying whether there is anything he missed and whether he has really placed all the cameras in the right places."[33] When he or she makes the right choices in camera location, shot selection, and pictorial pacing, the sporting event director creates seamless, real-time storytelling that can rival the intensity of even the best prescripted dramas.

In the case of prescripted entertainment shows, some situations call for close to real-time decision making on the part of the director; others allow for more studied show assemblage in postproduction editing. If the director is recording before a live audience, the aim is to capture as successful a take as possible the first time around. Unlike sports coverage, however, there is the possibility for shot retakes and fixing small deficiencies through postproduction editing. Nevertheless, series production works on very tight deadlines, so time is limited. A situation comedy director must normally deliver an episode a week, his or her soap opera counterpart one or even two shows a day, and the game show director perhaps three to five episodes in a single extended shooting session! Under such conditions, the more footage that can be captured in final form the first time, the better.

Although game show and daytime soap opera directors are recording their shows, they try to achieve as much of the final shot progression as possible in the initial run, operating much as a newscast director would work. A taped situation comedy director may function in a similar manner but with greater opportunity for selected retakes or for shooting corrective inserts that can be added in postproduction editing. For filmed (as opposed to taped) sitcoms, there tends to be a greater amount of postproduction work, and the director spends much more time on the studio floor with the cast than in the control booth. Filmed shows especially are shot a little long so that an editor always has more than the absolute minimum footage from which to trim the final product.

A typical five-day schedule for the completion of a single situation comedy episode is described below by John Finger, director of photography for *Cheers*. *Cheers* was a filmed show, but the progression remains much the same today for sitcoms shot on tape and film.

> During the first day, there is a production meeting involving all department heads, including props, lighting effects, wardrobe and special effects. We also determine our need for extras for this episode. Later that same day, the cast reads the script to see how it sounds. . . . After the reading, they have the first rehearsal on the stage. Meanwhile, the writers might be busy late into the night making revisions.
>
> Day two, changes are reviewed, and rehearsals begin. Later in the day, there is a complete run-through. . . . There are generally more rewrites. Day three is more or less a repetition of day two.
>
> On day four, the cameras are brought on stage, and they are used during rehearsals. Afterwards, stand-ins are used while [director] Jim Burrows and I discuss positioning [blocking] of the cameras. There is one last run-through without the cameras, since staging might have changed. Meanwhile, the writers are still fine-tuning.
>
> During the fifth day, most of the cast and crew come in around noon. They rehearse a couple of times. . . . Late in the afternoon, there is a dress rehearsal with everybody present. . . . Many times there are still changes being made in the script. After dinner, the show is filmed in front of a live audience.[34]

Throughout the entire process, the director's coordinating input is crucial for making quick decisions and communicating them to everyone involved.

Directors of hour-long series, which are virtually always shot on film, work much more like their movie than their situation comedy counterparts. Hour projects combine both soundstage and location shooting. They are shot out of order and then completely assembled in postproduction just as is a full-length movie, a process that expends several weeks. Because the director must be involved throughout the process, more than one director works on a given series, with the production of episodes assigned to each director frequently overlapping. The complexity and cost of hour shows (and of made-for-TV movies) invest their directors with greater overall creative responsibility. Unlike the situation comedy production environment cited above, industry watcher Neal Koch reports: "Once long-form productions start, the scripts are set and writers are often not even allowed into the filming sessions, which puts the director in control."[35]

We cannot leave the subject of electronic media directors without pointing out that such professionals are found on the radio side, too. Certainly, there are fewer opportunities for U.S. radio directors than in the thirties and forties when radio drama ruled the networks. Nevertheless, some skilled audio professionals still work as directors in two separate areas: syndicated program production and commercials.

Today's national syndicated radio shows are often music-related, like *American Top 40,* and may combine the playing of tunes with host commentary and interview segments. Frequently, the programs require extensive background research into the featured musicians, songs, and environment out of which they came. A director coordinates the consolidation of all of this material into a well-paced program that will play effectively in every area of the country and perhaps overseas as well. These syndicated projects must be consistent in their style and pacing from week to week and must develop a flowing format that is always maintained. Either live or, more frequently, in postproduction, the radio director's job is to ensure this continuity and the comfortable predictability that the show's listeners and affiliated stations expect.

Major-client radio commercials also employ directors to deliver spots that are as well engineered and enacted as anything heard on the air today. A talented radio copywriter can create a wonderfully involving and image-rich script, but it won't play well if the vocal casting or pacing is off the mark. Good radio spot directors (sometimes called producers) hear the optimal product in their heads and are able to motivate and navigate their talent to achieve that sound vision. Award-winning radio writer/director Joy Golden sees her job this way:

> The key to making a good radio commercial is to know how to direct the people.... What I often do, especially with some of the new people, is set up a visual scenario for them. I'll say, "You're a husband and wife in bed, right? And it's early in the morning, and he's feeling lousy, and she's feeling ___." I'll give them a whole physical and personality scenario, so they can perform, in their minds, in a setting that isn't so audio. Because I want a bigger thing to come out.... In the studio, most of the time it requires ten, twelve, fifteen takes for the talent to get it right, especially if it's a monologue. We do and we re-do and we re-do.[36]

Golden's description of the radio commercial director's job is probably an apt illumination of the key responsibility of *any* electronic media director. Even though the equipment varies (and there's lots of it), a successful director is still communicating with people (the audience) through the skillful guidance of other people. This guidance, or the lack of it, is most noticeable in how a director handles performers. "All actors," observed veteran television director Peter Dews, "are to some degree nervous, and the director can help them to overcome this by creating the right kind of environment to work in, so that they know where they are with the medium, come to working terms with it."[37] Fellow director Franklin Schaffner adds, "Simply defined, I suppose the director's function really is just staying one jump ahead of the cast."[38]

The special constraints of corporate video present a director with additional challenges, observes industry authority Stephen Barr:

> Working with tighter budgets, shorter time frames, smaller crews and less-extensive arsenals of equipment, they don't have the same creative clout; they find themselves walking a tightrope between aesthetic expression and economic reality. The employer may pull the plug before the ideal lighting setup is achieved; time constraints may keep flubbed lines from being reshot; there may not be enough money to rent a dolly—forcing the use of a zoom where a tracking shot would have worked better.[39]

Understandably, it is easy for a television or radio director to lose sight of the human function and to become preoccupied with all the technical tools and procedures of our industry. But like any other behind-the-scenes professionals in the electronic media, accomplished directors realize that hardware is only their avenue, not their destination. Equipment and its manipulation exist to package performances in such a way that these performances creatively strike responsive chords with the target audience. "Most of what you bring to this task is inherent in you," director James Gartner reveals. "Author H. L. Mencken said, 'Nothing can come out of the artist that isn't in the man.' But creativity is also a process that evolves. It is surrounding yourself with good art. Good film. Good literature. I believe creativity, largely, is having good taste."[40] This holds true whether we are directing a radio commercial, a television newscast, a prime-time entertainment series, or a corporate video.

Production Assistants

Last, but decidedly not least, are the production assistants who facilitate the work of the other professionals we have previously discussed in this chapter. Most production assistant jobs are entry level; providing a first extended exposure to the many phases of the business. PAs pull camera cable, help hang lights, locate and watch over props and costume parts, conduct news or music research, distribute promotional materials, keep track of financial records and receipts, procure project supplies, transport equipment and personnel, and provide a myriad of other services to keep the production on track.

A production assistant job is seldom glamourous but does serve as both introduction and gateway to any number of higher-level appointments. Sometimes the PA slot is a formal apprenticeship structured to help meet requirements for admission to technical craft unions. At other times it is a job in its own right that may cross several production areas. Either way, PA labors constitute an opportunity to observe closely the interdependence of a production's many facets. They also help isolate those positions that may be worthy of greater exploration in the future. In other words, PA experience can differentiate which career paths one wishes to follow from those of lesser interest.

Perhaps the greatest advantage to production assistant work is its inherently networking nature. Over the course of several assignments, a PA can meet professionals at all levels of the business: people who may hold the key to career advancement. One of the author's students began his career, while still in school, by signing on as a weekend crew member whenever ABC Sports was covering any kind of game within driving distance. These minimum-wage, one- to three-day jobs didn't net him much cash. But they did bring him, his skills, and his industriousness to the attention of crew chiefs who were happy to hire him as a full-time PA upon graduation. Ultimately, this led to successive positions as unit head and supervising sports producer.

If you are interested in a production career, it is likely your first role will be that of PA. Where

Tohry V. Petty
Producer, BASE Productions, Inc.

The role of a production assistant is very loosely defined. As the name implies, a PA provides assistance to the realization of a project. The company and/or project define the duties for which you, the PA, would be responsible. The nature of a PA position differs from a majority of other, more defined, jobs in the broadcast industry. Because as a production assistant, you are granted an all-access pass to observe and participate in multiple parts of an enterprise.

Key objectives that would describe an excellent production assistant are: resourceful, positive, creative, and responsible. A good PA must be an eager and fast learner—because your duties could change instantaneously.

In one minute you could be assisting the producer or director, the next you could be helping the lighting director or lending a hand with office work. Whether it's phoning crew members about availability and call times or picking up craft services and meals, you are an integral part of all stages of the production. Granted, you may not be calling the shots, but you will be receiving a variety of experiences during each production.

As tape is rolling, a production assistant could be asked to maintain *log notes* of the tape or film during production. What this means is to sit near the field monitor and jot down any action that is happening along with the corresponding time code and director/producer comments. These notes may consist of entries referring to good or bad takes, director and producer evaluative remarks, and/or a brief summary of an interview or scene. Whatever the venue, these notes are extremely important for the editor in postproduction so that he/she can know at a glance which takes can and should be integrated into the final product.

In a postproduction facility, a production assistant could be asked to take notes on the entire tape ("log" it), marking the time code of a segment and writing brief descriptions of the action, dialogue, and/or camera shots. Depending on the facility, some editors have specific key words that should be used when logging tapes. Descriptive words used to sift through footage could be yell, dance, crowd, or CU (close-up). In sports, key words like goal, save, fight, face-off, or dunk can be of tremendous help when editing specific packages. For interviews, the interviewee's name and a brief description and/or summarization of their comments are key.

On a shoot, a PA could be asked to pick up the meals for the crew from a nearby eatery or find miscellaneous props or makeup that are needed at the last minute. A production assistant could also act as a grip (an assistant to the lighting director) by helping set up and move lights around the set. Production assistants can also aid the production as a stand-in, filling in for talent as lights are set up and placed so that the camera operator and lighting director have a subject on which to focus. All the while the PA is given a unique opportunity to learn through observation and participation about the preparation and equipment needed for a production.

A production assistant can work in an office or in the field. A PA can be on the regular payroll as a full- or part-timer or can be freelance and hired on an as-needed basis. A great production assistant treats each task as a learning experience and takes great pride in their work. While on a shoot, make conversation with the crew in any downtime, ask each person what they do, how they got started. Don't be a nuisance, but make general conversation—especially if that specific job seems of particular interest to you. Genuinely ask if you can assist in any way—an extra set of hands is always helpful, and you can learn a new skill in the process. In the long run, it is your attitude and work ethic that make an impression on the crew. If a pleasant demeanor and willingness to learn shine through, a production assistant position can open the door to a wide range of professional opportunities.

Production Functions

It's said that it takes a whole village to raise a child. Well, a production assistant is as much a part of the village as the village elders. Because once the project is complete, you can know that you contributed meaningfully to a quality production.

---

you go from there depends on a number of factors—but most importantly the two mentioned above: the skills you bring to "the party" and the industriousness you demonstrate while you're there. Talent is indispensible. But talent without initiative is an unnoticed waste. Conversely, initiative is essential if a production team is to finish a project and finish it on time. But initiative without talent just gets in everybody else's way.

It is impossible to compile a complete list of all the types of PA positions found in our business. Some of them don't even have the words "production" or "assistant" in the job title. But all of them exist because they bring the audio, video, or online project to completion in as efficient and effective a matter as possible. The availability of skilled PAs means that the time of higher-priced, more veteran production staff is not wasted. PAs exist so that more senior people can place all of their focus on the crafts in which they are expert. Delegating supporting tasks to PAs keeps projects on time and on budget. And it creates a labor pool from which all craft areas can be replenished by promotion of the best people from within.

A production person doesn't set out "to be" a PA any more than a neurosurgeon sets out "to be" a first-year resident. But in each case, this early step is an essential prerequisite to, and clarifier of, the ultimate goal.

Chapter Flashback

Production professionals create the stages on which their performance and conceptual counterparts can display their talents. Technology has automated several operations that formerly required the constant ministration of technicians and engineers. This automation has reduced the number of lower-level jobs but has opened up entire new sets of opportunities for people who can manipulate state-of-the-art audio and video tools that are often computer assisted. Some television engineers are responsible for lighting as well as for video equipment. Larger facilities employ specialists who can manipulate illumination to accomplish sophisticated scene sculpting that transcends the two-dimensionality of the television screen. Much of the work of the lighting director/designer revolves around the meticulous construction of a *lighting plot.*

The individuals who run the cameras can generally be divided into three categories: *camera operators, videographers,* and *cinematographers.* Some camera operators have been replaced by robotic equipment, and videographers have supplanted cinematographers in television news gathering. Both film and videotape are widely shot in commercial, documentary, and entertainment productions, however, with the medium of choice depending mainly on the look desired for the final product. HDTV is blurring the distinctions between the two media, but it is unlikely to eliminate these distinctions in the foreseeable future. Both film and video *editors* (people who piece together isolated sequences into cohesive wholes) usually perform their tasks electronically for greater efficiency and element control. A skilled editor's work will not call attention to itself—so editing is referred to as "an invisible art."

Directors also can be divided into three main classifications: *technical, assistant,* and *main,* although in smaller facilities, all three functions can be exercised by a single individual. Technical directors assemble multiple video sources in real time and may oversee the entire technical crew. In more complex settings, an assistant director helps with preproduction planning and keeping track of shot sequences and running times during shooting and postproduction. The main director exercises the ultimate responsibility for bringing together all the technical and performance elements into a cogent whole; this responsibility holds true for both television and radio projects.

Production assistants function in a wide variety of capacities to keep projects on track and on time. PA work is the key gateway to higher-level production positions throughout the industry.

Review Probes

1. Define the difference between a "translator" technician and a "creative value" technician.
2. Compare and contrast Notan and Chiaroscuro lighting.
3. How has the replacement of cinematography with videography changed the practice of television news gathering?
4. What types of technical activities take place in *pre*production compared with *post*production?
5. What are the differences and similarities between the way news, sports, and scripted entertainment directors work?
6. In what ways do production assistants facilitate projects and advance their own careers?

Suggested Background Explorations

Alten, Stanley. *Audio in Media.* 6th ed. Belmont, CA: Wadsworth, 2002.
Bailou, Glen. *Handbook for Sound Engineers.* 3rd ed. Woburn, MA: Focal Press, 2001.
Caulfield, Louie. *Unit Runner and Assistant Camera Operator.* Woburn, MA: Focal Press, 2001.
Compesi, Ronald. *Video Field Production.* 6th ed. Boston: Allyn & Bacon, 2004.
Cury, Ivan. *Directing and Producing for Television: A Format Approach.* 2nd ed. Woburn, MA: Focal Press, 2001.
Dancyger, Ken. *Technique of Film and Video Editing: Theory and Practice.* 3rd ed. Woburn, MA: Focal Press, 2001.
DiZazzo, Ray. *Corporate Media Production.* Woburn, MA: Focal Press, 2000.
Donald, Ralph, *Fundamentals of Television Production.* Ames: Iowa State University Press, 2000.
Eargle, John. *The Microphone Book.* Woburn, MA: Focal Press, 2001.
Fauer, Jon. *Shooting Digital Video.* Woburn, MA: Focal Press, 2001.
Gross, Lynne, and Larry Ward. *Electronic Movie-Making.* 4th ed. Belmont, CA: Wadsworth, 2000.
Huber, David. *Modern Recording Techniques.* 5th ed. Woburn, MA: Focal Press, 2001.
Kindem, Gorham, and Robert Musberger. *Introduction to Media Production: From Analog to Digital.* 2nd ed. Woburn, MA: Focal Press, 2001.
Laycock, Roger. *Audio Techniques for Television Production.* Woburn, MA: Focal Press, 2001.
Millerson, Gerald. *Lighting for Television and Film.* 3rd ed. Woburn, MA: Focal Press, 1999.
___. *Television Production.* 13th ed. Woburn, MA: Focal Press, 1999.
Reese, David, and Lynne Gross. *Radio Production Worktext: Studio and Equipment.* 4th ed. Woburn, MA: Focal Press, 2001.
Rose, Brian. *Directing for Television: Conversations with American TV Directors.* Lanham, MD: Scarecrow Press, 1999.
Uva, Michael, and Sabrina Uva. *Uva's Rigging Guide for Studio and Location.* Woburn, MA: Focal Press, 2000.
Ward, Peter, Alan Birmingham, and Chris Wherry. *Multiskilling for Television Production.* Woburn, MA: Focal Press, 2000.
Zettl, Herbert. *Television Production Handbook.* 8th ed. Belmont, CA: Wadsworth, 2003.

CHAPTER 4

Sales Functions

This chapter focuses on the sales and promotion executives who generate the dollars and public or industry awareness that make it fiscally possible for electronic media businesses to continue to operate. These marketing and public relations functions are indispensible and, as a class, generally command higher salaries than most other aspects of our profession. The names of sales and promotions people are not found on program credits and are not read over the air. But it is these individuals who can turn intangibles of "time," "air," and "space" into highly profitable products.

Station/System Salespersons

Local commercial stations, cable systems, and most of the networks that service them all employ sales staff whose job it is to sell airtime to advertisers. For stations and broadcast networks, revenue from the sale of their commercial slots (called *avails*) constitutes virtually the sole source of income. For a cable system, on the other hand, this revenue stream is complemented by monthly subscriber fees from consumers as well as rights fees paid by smaller and newer cable networks that do not yet have the audience appeal to demand that the cable systems pay them for carriage rights.

Whatever the medium, the time salesperson (called an *account executive* at some outlets) makes a living by offering advertisers an intangible commodity (time) that these advertisers can use to attract prospects and build brand recognition.

In many ways, electronic media salespeople work in a more difficult arena than do their print counterparts. First, the availability of on-air time is fixed. There are only twenty-four hours in a day (assuming the outlet programs all twenty-four). The air schedule cannot be contracted when advertiser demand is low or expanded when it is high. *Unsold* time is lost revenue; the lack of *enough* time is lost business. Print media, conversely, can add or subtract pages at will to mirror advertiser needs. (Many local papers are much thicker on Wednesday, for example, when grocers are seeking the end-of-week food buyer, than they are on Tuesday.) So the air salesperson must not only secure advertising dollars but must also secure them in such a way that the resulting commercials will be more or less equally spread throughout the course of the week, the month, and the year.

Also unlike the case with print, the cost of operating a station bears no assured relationship to the number of people reached. A newspaper expecting to sell ten thousand copies will print ten thousand copies (with a few extra to meet unanticipated requests). As the paper's circulation expands, it can increase the press run incrementally to match actual demand. A station, on the other hand, must transmit a full program day whether one hundred or one hundred thousand people tune in. Its program costs are mainly determined by its market and signal dimensions and cannot easily be scaled down because it is failing to attract enough listeners. Advertisers pay rates based on the number and type of people the outlet is delivering. These rates fall as audience levels fall, but operating and programming costs stay relatively the same. If the on-air product is unattractive, the time salesperson is put in a double bind, being forced to sell more inventory at lower rates on a less desirable outlet.

Paul Boscarino
Station Manager and Director of Sales, Clear Channel Radio of Grand Rapids, Michigan

When I was a college student majoring in broadcasting, my career goal was to become a famous, highly paid, major-market disc jockey. Beyond my required coursework, I performed on-air shifts every weekday at one of the campus radio stations and on weekends worked for a small-market AM/FM/TV facility fifty-five miles away. Eight weeks prior to graduation, I came to the realization I needed to do *more* with my degree than what an air personality career would likely produce. A friend of mine who was aware of my epiphany encouraged me to speak with the faculty adviser overseeing the commercial student station. The adviser listened patiently as I told him about my lost interest in pursuing any future on-air or production work. His memorable response was, "Well, I guess you have no choice but to try doing sales for us *if* you want to remain in radio after you graduate."

And that is how I unceremoniously began my career in radio sales.

To my surprise, I immediately enjoyed working with area business owners. I learned about their businesses and then created written proposals addressing the logical reasons why they should advertise on the campus station to target the student population and sell more of the products and services their businesses offered.

Two days after I graduated from college, I eagerly began my professional career in radio advertising sales. The things I enjoyed about selling advertising for the college radio station expanded as my professional radio sales career unfolded. The daily opportunity to meet with people who owned and/or managed different businesses (furniture stores, car dealers, clothing stores, insurance companies, appliance stores, restaurants, shoe stores, banks, jewelry stores, etc.) and learn about their marketing challenges and/or successes was enlightening. Gaining approval from them to convert the information they shared with me into proposal form required me to put the research and writing skills I developed in college to practical use. Knowing that the marketing campaign recommendation I would present would be scrutinized against what area TV stations, newspaper, magazine, billboard, transit, and competing radio station advertising representatives were simultaneously presenting unleashed my competitive juices. The sense of victory when the client said, "I like what you've put together here. Let's go with it" was very satisfying. And the best part of all? Hearing the client later say, "My customer traffic and sales increased and my return on investment from that ad program you recommended was great! Let's talk about doing a similar program next month."

I am pleased to report these same feelings of

accomplishment continue on a daily basis with the thirty-three marketing professionals I oversee on behalf of a multimillion-dollar sales operation encompassing seven stations owned by the largest radio broadcasting company in the world. Being a broadcast salesperson is oftentimes not an easy job. But there truly is a great amount of gratification that can be realized—and hey, the money's not bad, either.

While you are in the midst of your college experience, I encourage you to give some consideration to a career in broadcast sales much earlier than I did. Seize the opportunity to take more business and marketing courses than I did. Talk to professional broadcasting salespeople and ask them to share their career experiences with you. It may turn out that a position in broadcast sales *could* be just the right career choice for you as well.

Finally, whereas a print layout is a concrete entity that sits there as long as one wants to look at it, a radio or television spot is a real-time emission that exists only at the moment it is being transmitted. The amount of information that even a sixty-second commercial can carry is also significantly less than can be accommodated in a quarter-page or larger print ad. Thus, many local advertisers either avoid the electronic media or have unreasonable expectations as to how much content can be memorably conveyed in a single spot.

Skilled time sellers have learned to surmount these problems while carving out lucrative careers. The specific techniques they use are beyond the scope of this book, but a few key principles shaping their task can be illuminated here.

Fundamentally, sellers and buyers of airtime negotiate on the basis of *audience ratings, shares, cost-per-thousand* (abbreviated *CPM* with *M* being the Roman numeral designator for 1,000), *gross rating points,* and *frequency* (message repetition). A rating measures the number of people or households reached by a given station/channel as compared with the total number of people or households in the market. Historically, radio has been measured in *persons* and television in *households.* However, the expanded use of more sophisticated electronic *peoplemeters* that count each person in a household is largely converting TV measurements to a person base as well. A rating then, is mathematically expressed this way:

$$\text{RATING} = \frac{\text{People/households using your station/channel}}{\text{Total surveyed people or homes}}$$

Thus, if we survey 280 persons and find that 42 of them are listening to our radio station, our station has attained a 15 rating:

$$\frac{42}{280} = .15$$

From this, (and assuming the measurement service's sample was representative of the composition of our market as a whole), we estimate that approximately 15% of the listeners in the market are tuned to our station (and its commercials!) at the time on which we are focusing.

A *share,* conversely, is the percentage of listeners/viewing households attained by our station/channel in relation to the number actually listening or watching at the time:

$$\text{SHARE} = \frac{\text{People/households using your station/channel}}{\text{Total surveyed people or homes } \textit{using the medium}}$$

Following up on our ratings example, let's assume that the number of people in the sample who report they are listening to the radio at the time is 126. Therefore, our equation is

$$\frac{42}{126} = .33$$

This means that we enjoy an audience share of 33 (in other words, one-third of the persons using radio at the time in the market are listening to us). These two numbers, the rating and the share, are often expressed together (15/33) in describing the performance of the station or program.

Shares are always larger than ratings, of course, because shares factor out persons/households not using the medium *before* station performance is computed. Even though shares are more important to the salesperson in selling against competing stations, ratings are also of interest to the advertiser because they indicate what percentage of the *total market* is tuned to the station—and, therefore (presumably), to that advertiser's commercial. The farther we get from prime listening (morning/afternoon drive time) or prime viewing (evening prime-time) periods, the wider is the discrepancy between ratings and share numbers because the greater is the number of people or households *not* using the medium.

In many markets, for example, *Late Night with David Letterman* pulls a 2 or 3 rating in the 11:45 to midnight quarter-hour. Even though this would be a failure by broadcast network prime-time standards, in its late-evening period it often means a 17 or higher share (depending on how many alternate broadcast and cable viewing choices are available in that locality). For advertisers who want to reach the *available* TV-viewing audience at that time, the station carrying *Letterman* is worthy of consideration.

Of course, an advertiser does not pay anywhere near as much for a commercial slot near midnight as for a spot in prime time. This is because commercial rates are figured on the basis of CPM. In other words, what does it cost to reach each group of 1,000 on the outlet under consideration for a time buy?

Let's go back to our original example and assume that the sample of 280 people is an accurate reflection of the entire market population of 98,000. We divide the total number listening (or viewing) by the total sample size to get our PUR (persons using radio) or, if we are measuring television, PUT (persons using television) level:

$$\frac{126}{280} = .45 \text{ (PUR level)}$$

In other words, our numbers indicate that 45% of the people in the market are listening to radio at the time in question. We then multiply the radio PUR (or the television PUT) level by the total number of actual people in that market to get the number using the medium. Continuing with our original example:

98,000 × .45 = 44,100 persons using the medium

As a salesperson for the radio station in our example, however, we know that the advertiser's cost-per-thousand expectations are based on the number of people our station reaches. We therefore multiply our rating (15) by the total market size to convert that rating to people reached:

98,000 × .15 = 14,700 persons listening to our station

Sales Functions

Now, let's say that advertisers are willing to pay a CPM for radio of about $8. We first divide our number of persons into units of 1,000:

$$\frac{14,700}{1,000} = 14.7 \text{ groups of } 1,000$$

and then we multiply the number of these thousand-people groups by what the advertiser is willing to pay to reach each group:

$$\$8 \times 14.7 = \$117.60$$

Thus, as a salesperson, we seemingly could be comfortable in trying to sell our spot positions in this particular time period at a price of somewhere between $115 and $120.

In actuality, things are much more complicated than this. The 14,700 people the salesperson is selling may not, in whole or part, be the demographic types of people this particular advertiser is most interested in reaching. Perhaps that advertiser does not think the content of the program is a suitable image vehicle for his or her company. Or perhaps this buy is made contingent on the availability of a number of other time slots in parts of the day in which the station is already sold out. Consequently, the advertiser may not be able to purchase the *gross rating points* (total of all rating points for a specific schedule of spots) or *frequency* (number of times a given listener/viewer will be exposed to the spot) required. These and a multiplicity of other factors enter into the negotiation, and the salesperson must anticipate them all before making a pitch.

Qualitative factors also come into consideration. Certain clients may be willing to pay a considerably higher CPM in order to reach a more specific/higher-quality audience. All-news radio salespersons, for example, can usually out-price their station's rating/share level because all-news tends to reach older, more affluent male professionals who are prime targets for luxury cars, stock/commodities offerings, travel services, and similar clients. On the other hand, certain formats/stations are perceived as delivering downscale listeners, and thus the salesperson faces a constant battle against pressures to underprice the rating/share numbers because they constitute less valuable prospects for many advertisers.

Most salespeople, whether they work for radio, television, or cable, operate from a *rate card* that sets forth the costs of airtime on different parts of the schedule, in different packages (*volume buys*), and in a variety of spot lengths. Standard rate cards (like the prototype described in figure 4-1) are usually *grids* that list several different pricing options. As advertiser demand increases, single and package asking prices move to a more costly grid on the card. Thus, the numbers in figure 4-1's boxes are dollars, with the price varying by time of day and advertiser demand.

Increasingly, rate cards exist in computer rather than preprinted form. This allows time selling to work like an airline ticket reservation supply-and-demand system (extending the comparison introduced on our prototype grid). As veteran radio sales executive Paul Boscarino says of such software, "Gone are the days when I would have to second-guess when and how much (or how little) to raise or lower our rates, based on (projected) inventory demand. Now, via our software-generated rate card, our rates are raised (or lowered) in exact proportion to our actual inventory demand."[1] Fourth-quarter (October to December) demand, for example, is normally much higher than first-quarter (January to March) because of the effects of the holiday season and the need for many businesses to reduce end-of-year stock for tax purposes. Of course, because of their prominence and audience penetration, certain singular television events like the Super Bowl and the Academy Awards (the Oscars)

Sample Grid Rate Card

GRID	6-10 a.m.	10 a.m.-3 p.m.	3-7 p.m.	7 p.m.-midnight
I	150	100	130	60
II	140	90	120	50
III	130	80	110	40
IV	120	70	100	30
V	110	60	90	20

The level of grid used by a station relates directly to the amount of inventory already sold. For instance, the station would be at Grid I if 90 percent of the inventory for that time period were sold. The grids match up with sell-out levels as follows:

Grid I	90 %
Grid II	80 %
Grid III	70 %
Grid IV	60 %
Grid V	Less than 60 %

As inventory is sold, the price goes up. This is the concept of "yield management" that was invented by the airlines. The farther out that you buy, because the demand for the inventory is less developed, the cheaper the rates will be. Like the airlines that sell seats six months out for less than what they sell a seat for one week before departure, advertisers who commit to long-term contracts can buy the same time periods for less than the advertisers who buy only a week or month at a time.

Figure 4-1. Layout of a "grid" pricing matrix.

command huge advertising dollars even though they are first-quarter occurrences. Keep in mind, however, that the wisdom of buying time around even such prestigious shows as these depends on your target audience. The Super Bowl delivers relatively reasonable CPMs for men eighteen to forty-nine, and the Academy Awards show provides almost as reasonable CPMs for women in that same age group. However, neither show is as efficient a buy in reaching the opposite demographic.

Broadcast or cable, network or local, television or radio, a salesperson's job is to package and deliver to advertisers an air schedule that meets their marketing needs. But local cable sellers must contend with two special conditions. First, the extensive marketing of advertising time on cable systems is still a relatively new enterprise. Advertising agencies and local businesses alike are still attempting to ascertain the real value of local cable buys because audience measurement is not yet able to fully track the effectiveness (exposure) of commercials placed on this multichannel medium. Consequently, cable sales executives must sell the value/prestige of an advertiser's placement on a given channel(s) rather than documented rating and share points.

Second, a local cable spot normally is inserted on a number (package) of basic cable channels. Virtually all the national basic cable networks (CNN, Nickelodeon, USA, etc.) make local insert slots available within their programming into which each cable system can inject commercials for advertisers in its market. Consequently, even though many of these individual cable networks may not garner enough audience numbers to appear in a market's ratings book, cable salespersons do have great flexibility in packaging a buy. They can sell the advertiser placement on not just one but a cluster of several programming channels. Local clients seeking upscale viewers might buy a package on CNN, the History Channel, and Discovery, for instance, whereas clients interested in being associated with sports might select a package consisting of placements on ESPN, a regional sports service, and the Golf Channel. In a similar manner, radio salespersons working for stations combined into multistation clusters now can offer "multiple-vehicle" buys to their clients, too.

As compared with local outlet sales, *network* selling—both broadcast and cable—involves much bigger decisions and, as we have seen, much bigger dollar values. (Figure 4-2 presents a snapshot history of broadcast network advertising costs.) The fundamental selling dynamic remains the same, but the network sales executive deals almost exclusively with *media buyers* (see chapter 6) at advertising agencies and media service firms rather than with clients themselves.

Dixie Gostola
Sales Executive, Charter Media

The next time you're watching cable television you'll probably see commercials from your local community. If you see spots from a local carpet store, local car dealer, or hair salon, then you've seen an example of the work of a cable advertising salesperson.

Selling cable television advertising is a unique enterprise. Yes, you need to have the ability to sell to others. But with this career you'll need a blend of versatility combined with flair for creativity. And income opportunities that really have no limits.

A smart business owner understands the value of advertising but doesn't always know the best approach to take. To be successful in cable sales you must understand the needs of each client and how your product can help them. Are you a good listener? Can you communicate well with others? Any sales position goes way beyond *making* a sale. It's really about creating a relationship with each of your clients. If you enjoy working with people and have a genuine interest in helping them, then you possess the most important qualities.

As a cable sales representative you will be responsible for a sales region or area. My sales territory has 18,500 subscribers—the number of homes hooked up to cable. Almost every business is a potential client and there are very few limitations as to who can advertise. Since I'm paid on how much I sell, time is money. My boss tells me, "Work smarter, not harder." When I first started selling cable advertising eight years ago I was given a few small local accounts to help me get started. Today my client list is quite impressive. Almost every business needs to advertise so you'll find plenty of opportunities. Within my local market I have twenty-eight networks available for local commercials. Several years ago there were only four networks—ESPN, CNN, TNT, and USA. Now a business owner has more choices than ever before.

My average workweek is forty hours, which is spent calling and meeting with clients, developing ad copy ideas, preparing proposals, doing research, and various other projects. Most of my work is done on my laptop—thank heaven for technology.

What do I like best about my job? I really enjoy the variety of my work. Every day is different than the one before. Each client is a work in progress. Some days are spent doing paperwork, other days I'm off meeting with clients. Then there's the creative side. Clients get excited when they see themselves and their business on TV. If you know how to showcase them, that will be a great asset! It may take some time but if you enjoy this kind of client serving, cable sales can become a great career.

Prime Time's Rising Costs

Year	Program	Cost
1970	Bewitched	$5,200 per :60
1972	Bonanza	$26,000 per :30
1980	Dallas	$145,000 per :30
1987	Cheers	$307,000 per :30
1992	Murphy Brown	$310,000 per :30
1994	Seinfeld	$390,000 per :30
1997	Friends	$410,000 per :30
2003	Survivor	$425,000 per :30

Figure 4-2. The price of popular programs.

In general terms, network sales are divided into the *upfront* and *scatter* markets. As trade journalist Joe Mandese explains: "During the upfront, the networks typically sell upwards of 75% to 80% of their season's ad inventory—excluding slots held back for 'makegoods' in the event a show doesn't attain its projected audience. The remainder is sold in quarterly 'scatter' markets. Prices historically are more expensive during scatter because there's less inventory to meet demand."[2]

Upfront selling usually commences in early June, and as much as two-thirds of each network's prime-time inventory for the upcoming season may be sold in a two-week period or less. Other dayparts' time selling then moves rapidly or slowly depending on advertiser demand for prime-time placements.

Advertisers who buy during upfront normally receive ratings guarantees. If the network fails to deliver the promised rating levels, additional advertising time is provided free as *makegoods*. Meanwhile, clients who wait until after the upfront market closes must buy in the scatter market at presumably higher rates, with minimal or no ratings guarantees and with fewer choices as to the shows around which their spots may be placed. However, if the network salespersons have been unable to generate extensive enough sales in upfront, or if certain advertisers have pulled their commercials from a show that has become controversial, significant airtime bargains may surface during scatter sales.

More and more, networks are turning to demographic rather than to ratings-level sales. National advertisers are less willing to spend money for undifferentiated tonnage numbers. Instead, these marketers seek assurances that their advertising messages are being delivered to at least the broadly delineated audience segments that make up their prime consumer market. The overall household race, "which still gets most of the press attention, is largely irrelevant from the advertiser's perspective," asserts major advertising agency time buyer Paul Isaacsson. "TV time just isn't bought that way anymore."[3]

An effective salesperson, for whatever medium, knows exactly how time is being bought in his or her branch of the industry and proceeds accordingly. This is merely the most obvious facet of the salesperson's most important function: learning and serving the prospective time buyer's needs. "Before I walk into a door," reveals one successful time salesperson, "I try to bring to the table what's going to help this client's business. It's the creative ideas that distinguish us from our competitors."[4] "The salesperson today is an orchestrator and resource for their clients," chimes in Detroit radio executive Bruce Stoller.[5] And from the potential client's perspective, agency media executive Sondra Michaelson adds, "[m]ost salespeople are too interested in making sure that their pitch is heard, at the sacrifice of listening to [the] buyer's needs. Frankly this is a brand of a bad salesperson. Knowledge of client and agency needs creates opportunities."[6]

Station Representatives

To this point, we have focused on the sale of two kinds of airtime: *local* and *network* (national). There also is an important intermediate category known as *national spot,* or just *spot* for short. Simply put, *spot* is the sale of time on local media outlets to national advertisers. But it is not humanly possible for agency buyers to meet with salespersons from hundreds of stations and cable systems. And these stations and systems can't afford to send their sales force to meet with agency time buyers in all those major cities where top agency executives are headquartered. Consequently, a vitally important business has grown up to provide this liaison efficiently and make spot buying work. This is the business of the *station representative* or *rep firm.*

Every major radio or television station is affiliated with a rep firm. This firm may have dozens or even hundreds of stations among its clients. Normally, a given rep will handle only one company's stations in a market, although this rule may be broken in the case of outlets having widely divergent formats (and thus widely divergent and mutually exclusive listenerships). The rep salesperson compiles the avails from his or her client stations and packages them in a virtually endless variety of patterns to meet the specific demographic and CPM requirements of advertising agency clients. Using a rep firm, an agency time buyer can thus purchase avails on a number of stations throughout the country in one single business transaction. In this enterprise, "the agencies and the reps are both middlemen," points out Jim Greenwald, chairman of major rep firm Katz Media. "The agency represents the client and the reps serve the media. Both must work together to grow the industry."[7]

In selling spot, the rep is offering a much more flexible vehicle than network because the buyer can emphasize a given type of station, or region of the country, without having to purchase the network as a whole. "It is our responsibility not only to maximize our client stations' share of budget that agencies are planning, it is just as important—if not more important—for national reps to position the medium to attract more dollars to spot," declares Katz's James Beloyianis. "It's the only medium that can provide market-by-market pinpoint accuracy."[8] A snowmobile manufacturer, for example, can use a rep firm or firms to purchase an advertising schedule on the sports programming of stations located exclusively in the dozen or so northern states where snowmobiling is most popular.

In packaging a buy, the rep firm sometimes creates what is known as an *unwired network*—an assemblage of stations whose key commonalities are that they have the same representative and are airing the same commercial at about the same time. The program adjacent to that commercial may vary widely from station to station. And these outlets may be network affiliates, independents, or a blend of both. But the essential thing is that the rep salesperson has cleared simultaneous time on all of them on behalf of the same advertiser.

Rep firms earn their money via the commission they make for the time they sell. Let's say that a rep firm collects $25,000 from an advertising agency for spots it has sold the agency on a given station. The rep keeps its commission (usually 6% to 9%) and forwards the remainder to the outlet. Thus, in this case, the rep firm makes anywhere from $1,500 to $2,250, and the station receives the remainder. There are, of course, variations on this pattern depending on the size of the buy, whether it involves an entire station group, and the degree of importance that the parties assign to each other. A rep firm that really needs a powerhouse station to fill out its client list may thus be willing to take less commission than when it is selling time on behalf of an underperforming or small-market facility. Like any national distributor, a rep firm seeks strong local outlets that offer the dollar volume to generate lucrative commission levels.

As a means of strengthening their client stations (and thus their own revenues), rep firms frequently also offer researched advice on programming to those stations. Reps have access to more sophisticated audience data on shows and formats than do local facilities and also employ a research staff to analyze these data in formulating recommendations. Particularly in the case of television, the rep firm is also given the chance by syndicators to preview their new series. Client stations of the rep thereby enjoy advance information that can help them decide whether to purchase new program properties and how much to bid for them.

Web Sales Executives

This need for comprehensive data on a wide range of "product" is just as important for professionals who sell Internet advertising. In addition, the Web seller must be conversant in all of the huge variety of advertising products that the online environment has to offer. Banners, pop-ups, skyscrapers (vertical ads at the margins of the screen), and streamed audio and video "commercials" are just some of the advertising vehicles that populate the Web. Each of these vehicles has its own inherent advantages and disadvantages for the advertiser and the consumer that advertiser is trying to reach. The Web sales executive must juggle all of these factors in coming up with a plan that makes economic sense for the advertiser and that, in some way, can be validated. Firms like Nielsen/Net Ratings and Media Metrix now provide ratings for top Internet sites. But the interactive nature of the medium and the wide variety of advertising forms found on it do not make for easy translation of these ratings into measures of marketing effectiveness.

The principles of effective selling as practiced by broadcast and cable representatives apply to Web salespersons as well. It is just that the Internet seller has so many more products to keep track of. "I think the primary reason that advertisers haven't either gone online or made larger investments is that there's never been a clear and simple way to think about the medium," says online advertising executive Allie Savarino. She, like many of her colleagues, believes that it is the responsibility of Web salespeople to free advertisers from worrying about technology so they can focus on "what's the audience and what's the physical space you want to reach them in?"[9] Some audience-preference patterns about that space have emerged to guide sales activities. A 2003 survey by Arbitron and Edison Media Research, for example, found that pop-up ads were, by a large margin, the most annoying to consumers, with 65% of the survey so identifying them. On the other hand, only 2% of the survey felt "commercials" running before Web audio or video content were annoying. "For sites offering streaming content, a commercial is not as intrusive, since people are used to it from watching TV," asserts Arbitron Internet Broadcast Services general manager Bill Rose. "In the cases of media such as TV, there is a quid pro quo that consumers have accepted: 20 minutes of ads for 40 minutes of content," adds Greg Stuart, CEO of the Interactive Advertising Bureau (IAB). "That model holds a lot of promise for online advertising."[10]

As more and more homes and businesses acquire high-speed broadband hookups via telephone, cable, or wireless provider, "commercial-like" content is becoming increasingly viable. Sometimes, these spots may be very similar to those that have run on conventional television. Increasingly, however, broadband avenues make possible long-form commercials and commercials with which the viewer can interact, much like in a video game. It is the Web salesperson's job to keep abreast of these new opportunities and accurately calculate whether they make sense for a particular advertiser.

Mike Feltz
Director, National Account Development, Yahoo!

The customer said, "I don't know why you're here. I can't even believe I let you come into my office, and there's nothing for us to talk about!" This was the beginning of the most harrowing *and* satisfying sales call of my career.

Don't worry. This customer "greeting" is certainly the exception, not the norm. Taking on this type of challenge defines the courage of media salespeople who operate in pressurized environments. They are persistent in overcoming objections. They are positive in their approach. Media salespeople are passionate about their work and who they represent. They must be tireless in following up and following through on the sales mission. Each is a driven, goal-oriented individual who finds unique ways to create success for their customer, their business, and for themselves. They are adept at juggling multiple projects and personalities simultaneously. Most of all, salespeople are talented, dynamic individuals who make an exceptional difference for their organization. And they like to have fun in what they do!

Beyond these common traits, media salespeople are varied in their background, personality, and methods of doing business. A salesperson may have previous experience representing another media organization. Some come from advertising agencies. Others are drawn to sales from the client side. And some leverage their successes from other industries and apply them to the media environment. Personalities vary by salesperson as well. Some are highly aggressive, some very low-key. Most strike a balance between these extremes. It's important that you discover your "style" and be yourself as you present in the marketplace.

Salespeople must be intensely focused on the needs of the customer. They must have a good working knowledge of the advertisers and ad agencies they call on. They must be able to use that information in their discussions and proposals with clients. Salespeople create an environment of trust with their customers. The clients desire you to be their advocate within your organization. They expect you to be creative in proposing new and relevant ideas. They expect you to stay on top of all the details in your campaign and keep them informed of positive or negative news ahead. No client wants a surprise! Overall, there's a high expectation for you to deliver value to the customer. If you succeed, they'll likely come back!

As a relatively new medium, online involves unique sales challenges compared to other media. The first step is measuring a client's perspective on the Web as a marketing tool. Some customers may be reluctant or uncomfortable with their budget commitment to the Internet. For these clients, you first need to educate and sell the benefits of increasing overall online share (or the medium itself).

There's great creativity in the online world. You have the advantage of selling ideas using an assortment of ad placements (banners, text links, e-mails), targeting methods (by demography, context, relevance, and/or behaviors), creative options (rich media, streaming video), and other applications (Web pages, instant messaging, promotions).

This "variety" also creates some complexity for both buyer and seller. Expect to invest time to stay on top of ever-changing technologies; it is extremely important to your success.

Speed and measurement are key components of online sales. Clients expect lightning-quick delivery throughout the sales process, and the medium's flexibility enables fast and frequent changes during a campaign. Measurement is pervasive on the Web. Unlike print or broadcast, Web publishers provide nearly real-time tracking of metrics such as impressions served, clicks, and creative performance. With online, clients evaluate their return on investment by linking media performance to the desired marketing objective (increases in brand awareness, qualified leads, product sales). The impact of an efficient media placement in tandem with the client's compelling creative offer is seen in the data, keeping both parties accountable to each other for success of the partnership. A win-win for all is the key!

It should not be surprising, therefore, that an expanding number of Web sales executives are coming to the medium with previous broadcast, as opposed to print, marketing experience. As Internet rating systems are refined, selling in the online world is likely to become more and more similar to traditional broadcast selling—but with the added complexities of a multitude of print-like formats that will still be prominent features of the Web advertising scene for a long time to come. Ironically, once television stations are fully converted to multiple-stream digital providers, it may be that this process will reverse itself. The selling of broadcast advertising may come to be more like that of selling the Web, with TV sales executives having a multiplicity of marketing options that can be offered to clients individually or in packages.

Program Salespersons (Syndicators)

Up to this point we have been talking about the sale of advertising time and (on the Internet) advertising space. An equally important selling function is handled by those professionals who sell the programming around and within which much of this advertising appears. In both radio and television, *syndication* firms employ salespersons whose job it is to place on local outlets the shows and formats packaged by their producer clients.

In radio, because of the lower cost figures involved and the multiplicity of outlets, salespersons usually work directly for the company (radio network or syndicator) that has created the property. In television, on the other hand, the sales force is likely constituted as a separate company or division of a studio-owning corporation that probably owns a network as well. Over the last few years, the mergers of Disney with ABC, CBS with Viacom, and NBC with Universal have meant that all the broadcast networks now have a sister studio. While there is an obvious tendency to have that studio produce programming for its sibling network, the parent corporations will also sell programs they produce to competing networks if it is to their economic benefit. Meanwhile, other self-standing syndicators (like the one promoted in the figure 4-3 print ad) package programs from a variety of foreign, independent, and other unaffiliated studios for direct sale to stations, cable outlets, or satellite programmers.

TV syndicators sell a product that can be broadly classified as either *off network* or *first run*. Off-network material, as the name implies, has been originally exposed on one of the major broadcast networks, with episodes now made available for running on local stations or on cable. (Recently, an *off-cable* category has started to emerge as series launched on the wired medium are subsequently being sold to broadcasters, reversing the traditional progression.) Sometimes hit network series are first marketed as *futures*—for local exposure at least two years in advance and be-

Sales Functions   103

Figure 4-3. Promotional piece from a program syndicator. *Courtesy of Sophie Beaurpere, CABLEready Corporation.*

fore some of the later on-network episodes have even been produced. A station or cable network takes a risk that the show will still be popular two years or more down the road and may even have to commit to sharing production costs if the network cancels the project before a sufficient number of episodes have been made. On the other hand, a futures buy could be a bargain if the network popularity of the series stays stable or increases in the interim.

First-run syndicated projects, conversely, have never appeared on a broadcast network and will be exposed for the first time at the outlet level. Many times, the syndicator/salesperson will market a first-run series on a *barter* basis. This means that the station may secure the show for free or at a reduced rate by carrying at no charge the commercials that come within the program. What has actually happened is that the production company has presold advertising positions contained in the vehicle to certain clients via their advertising agencies. This money then helps the producer with the up-front financing for project production. (Some high-powered off-network properties are also sold on a partial barter basis to raise more revenue than would be available from stations on a straight-cash deal.) Rather than sell *all* of the advertising time within and around the show, local stations then give up some slots (avails) to the bartered spots on behalf of those national advertisers. In short, the station exchanges some of its commercial time for syndicated programming.

Gary Lico
President and Chief Executive Officer, CABLEready Corporation

As an industry evolves, and the electronic media business certainly does that, one element does not. Sales—the generation of revenue through the offering of a commodity somehow unique to the seller and attractive to the buyer—is still the key to an organization's survival. "Nothing happens until somebody sells something" is the adage. Yet many people avoid a selling career like the plague.

It's a pity. Because sales really is the epicenter of a media operation, be it the sale of *advertising time* on the local or national level or the sale of *television programming* to domestic and foreign networks, to cable, to individual stations through syndication, to airlines, and to home and/or educational video. There's no other job in our business that interacts with as many other departments within a media outlet as does sales, while providing it with its lifeblood—money. More upper-management jobs are filled by people with sales backgrounds than any other single area (except maybe the mail room at talent agency William Morris).

Program sales, which we do around here, are no longer a slick suit, gray shoe, fancy patter, and a "hi-how-are-ya" smile. Today's salesperson must be an expert on the format of the station, network, or channel, its program content, its strategic direction, its financial policies, the competition's strategies, and above all, the client's needs. Selling a sitcom to a news-and-talk formatted television station or an ad flight on a teen-oriented station to Geritol is a good way to lose credibility fast.

You must be organized. You must have a plan. You must keep track of your client's needs and your ability to meet them. In the age of the Internet, and so many buyers' websites, there is no longer an excuse for not knowing the client and the territory. You must know your inventory (programs, ad time, whatever it is you're selling) better than your mother's maiden name. You must be persistent and follow through. So many salespeople—most of 'em, in fact—make the pitch and then drop off the face of the earth.

Making the follow-up call puts you above the rest. Half of success is just showing up, says Woody Allen. Staying on top of the client to conclusion is a rarity. So our credo at CABLE*ready* is we don't stop the effort until the client tells us to go to hell and draws us a map.

Travel is a function of many sales jobs and was a lot more glamorous before airline deregulation, tougher IRS rules, and the fears of terrorism. Still, it's a great opportunity for a beginner to pay the early dues, really immerse him/herself in a business, and meet a *lot* of people. Domestic syndication selling has been famous for its travel demands. But those jobs are drying up as more companies merge, more networks become involved in program production for their affiliates, and local news expands. Nevertheless, though the number of syndicators is being drastically reduced here in the U.S., new opportunities abound as international television grows.

The program salesperson of tomorrow will have to function as an entrepreneur, a self-starter, a hired gun. You'll never go hungry if you develop and maintain good organizational habits, a high level of energy, a positive mental attitude, and a network of solid contacts. You must be able to read a rating book and possess the ability to separate "no" from feelings of rejection. Staying well read on your industry and your clients' industries will make you a source of information. And information is power. The *Wall Street Journal* and a solid national newspaper (my favorite is the *New York Times*) will pretty much keep you timely on a daily basis. Read all the trades and industry websites and e-newsletters you can afford. There's nothing like sending an article that pertains to your client's business with a little note that says, "Saw this—thought of you" for keeping your name and company uppermost in the client's mind.

Sales Functions

None of this is insincere or phony. The care and feeding of clients is crucial to your company's growth and your own income, particularly if you work on commission, which is the norm. Staying up to speed and up-to-date and respecting where you *and* your company's next dollar comes from will make you a well-paid and employable person until the day you die.

For a barter project to work, the program salespeople must be able to place the series with stations covering at least 70 to 80% of the country. Anything less will not give the participating national advertisers the commercial penetration they need to justify their barter costs. Stations, for their part, must be careful not to contract for too much barter programming. For one thing, they thereby incur the obligation to air the bartered spots—even if they pull the program from the air before the contract expires. Furthermore, every minute of bartered time means one less minute that the station's own sales force can sell, one less chance to obtain revenue from local or spot sources. Barter syndication is thus a commodity that both program and time sellers must analyze carefully before offering or making binding commitments.

Other syndicated shows are made available on a straight-cash basis. Among these are usually the *evergreens*—series that have been so popular for so long that it is hard to remember whether they began life as network or first-run entities. *I Love Lucy, The Brady Bunch, The Andy Griffith Show,* and *M*A*S*H.* are a few such recurrent properties that continue to draw salable audiences even after years of rerunning. Occasionally, an evergreen concept like *Lassie* will even be brought back into production for release as a first-run syndication project.

Off network or first run, evergreen or new, barter or cash, a program salesperson's job is to get the widest possible placement for his or her properties, on the best possible outlets, and at the most favorable price. For example, the broadcast market for off-network hour-long shows has been soft because local stations have found it difficult to accommodate them in schedules already committed to network, local, and shorter-form fare. Thus, program salespeople have sold several such properties to cable networks on an exclusivity basis. The cable network sale may bring in less money than will peddling to individual stations, but this loss is balanced by the fact that the syndicator's administrative costs are much lower when dealing with one cable net than with a multiplicity of individual television stations.

Promotions People

Once a program is placed on a station, network, or even local cable channel, it becomes the job of promotions people to make certain that the public is made positively aware of the show. Using a combination of *on-* and *off-air* promotions, these professionals do whatever they can to ensure that their outlet is *sampled*—in other words, that sufficient numbers of listeners or viewers give it a try. When the product is appealing, this sampling can translate into a measurable and loyal audience whose predicted attention can be sold to advertisers. "If you get reasonable sampling of a good program, you get relatively strong repeat rates [consumers returning for a subsequent episode] in this industry," maintains CBS chairman Michael Jordan. "The repeat rates are probably the same as 15 to 20 years ago, but the sampling is so much lower because of all the choices."[11] In a multichannel world, sampling enhancement thus becomes an ever more critical task for the promotions professional.

On-air promotion activities consist of the messages carried by the outlet itself or placed on other noncompeting electronic media vehicles. These messages range from simple identification

(ID) slogans, jingles, and graphics to full-scale commercials (promos) on behalf of the facility and its programming. These longer promos are often subdivided into three categories: image pieces, generics, and topicals. *Image pieces*—sometimes referred to as *brand enhancers*—try to position the station or network as a whole in the consumer's mind. We are the fun station, the station where news comes first, the station where you always hear your favorites, the place to turn for family entertainment, or the people who know your community best. The best image pieces articulate a clear and attainable benefit/promise. The promotions person stakes out a personality for the outlet that is both memorable and appealingly in line with what the target audience is looking for in a radio or television service.

Generics are messages designed to publicize a particular show or series. On radio as a whole, or in television news, these generics often focus on an air personality or newscaster because this on-air talent is the outlet's most consistent and distinguishable element. The individual musical or journalistic elements change from day to day, but the talent stays the same. In the case of an entertainment series, the focus may still be on the personalities, but the environment and central premise of the program must also be established to aid viewer recall and to pique curiosity.

Topicals (sometimes called *episodics*) then stress what is to happen on a particular segment (episode) of the show. Often a group of topicals will be grouped together to form a *block promo*, which teases the happenings the listener or viewer can enjoy on several successive programs.

Promotions or *creative services* staff (as they are also known) balance these on-air efforts with a wealth of off-air activities. *Client (advertiser) promotions* endeavors might include printed sales kits and rate cards and also extend to more exotic activities such as recorded sampler tapes touting the outlet's programming, or hosting full-scale client theme parties tied to a new station campaign or show concept. *Audience promotions* might involve anything from basic newspaper ads (like the one in figure 4-4) to bumper stickers to cosponsorship of live concerts or community health fairs where pocket blood pressure charts come emblazoned with the station's logo. A whole range of contests might combine both on- and off-air activities to enhance consumer involvement with the station.

Value-added events carry the practice even further by creating joint station/client promotions to give advertisers bonus exposure. However, though value-added is becoming increasingly popular with advertisers seeking to extend the worth of their regular spot buys, it presents the station marketing director with special pitfalls. Remotes, giveaways, and air staff personal appearances that are dictated by the advertiser without the considered input of station promotion executives can serve to clutter or distort programming and thereby drive the audience away.

"For a marketing director, the goal is to provide a promotional or merchandising stage that exposes the client's products or services yet does not compromise the integrity of the station," cautions trade publisher Dan Acree. "If you consider that a station's call letters are a *brand name* in the same way that *Tide* is a Proctor & Gamble brand name, then you begin to understand the importance of protecting how that brand is used promotionally."[12] Adds Pulitzer Broadcasting marketing vice president Peter Smith: "[Value-added] deals the sales department comes up with are not always designed to enhance the image of the station, but rather the image of the advertiser."[13] In short, promotions professionals must market their outlet so that their sales colleagues can more easily peddle avails on it—while steering these colleagues away from short-sighted value-added gimmicks that will dilute the station's positive long-term image.

Effective promotion is a key enterprise in noncommercial broadcasting, too: attracting the audience attention that justifies and expedites grants, donations, and pledges. Thus, poorly conceived auctions and boorish on-air fund-raising festivals are just as counterproductive to public outlet branding as client-strident, value-added schemes are to the image of commercial stations.

Newsradio 78 Tastes Good, Too!

From Monday, April 16 through Friday, April 20, Sherman Kaplan's In the Kitchen will feature recipes for Asparagus Parmigiano, Fish Etoufeé, Sally Lunn Bread, Dixie Peanut Brittle (yum!) and Easy Skillet Gumbo.
Listen every weekday at 10:38 am as culinary experts join Sherman on Newsradio 78 to share their cooking secrets and recipes.
Know what's cooking.
Turn to us first.
WBBM
Newsradio 78.

Figure 4-4. Newspaper layouts have long been a staple of off-air promotion. *Courtesy of Barbara DiGuido, WBBM.*

With the continued expansion in the number of electronic media vehicles competing for consumer attention, the job of the promotions or creative services director continues to grow. Nancy Smith, former chair of PROMAX (the professional association for promotions executives) points out that as these executives "evolve into 'brand managers,' they will continue to play a greater role in programming. In fact, a PROMAX survey revealed that 42% of all promotion managers carry significant programming duties."[14] Looking ahead, promotions expert Vince Manzi predicts that in the coming century, the mandate to "get noticed" will be as applicable as ever, so the need for skilled promotions people can only increase. "You'll have to pay them more and you have to give them some opportunity to do good creative," Manzi advised station managers.[15] "Getting noticed" in our business does not come cheap or easy. But not getting noticed because of deficient promotion means almost certain bankruptcy. Promotions is highly demanding work—but it is also work in high demand.

Production House Marketers

The last sales function to be discussed combines aspects of direct selling and promotions. The job of *production house marketer* is thus a fitting subject through which to summarize this chapter as a whole. In essence, these marketers explain and publicize their firms' technical capabilities in order to sell those capabilities to other electronic media players.

With state-of-the-art equipment becoming more sophisticated and multifaceted every day, few

Judy Paluso
*Director of Creative Services,
CBS/UPN, Detroit, Michigan*

Do you want to effect change? Do you want to influence how your station is perceived in a market? Then you might want to consider a career in television as a creative services director.

What does one do as a creative services director? This is a loaded question. Each day is slightly different. Your number one priority is to increase the ratings and shares of your programming. The higher the numbers, the more your sales department can charge. The more money the station makes. The better your chances are of keeping your job. Sound simple? Not so. You do the above in a number of ways. You can work with your writer/producers to create earth-shattering promo spots, sure to catch viewers' attention. You can purchase outside media to capture those people who may or may not be your viewers. You can implement guerilla-marketing tactics that will create buzz in your market. The key is to be creative, ever changing, and inventive.

So what skills will you need? You will need a knowledge of advertising and media buying. You will need to understand audience trends. You should be able to analyze ratings. You should be familiar with your audience and know where to plot promotional spots to effectively maximize your inventory. You should be creative. You should be able to write and produce promotional spots. You need to know how to lead a creative team. (Not so easy a task.) And lastly, think way, way outside the box.

Another role of the creative services director is to work with the sales team to develop exciting, revenue-generating sales promotions. These are campaigns that the sales force can take to clients in exchange for a media buy on the station. The key in working with the sales team is to develop a win-win situation for the station and the client. The most important goal is to ensure that the promotion fits your brand and image.

Event marketing is another means of growing your audience. Getting out in front of the very people who watch your station is important. Again, you must select events that fit your brand and image. And you must make the station look its best. This means working evenings and weekends. It means setting up kiosks and tents and ordering giveaway items. You'll need to work the event to make sure the station is presented in the best light. Often you will tie in with an advertiser to develop an event. For example, as a CBS affiliate, our station participated in the *Survivor* casting calls. People came to audition for a part on an upcoming *Survivor* episode. We tied in an advertiser for a cash schedule. The casting call took place at the client's business. The station wins because money is made and publicity is generated for *Survivor*. The client wins because people have been made aware of their business.

Where do you start? I recommend becoming proficient in nonlinear editing. Know how to write eye-catching promotional spots. Take marketing classes to learn advertising and audience trends. Take a television performance class to understand what it is like to be in front of the camera. You will need to work with news personnel. Knowing what it is like to be "talent" is important if you are going to direct them. My number one advice is to acquire an internship at a television station *and* an ad agency. Get the experience before you start full-time job interviewing. I have personally hired many of my past interns. Get your foot in the door. And then become indispensable.

Sales Functions

media companies are in a position to produce their messages entirely in house. Advertising agencies, public relations firms, industrial media units, and even stations and networks must often turn to outside companies to obtain cutting-edge audio and video production services. Because their only business is to produce projects brought to them by others, production houses can concentrate exclusively on technical rather than conceptual processes. By booking projects from a variety of clients, they can keep their expensive physical plants and expert staffs in continual use. The marketing managers for these shops therefore have as their top priority the booking of production business so that costly facilities are never standing idle, on the one hand, and never overbooked, on the other.

Ron Herman
President, The Production Café

In the 1997 version of this text I stated, "For people to be successful in the communications industry, they must be able to thrive on change." It is still the case now. As a matter of fact, the pace of change has gone into (digital) warp speed! For the production house sales and marketing pro it is imperative that we keep up with what I call the three "T's." The first is *technology*, the second is *talent*, and the third is *trends*.

Technology

Technology is cool and it helps sell! The effects, the mixes, the sounds, the speed, are all parts of the technological package that make our jobs as sales people easier. As a young sales representative in the Los Angeles market in the early seventies, I had to convince network producers and directors to use our studios and editing facilities. This task was not very daunting due to the fact that we were one of a handful of "state-of-the-art" facilities. Our shop, the Vidtronics Company, was responsible for designing SMPTE time code and the Technimatte (the forerunner of the Ultimatte). We also had the first scene-to-scene color corrector system and we invented Pan and Scan technology. With this type of engineering prowess selling our services was not difficult.

Back in the seventies, as more and more facilities came on line, the competition grew. We always had to find unique ways to attract new customers and maintain our existing ones. We did this in a number of ways. We created exciting campaigns with new demos, advertising, brand extensions, public relations, and events. We were also able to drive new business to our company as our service acumen and reputation became more recognized in the marketplace.

In production sales and marketing, it is imperative to have a thorough understanding of the operation of the technology. This operational understanding allows one to meet with prospective clients, break down their projects, and make recommendations on how to achieve their visions. As reps we often find ourselves as extensions of our clients' companies. When a director requires a certain shot, we step in and make recommendations of lens compliments or the camera-mounting unit. When a project reaches the postproduction stage, we prep the edit staff so that they are prepared to complete the project.

The digital technology continues to impact the market place. Symphonies, Flints, Flames, Mayas, Smoke, and high-definition systems make up just a portion of what a rep needs to understand in order to make a case for his or her company. If you're in facility sales and marketing, having access to and knowing the capabilities of the newest type of editing hardware is critical. One of my former editors built a strong business based on his talent and Avid editing equipment. When the Final

Cut Pro system became viable, he refused to commit to the new technology. He sat back and didn't want to invest his money. He is no longer in business!

Talent

The second "T" is talent. If all of the competition has an equal investment in equipment, then the talent that operates it will make the difference. A strong creative editor is worth his or her weight in gold.

In the past ten years we transitioned from being a boxes and wires facility to being a creative production shop. My experience had taught me that being associated with a high-end talent is the key to success. With the right talent and the right style, you can create a client base that will last for years. With a talented staff of creatives, our sales efforts extend past the borders of our home state of Michigan. We package projects in California, Florida, or even overseas. Once again, reputation goes a long way in being able to bring home a particular project.

When I was in Chicago in the early eighties, we had a director who was so popular we would receive eight to ten scripts or storyboards to bid a week. He would pick and choose the project that he would consider directing. This wasn't always the best for the reps. But with the dollar volume he generated, it didn't particularly matter.

Trends

The final "T" is trends. What is in vogue today? What is the shooting style that your clients will buy? It is critical that you stay on top of what is hot today. If you're in corporate video, make sure you know what is "hitting the hot buttons" of your customers. If you are in spot (commercial) production sales, make sure that your creative team is able to deliver a unique style that is marketable. Having the right directors and the right director of photography will make selling your wares a lot easier at the advertising agencies.

Effects are everywhere. We see them on the big screen, we see them in spots, and we see them in corporate communications. Learning the techniques to accomplish these effects will go a long way when a customer asks, "How did they do that?"

Finally, successful selling in our business is based on relationships. Sometimes we spend as much time nurturing these relationships as we do nurturing the relationships within our own families. If you provide this nurturing and possess the passion, the attitude, and the work ethic, opportunity for entry into sales and marketing of production services has never been greater.

The real key to success is clearly defined by one's ability to adapt to change. The pace of change is continuing to accelerate. Don't stand at the corner and let it pass you by. Grab it, hold on tight, and have a great ride!

Some production houses specialize in multitrack audio recordings. Others concentrate on in-studio or remote video shoots or on the postproduction editing and sophisticated assemblage of footage shot by other professionals. A few houses focus exclusively on computer effects and graphics or even on narrow specialties such as Claymation (animation derived from manipulating clay figures such as the famous California Raisins). Whatever the case, the reputations of these firms are only as good as their last project. Thus, their marketing executives must keep abreast of the triumphs that have just come through their shops and must use this information to attract new and repeat business. Having one foot on the technical side and the other on the sales side, these marketers facilitate on two levels: they ensure that the communication projects from other companies are successfully completed while making certain that their own companies' client volume goals are sustained.

Because it is exclusively devoted to the optimal consummation of projects brought to it, the production house is the best positioned of any branch of our industry to probe and experiment with new techniques and ways of packaging an audio/visual message. These firms thus constitute the frontier for technology and the application of it to new conceptual challenges. The marketers of such services realize, however, that their clients are not interested in these techniques for themselves—but only in how such abilities can create more effective commercials, programs, or training videos. Like any electronic media seller, the production house marketer must reconcile his or her services with the communication objective of whoever is paying the bill.

Chapter Flashback

Salespersons generate the revenue that keeps electronic media companies operating. At stations and cable systems, they sell airtime to advertisers. This commodity is fixed. Unsold airtime cannot be retrieved at some later date, and additional airtime cannot be created—at least not without cutting into program content. Time sales are negotiated based on a number of considerations, including *ratings, shares, cost-per-thousand, gross rating points,* and *frequency.* Qualitative factors such as the type of programming being aired and the type of audience it attracts may sometimes outweigh purely quantitative considerations. Local salespersons often deal directly with advertisers, but network selling is usually done through media buyers at advertising agencies. *Station reps* also pitch to these buyers by offering spot packages on a number of their client outlets as a single transaction. Reps often also provide their stations with research-based programming advice.

Web sales executives must stay on top of an Internet environment that is in a constant state of technological flux. They must know how to compare and sell a wide variety of advertising tools that range from static banners to broadband streamed video. Increasingly, their ranks are drawn from broadcast sales. Program sellers also have a wide variety of products to offer clients and must be familiar with the differing needs of the outlets with which they deal. Program salespersons represent the syndication firms that sell *off-net, off-cable, first-run,* and *evergreen* properties on a cash and/or barter basis.

Whatever programming is selected, it is the job of promotions or creative services staff to see that this programming gets sampled by listeners or viewers. *Audience promotion* activities take place both on air and off air, the latter encompassing everything from bumper stickers to concerts. Simultaneously, off-air *client promotions* are developed to court current and potential advertisers. Finally, production house marketers must exercise the entire range of sales and promotion functions. They explain and publicize their firms' technical capabilities in order to market these services to other electronic media enterprises.

Review Probes

1. What are the differences between selling print layouts and selling airtime?
2. How are ratings and shares computed? Why will a share always be larger than a rating?
3. What is wrong with using CPM as the sole measure of time buys?
4. Differentiate *local, network, barter,* and *spot* airtime sales.
5. What are the key challenges facing the Web sales executive?
6. In what ways might the work of the program seller and promotions executive intersect?

Suggested Background Explorations

Albarran, Alan, and Angel Arrese. *Time and Media Markets*. Mahwah, NJ: Lawrence Erlbaum, 2003.
Alexander, Alison, et al. *Media Economics: Theory and Practice*. 3rd ed. Mahwah, NJ: Lawrence Erlbaum, 2003.

Asher, Spring, and Wicke Chambers. *Wooing and Winning Business: The Foolproof Formula for Making Persuasive Business Presentations.* New York: John Wiley, 1997.

Blumenthal, Howard, and Oliver Goodenough. *The Business of Television.* New York: Billboard Books, 1998.

Eastman, Susan. *Research in Media Promotion.* Mahwah, NJ: Lawrence Erlbaum, 2000.

Eastman, Susan, et al. *Promotion and Marketing for Broadcasting, Cable, and the Web.* 4th ed. Woburn, MA: Focal Press, 2001.

Herweg, Godfrey, and Ashley Herweg. *Radio's Niche Marketing Revolution: Futuresell.* Woburn, MA: Butterworth-Heinemann, 1997.

Holtz, Herman. *Getting Started in Sales Consulting.* New York: John Wiley, 2000.

Katz, Helen. *The Media Handbook: A Complete Guide to Advertising Media Selection, Planning, Research, and Buying.* 2nd ed. Mahwah, NJ: Lawrence Erlbaum, 2003.

Morrison, Margaret, et al. *Using Qualitative Research in Advertising.* Thousands Oaks, CA: Sage, 2002.

Plum, Shyrl. *Underwriting 101: Selling College Radio.* Mahwah, NJ: Lawrence Erlbaum, 2003.

Twitchell, James. *Adcult USA: The Triumph of Advertising in American Culture.* New York: Columbia University Press, 1997.

Warner, Charles, and Jack Buchman. *Broadcast and Cable Selling.* 2nd ed. Belmont, CA: Wadsworth, 1993.

White, Barton, and N. Doyle Satterthwaite. *But First, These Messages: The Selling of Broadcast Advertising.* Boston: Allyn &Bacon, 1989.

CHAPTER 5

Directive Functions

Up to this point, we've looked at the performers, who are the most noticeable aspect of our industry, and the conceptualizers, who formulate much of the material these performers deliver. We've also examined the sales and promotion specialists who bring in the revenues necessary to sustain the media systems that carry the work of these other professionals. Now it is time to focus attention on the executives who provide the managerial guidance and decision making that determine how all of these other people and resources are used in pursuing the needs of clients and intended audiences. To be effective in this task, these "directors" must move beyond being *managers* to become *leaders*. As famed business consultant Tom Peters draws the distinction, "Leadership is not the same as management. Management is mostly about 'to do' lists. Leadership is about tapping the wellspring of human motivation."[1]

Program Directors

At both radio and television outlets, the program director (PD) or program manager is responsible for stocking a media content store that the target audience will want to patronize. As professor Susan Tyler Eastman describes it, "[T]he classic programming paradigm consists of four major parts: *the selection of content, the evaluation of content and audiences, the scheduling of content, and the promotion of viewing/listening/using that content.*"[2]

In radio, this intrinsically involves: (1) choosing a particular format that is calculated to appeal to an attainable demographic; and then (2) fine-tuning that format to improve and solidify this appeal. The owner and/or general manager may play an important role in the first task, but the second task is the fundamental responsibility of the program director. Today, even a relatively small market may have a dozen or more listenable radio signals from which to pick. The radio dial thus becomes much like a shopping mall. A number of competing and contrasting alternatives are set adjacent to each other, and from these, consumers make a decision as to which ones to visit within the time they have to spend. How long or how often each consumer "visits" a given station depends on a number of factors. But the on-air programming constitutes the outlet's essence.

In selecting, refining, and delivering an on-air "sound," successful program directors realize that they are not in the business of manufacturing programming itself. Rather, they are in the business of *manufacturing an audience* that is salable to advertisers. Individual elements are selected for airing because they seem to have appeal for the type of listener the station is seeking. Once these individual listeners are assembled into an audience by the program director's product, this audience is then rented out to advertisers—usually for thirty or sixty seconds at a time. The more similar these listeners are to each other in demographic or lifestyle characteristics, the more appealing the station's audience is to clients seeking to reach this consumer type. If, on the other hand, the listenership consists of many dissimilar groups of people, it may be more difficult to "rent" it to advertisers who do not wish to pay to reach people who aren't prime prospects for their business. Such unwanted listeners constitute *waste circulation* as far as an advertiser is concerned. They inflate the cost-per-thousand (see chapter 4) through inclusion of irrelevant ears.

The radio program director therefore seeks to attract not only raw numbers but also numbers that cluster into population groupings that are desirable to available advertisers. By manipulating music selection, amount of news, prominence or de-emphasis of deejay patter, commercial placement, and other formatic factors, programmers hope to find the key mixture that will attract an easily identifiable audience segment and also clearly distinguish their station's air sound from that of other stations.

As the number of radio facilities has increased, formats have been slivered into smaller and smaller segments. What was once just a "country" format now might be "traditional country," "young country," "pop country," "soft country," or some other variant derived by an enterprising program director or consultant. Duopolies and multistation clusters have solidified and even furthered this fragmentation trend as companies seek to develop mutually supportive stations in the same market. All of these niche possibilities place radio program directors on a tightrope. On the one hand, if they try to buck the trend and appeal to a broader listener base, their station image may be blurred, with many advertisers avoiding the outlet because it delivers too much waste circulation. On the other hand, if the format is too narrowly drawn, it will not attract a large enough listenership to be salable to advertisers at rates high enough to keep the station financially solvent. Maintaining a balance between these two extremes makes a PD job in the larger markets especially challenging.

Even in smaller towns, where there are fewer competing signals, program directors still must wrestle with these image and audience-targeting issues. They must set their sights on attracting a larger slice of a smaller audience pie while still making certain that each station for which they are responsible is projecting a distinct personality. Sometimes, formatic techniques developed in one market can be applied elsewhere, but a sensitive programmer knows that local conditions must still be accommodated. Programming consultant Bill Hennes therefore advises that

> programmers need to understand what makes a radio format work, then translate that to their market. Go ahead and monitor successful stations, but do it with a grain of salt. A certain percentage of the things you monitor will remain the same, but it's the expansion of those basics that will make the radio station successful. Expand on these basics by applying them to the unique aspects of your own market.
>
> Before you monitor any station, analyze the market. Don't just look at the demographics either. The mind-set of the market is important. What makes the market tick? Getting a handle on the real pulse of the listener base is the first step to understanding why the station you're monitoring is successful. . . .
>
> Once you have found out what makes that radio station tick, then you can come back to your market and take the bits and pieces that apply to your station. I caution you in taking a cookie cutter approach though. Dissecting your market and refining those bits and pieces to fit the uniqueness of your audience is what will work.[3]

Anyone can play the hits. But it's what you do with them or with any other on-air element that distinguishes innovative program directors from drone formula followers.

Directive Functions

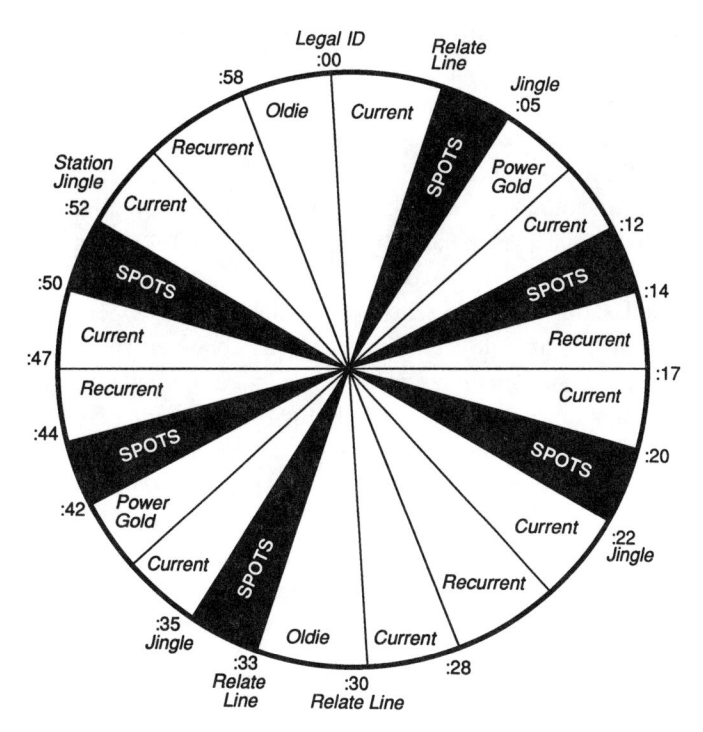

Figure 5-1. Hot clock developed by Broadcast Programming, Inc., for its Z format. *Courtesy of Dennis Soapes, Broadcast Programming.*

Even the best radio format design will fail if it is not properly executed. Program directors thus must clearly communicate their managerial plan to the staff in their department and must continually monitor performance to ensure that these plans are carried out. In larger markets, PDs are often assisted in these endeavors by *production managers, music directors,* and *traffic directors.* At smaller stations, these functions may be combined.

The production manager is most directly responsible for the sheen of the on-air sound. This includes overseeing of in-studio, remote, and recorded programming to make certain that technical quality and style remain consistent. The production manager must also see that all appropriate personnel are competent in equipment operation. (At the largest outlets, this duty may be delegated to an *operations manager*.) Just as important, the production manager serves as the quality control executive for all commercials, promos, and other recorded messages prepared at the station. Because the commercials pay all the bills, their production standards become a top priority.

The music director helps translate the PD's overall sound goals into specific selections. Often, the music director screens all incoming records (frequently referred to as *product*) to ascertain which songs should be added to the station playlist and with what frequency. Record *rotation* is a key component of any music-driven format, and so the music director's categorization of a tune is a key factor in how much airplay it will receive. A "hot clock" like that in figure 5-1 determines the placement of these categories and other on-air elements within a given hour. Increasingly, computer programs operationalize such decisions.

Focus-group and call-out telephone research can also be used to gauge listener perceptions and preferences. Focus groups are comprised of up to a dozen people whose recollections and attitudes are probed by a trained moderator. Call-out research interviews people individually via the telephone. To help keep track of new product, the music director also reads trade publications such as *Radio and Records* and *Billboard* and serves as the contact person with record company representatives. This last task is a sensitive one given the *payola* possibilities that can occasionally arise. Undisclosed payments and other inducements of value provided by a record distributor to get songs played on the air are a federal offense. Conviction on payola charges can result in fines, imprisonment, and even loss of a station's license. Many large stations, therefore, have created a group screening process so that no one person has complete control in specifying the individual tunes that will receive heavy play.

As their name implies, traffic directors are the station schedulers. Operating within the patterns and flow developed by the program director (as graphically conveyed via daypart hot clocks), traffic people arrange the specific commercials, promos, and programs and prepare the master log that sets forth everything that is to air on a given day. In final form, this log becomes a validation that contracted-for spots were played. At all but the smallest stations, the traffic director prepares the log via a computer that automatically generates advertiser invoices and confirmations as the various advertisements are run.

However many or few personnel are available to assist the program director, this executive must remember that these off-air people, together with the on-air talent, are what will make or break format execution. How they are treated by the PD is the key to success. As programming consultant John Silliman Dodge cautions:

> We think we're in the music business. We're not. We're in the relationship business. Relationships between listeners and air personalities, between leaders and staff, between sales people and clients, between labels, program directors and music directors—this is the glue that holds the Big Show together. And who are your outreach experts, your relationship managers, your most valuable customer service reps? The announcers. Their presentation, the way they deliver your carefully crafted package is more important now than ever before. Because the songs you play aren't exclusive to you. Your playlist isn't defensible. Even your promotions can be scooped or one-upped. We should post a sign that says: It's the PEOPLE, stupid.[4]

This people and product management task becomes more difficult in programming a multistation cluster. The vast consolidation in the radio industry that was set off by the Telecommunications Act of 1996 has meant that a single company can now own several stations in the same locale—up to eight in the largest markets. Each facility may have its own program director, who now must not only make sure his or her own outlet's format is market attuned and well executed, but also that it complements the sister stations' formats without replicating them. In many such clusters a new *director of programming* position has been installed. "This position parallels the newly configured 'director of sales' job that coordinates all the sales staffs in a local cluster," reports radio consultant Ed Shane. "The person most qualified to be director of programming has experience in many formats and an ability to manage highly charged, ego-driven talents. Answering to the director of programming, individual station program directors are specialists in their formats."[5]

In smaller-market clusters, the director of programming may just be a single program director who is solely responsible for designing what goes out over the air of each of the several facilities in the group. This operational burden is lessened when one or more of these stations airs satellite-delivered content from a distant supplier for all or part of the programming day. Of course, once a satellite service is selected, the program director has little subsequent control over the sculpting of that content and no meaningful interaction with or influence over the *voice tracking* (see chapter 1) talent who announce within it. The professional and public service drawbacks to such a system are

Figure 5-2. Bing Crosby's use of new audiotape technology overcame network and radio industry opposition to transcribed (recorded) programs. *Courtesy of 3M Corporate Marketing Services.*

being widely debated—but this operational by-product of industry consolidation is a fact of life with which many of today's radio program directors must deal. The radio industry has come a long way from 1947 when crooner Bing Crosby had to battle with ABC to allow him to record his show so he didn't have to repeat his performances for western time zones (see figure 5-2). The question is whether the recording of air talent has gone *too* far, depriving program directors of an important tool for spontaneous, local-attuned presentation.

Up to this point, we have been discussing the span of control exercised by radio program directors. Television station PDs (sometimes called *program managers* or *operations* directors) play a similar role but with modified emphases. Instead of concentrating on the design and maintenance of a format, television program directors must concentrate on the individual and often highly divergent *programs* that make up a given broadcast day. The task here is to encourage audience flow from one show to the next while keeping an eye on the programs being scheduled by competing outlets. If other major local stations are airing material with heavy male appeal, such as a sporting event followed by an action/adventure cop show, the wise TV programmer may choose a property like a nighttime soap that will skew more to female viewers. If a station is a network affiliate, of course, some of these options are precluded by what the net has decided to slot during that time.

Regardless of network affiliation, it is important to recognize that no network can dictate what an affiliate station must air. Over the years and via several regulations and policy statements, the FCC has made it clear that a station is licensed to serve a local community. Program decisions, therefore, are to be made at the local level. With the backing of the general manager, any program director can refuse to clear a given network show in the station's market, though this occurrence is quite rare due to a variety of network disincentives. The reason for preemption could be anything from a time conflict with a local hockey broadcast to a concern that the net show's content will offend local community values. Of course, if the net show is not run, the station will not receive compensation (*net comp*) for the network commercials within the preempted show that were also not run. The larger the market, however, the less a station depends on net comp, so this factor is not the major deterrent—at least in the top fifty population centers.

In today's world, in fact, some stations have been forced to pay *reverse compensation* to the network to avoid the net taking its affiliation elsewhere. Clearly networks are no longer at the mercy of their affiliates—if they ever were. A station that consistently fails to clear a significant number of net programs may find the network threatening to move its affiliation to another outlet. This, of course, assumes that there is a local outlet with equal technical facilities (signal reach) in search of such an affiliation. To assure stability for both parties, all the broadcast networks have set about signing long-term contracts with stations—some as long as ten years. These contracts also seek to improve station clearance levels by offering such incentives as cash bonuses and more local avails adjoining net programs. Nevertheless, on a week-to-week basis, individual program directors still must fashion a series of compromises between the optimum local schedule and the offerings being fed from the net.

Though relatively few independent (non-network-affiliated) stations remain, program directors working at them are free from the limitations posed by network feeds. However, the pressure to fill an entire program day generates significant administrative and financial problems in its own right. Purchasing a spate of low-cost properties will make the station look cheap and unattractive to audiences and advertisers alike. But spending vast sums on consistently high-appeal programs may put untenable pressures on the sales staff to sell at rates high enough to pay for the on-air product. To lessen this dilemma, many independents affiliated with the young Fox network in the late 1980s and early 1990s. Fox's comparatively limited schedule of offerings allowed these formerly independent outlets to maintain much of their flexibility while still freeing up station resources to purchase better programming for remaining locally determined dayparts. This process repeated itself in the mid-1990s as yet other independents signed up to carry the comparatively limited amount of programming mounted by new WB and UPN nets.

Television station program managers, like their radio counterparts, have production/operations executives and traffic directors to assist in the administration of their departments. Given the requirements of a visual medium, they are also responsible for art/graphics departments and for the larger number of technical personnel required to produce local pictorial content. With cable and DBS (direct-to-consumer satellite services) bringing more and more viewing options to town, television station program directors must use these resources more deftly than ever in delineating their outlets from the expanding list of alternatives. "With the average household having 100 channels of broadcast and cable," says Initiative Media's executive vice president Tim Spengler, "you've got to be spectacular to break through."[6] As has been the case in radio since the 1950s, television PDs now must labor to find and fill what broadcast consultant Clark Smidt labels the "profitable market niche." As Smidt sees it,

> Proper positioning results from understanding a constantly changing competitive field within a defined area. . . . The total packaging of the station is the amalgam of all elements reaching the public, and doing so in synch. It includes a properly tuned presentation, appropriate style and meaningful services. The total package is your station's identity and personality. . . .

Station development is a deliberate and gradual process, requiring step-by-step building and minimal error. A well-put-together product invites many vehicles for reinforcing ongoing presence in your competitive arena.

Management must respect its own staff capabilities while carefully monitoring the competition. Ratings are the product of not only what's been on your air but also what everyone else in the market has or hasn't been doing.[7]

Beyond the individual formatic and show scheduling elements, today's station program director also must be a marketer with a marketer's sensitivity to audience targeting and audience expectations.

In fact, the preeminence of marketing in today's television business has caused a fundamental rethinking of the job of program director. More and more PD jobs are being eliminated, points out trade reporter Steve McClellan, "as program costs have risen and as promotion has come to be seen as important to a show's success as the show itself."[8] Program and promotion departments thus are being combined into a high-powered *marketing* unit that either makes the program decisions itself or executes program actions flowing from the station's general manager. Alternatively, programming duties are now placed under a strengthened *operations manager* or *operations director* responsible for all personnel, promotion, production, and program activities. In short, program director-type tasks are still being exercised—but in different offices, intertwined with broader marketing and operational concerns.

Jon Bengtson
Operations Director, WEYI-TV,
Flint/Saginaw/Bay City, Michigan

Consolidation, ownership caps, hubbing, duopolies, and cross-ownership are the current broadcast industry buzzwords. Like other broadcasters, I have found that, in order to maintain my value as an employee, I need to diversify my skills and adapt to changing job requirements. My job as operations director is constantly evolving and today covers a number of different station tasks.

A major task is to acquire programming that will attract viewers. The value of our media and the cost of our commercial time is determined by our viewers, whose changing numbers and demographics are recorded in paper diaries—a very "low-tech" method. Some of the television markets enjoy electronic information-gathering technology provided by Nielsen Station Index. However, the Flint/Saginaw/Bay City market continues to use the manual diary method because electronic set top boxes are too costly for a market our size.

Each year, studios produce thousands of hours of programming intended for broadcast by the networks as well as by individual television stations like WEYI-TV. "Off-network" programming would be programs that were originally aired on a network and are now up for sale in each of the television markets. These programs have normally been dramas, sitcoms, and now—reality shows. A second program category is what the industry calls "first run." This is a television program that gets its first telecast on stations selected individually from each of the television markets across the country. There are more than two hundred separate television markets in the United States.

The purchase of a particular show is a complicated negotiation often involving multiple stations. Each show is unique, with "deals" often covering multiple years, specific time periods in which it can be broadcast, additional episode availability, and promotional spot guarantees. Sometimes the value of the show is diluted by being broadcast on cable superstations during the same contract years that it is being offered for telecast by local stations. This fractionalization changes the potential value of the program and makes it more difficult to anticipate the audience attraction in the various time periods.

The syndication company will provide the basic contract terms—typically in some form of offering document. From that point, each segment of the contract is negotiated between the station and the syndication company. The station will provide a cash offer after all the contractual obligations are evaluated.

Program selection at WEYI-TV is a careful calculation of cost versus profit. We analyze the profit by calculating estimates of potential ratings that the program is likely to attract. This estimate will use historical time period ratings as well as data provided by research companies and Nielsen station-rating information. The syndicator (the seller) often comes to the station with a very detailed "pitch" indicating the program's potential audience. It is vitally important that all this information be fully evaluated so that a realistic picture of the profitability of the show can be achieved.

In past years, it was common for the individual station (WEYI-TV) to negotiate directly with the syndicator when a program was up for sale. Today, as larger and larger television groups are formed, the groups are using their clout to make combined program offers. A combined offer can result in lower programming cost per market, as well as serving to eliminate some individual stations from competing for the television program. Regardless of whether it is a "group purchase" or an individual purchase by WEYI-TV alone, we need to craft bid amounts and contract terms that make sense for our station's business plan.

As operations director, I am also the primary computer networking or information technology person at NewsCenter 25. I work in conjunction with our engineering staff to design and maintain over seventy-six personal computers and five servers on multiple operating systems. A typical work week will find me fixing a number of operator errors, repairing one or two hardware failures, and providing training and instruction on how to properly use the various systems and the software packages.

I am further responsible for the operation of the production department. This includes the directors, the audio operators, character-generator operators, camera operators, writers, and editors. Scheduling of these persons, as well as conducting their performance reviews, are essential parts of operations management.

The production department serves a variety of support roles within the station. Our photographer, editors, and writers work closely with our sales department to provide creative local commercials for advertisers. This allows our sales executives to forge a relationship with the client that will evolve and grow into a mutually rewarding partnership. My involvement is to make sure that commercial spot announcements are legal and in compliance with Federal Trade Commission and other federal and state regulations.

In today's segmented television marketplace, most television stations have chosen to have a local news presence in their community. NewsCenter 25 (WEYI-TV and WEYI-DT) is owned by LIN Television, which has made the marketing decision, in virtually all of their stations, to adopt a very strong news presence. Our production department and personnel all serve in a supporting role to our news operation. We currently telecast *NewsCenter 25 Today*, *NewsCenter 25 at Noon*, and evening newscasts at 5:30, at 6:00, and at 11:00.

Programming tasks are significantly impacted by Federal Communications Commission (FCC) rules and regulations. Our broadcast license is granted by the FCC for a period of five years. WEYI-TV must provide materials in our Public File and various periodic reports that demonstrate that we operate within FCC guidelines and fulfill the promises we made in our license renewal request. FCC regulations require the continual monitoring of commercial content within shows specifically broadcast to children ages two to eleven. This information is submitted in quarterly program reports. Similar station reporting is required for closed captioning for the hearing impaired, with a specific minimum number of hours required. The completeness of our Public File is an important focus of any FCC inspection, and maintaining the proper records and station information is essential for the protection of our broadcast license.

Over the years I have assisted in the negotiations of a number of labor contracts and, as a result, part of my day is often filled with various labor issues and interpretation of contract language as it impacts our day-to-day operation. Labor issues, Equal Employment Opportunity (EEO) compliance, and the ability to provide a safe and effective working environment for all employees are time-consuming concerns.

Station sales personnel face a much more competitive market then ever before. To meet advertiser demands, NewsCenter 25 sales executives must have the latest long-range programming information from which they may choose the most effective advertising opportunities. Our program schedules are created many weeks in advance and provided to our sales staff. The schedules are updated on a weekly basis and contain essential information such as episode synopsis information. Additional information on NBC programming is available via special secured Internet sites that can be accessed by station personnel.

I monitor the station's participation in the SHVA (the Satellite Home Viewer Improvement Act). This is the compliance of satellite program suppliers with the requirement that they restrict the sales of network television signals to areas where there are no local broadcast affiliate signals.

Similarly, every three years, broadcast affiliates and cable operators must agree on the terms under which the broadcast affiliate is carried on the cable system. Television operators must choose either "must carry" or "retransmission consent." The latter often invokes financial payments or other consideration by the cable system to be paid to the broadcast station. Each station must choose and carefully calculate the "value" of their television signal before making the choice. If stations elect retransmission consent, it is a negotiated outcome. There is no guarantee or requirement that the cable operator carry the specific television station should the station demand terms of carriage that the cable system feels are too expensive. I negotiate these "must carry" or "retransmission consent" contracts with each of the cable systems that carry the broadcast signal of WEYI-TV or WEYI-DT (our high-definition digital signal).

As a programming executive, it has been my choice always to continue to focus on new technologies and attempt to gain an understanding of how these might be applied in WEYI-TV. The Internet is one area where stations struggle. Some see it as only a promotional arm of television rather than a distinct media path or competitor. It may well be all of these.

Probably the biggest trend will be the practice by large media operations to "hub" their stations. This centralization process reduces costs and increases profits for the industry by remotely controlling stations from a central location in a city many miles from the affected stations. As a result, each station need only maintain a small engineering, news, and sales component. My group, LIN Television, has pioneered this concept at our stations in Michigan, Indiana, and New York.

I expect my daily tasks will continue to change and that technology will affect more and more of my managerial duties at WEYI-TV. I embrace these changes and attempt to anticipate those areas in which I can apply them for the benefit of our daily station operation.

Whoever makes the program decisions, these decisions continue to reflect a reliance on certain basic strategies that have proven successful over the years. Because this is not a programming textbook, we can mention only a few of these strategies. *Strip* programming means airing the same show at the same time five or more days a week. The alternative is *checkerboarding,* where several different shows appear on succeeding days. In prime-time access (the hour before network evening programming begins), most affiliate and independent stations alike have gone to stripping for consistency and easier promotability. Checkerboarding is, of course, the common network prime-time strategy—which some independent outlets try to counter via a strip with high appeal to a certain demographic.

Vertical programming is the airing of several same-category shows in succession, such as a block of situation comedies, one-hour action/adventure series, or game shows. The assumption is that viewers who enjoy this type of offering are more likely to stay with the channel as long as their preference is available.

Another pair of strategies relates to the strength or initial weakness of properties the programmer has available to schedule. *Hammocking* is the placement of a new or untried series between two established successes. The objective is to support the new offering between the two proven properties to ensure it gets sampled—and, hopefully, builds a following. However, if the new program is a dud, it can harm the established show that follows it by driving viewers to other channels. *Tentpoling,* conversely, is using a proven series to prop up newer and/or weaker properties scheduled before and after it. To a degree, tentpoling is a desperation strategy that, if it fails, can torpedo the established show as well and sink the whole program block.

In a viewing world of remotes, PVRs (personal video recorders), and streaming video on the Internet, some feel that such program director strategies no longer matter—that consumers now will make up their own program schedules and defeat the best efforts of programmers to keep them tuned to a particular station or network. However, as programming researcher Susan Tyler Eastman has found: "One inescapable facet of television programming is that most viewers, and most listeners, are passive most of the time. The fuller one's life, the more complex the world, the more tired one is, and the more one uses computers for professional work, the less energy one has left to devote to making choices about how one is going to be entertained. Consuming television is, for most people most of the time, an activity that must be relaxing and easy, not interactive and challenging."[9]

Before concluding our assessment of program directors, we need briefly to explore how this position functions on the cable side of our business. In the wired environment, PDs are not concerned with individual programs but rather with the total arrangement of entire program channels on their system or systems.

What broadcast signals will be carried and on what system channel? Will distant broadcast signals, including superstations, be offered? What basic cable networks should be run? How many and which pay-cable services are to be provided? And should these services be grouped into packages, or *tiers,* for which subscribers pay additional fees? Should certain basic cable networks be tiered as well? What about local origination, public-access channels for governmental units, or leased access channels to be rented to realtors, home-shopping services, or other enterprises? Together with the system manager, a cable program director deals with all of these questions.

For the most part, however, most local cable systems are now part of *MSOs—multiple-system operators* who own a number of individual systems that collectively might serve millions of subscribers. In these cases, the majority of decisions are made by a program director at MSO headquarters, leaving the local staffer with little more than traffic-director duties. In addition, certain other determinations (such as the number and use of access channels) may have been specified in the franchise agreement that the system has signed with the local government in whose jurisdiction it is operating. Unlike the situation found at stations, therefore, a local cable system programmer may have limited decision-making powers and may serve more a facilitative than a directive function.

The MSO program director, conversely, is burdened with vast decision-making responsibility. As trade reporter Deborah McAdams describes the situation:

> Talk about being in the hot seat. Imagine deciding what gets pumped into millions of cable homes. Imagine dealing with scores of executives, from entrepreneurs to billionaires' emissaries, all of whom want their product to be a part of that payload. Imagine doing it in an atmosphere where programming costs outpace inflation, but consumers resist paying more while stockholders expect spec-

tacular results, which, when realized, can mean getting merged out of a job. That, in case you haven't figured it out, is a fairly accurate job description for a cable [MSO] programming executive.[10]

The same description could also be applied to the programming executives working for DBS companies.

Whatever the electronic medium and whatever the job title, professionals exercising programming responsibilities must constantly be focused on what their target audiences will be seeking in the future. In many instances, program properties purchased today will not be available to air for months or even years—and the multiyear nature of most syndication contracts means that the show must be appropriate for several seasons after that. The programmer who is fixated only on today will not be around to fill that position tomorrow. Robert Pittman, the founder of MTV, once put it this way: "A programmer is really a sociologist—a reader of the trends in society."[11] And Saga Communications' Steven Goldstein adds, "The program director's job has become complicated over time. The PD has, in many ways, become a brand manager. Knowledge of research, marketing, people management, and business skills are all essential ingredients."[12]

Sales Managers

The sales-side counterpart of the program director is the sales manager. The time sellers discussed in chapter 4 report to this executive. The sales manager directs and monitors the efforts of the individual sellers to meet the projected revenue goals of the station, system, network, or Web service. This executive must also set the rates specified in the grid card (review figure 4-1) or similar pricing schedule. If the outlet for which the sales manager is responsible does not perform to expected capacity, the future of the entire operation and everyone who works for it is in jeopardy.

In setting goals for subordinates, sales executives stress one of two things: *share-of-market* or *budget*. In *share-of-market,* a computation is made of the total advertising dollars being spent in the particular medium for that geographic area. Then, in light of the outlet's programming performance, the sales manager targets a certain percentage of the total ad dollars that his or her sales staff should be able to obtain. If increased money is being spent in radio advertising in the market, for example, the radio sales manager using a share-of-market approach expects that the sales force will bring a proportionate increase in revenue into the station. Conversely, *budget* targeting looks at the revenues needed to operate at the expected profit margin and then sets the individual salesperson's dollar goals accordingly. If market expenditures increase, it will be easier for the staff to deliver their goals. If, on the other hand, advertisers are spending fewer dollars in the medium, time sellers must actually increase their share-of-market in order to meet the budget mandates.

Sales managers must be adept forecasters as well as skilled motivators. They must project what performance standard and level to set and then activate their staff to reach these objectives. By themselves, goals and targets are worthless if the people who must attain them are not given supportive direction and organized assistance by their sales boss. Broadcast sales consultant Martin Antonelli puts it this way:

> Managing at a station or rep firm must be approached with a definite plan of action; one that involves discipline and organization. Without a plan we see nothing more than activity without action; motion without movement. Salespeople have a right to be managed with competence and dedication. Anything less will lead to breakdown and chaos. . . .

> Perhaps the single most important responsibility of the sales manager involves spending time with salespeople. Daily sessions should be scheduled at which the following subjects are covered: pending business; establishing target shares for specific buys; what salespeople are saying to clients; how clients are responding; whether the appropriate areas are being pitched on specific buys; whether specials are being emphasized; what the client requirements are (cost per point, target demo, number of spots, reach, frequency, traffic building, etc.); what type of order it looks like the station will get; determining the problems that exist and possible solutions. Time spent on these areas can be invaluable in determining strengths and weaknesses of the sales staff as well as providing direction and leadership for the team.[13]

In other words, leaders who don't lead are as damaging to a sales effort as they are to programming, engineering, and any other aspect of our industry.

While directing their own staffs, sales managers at electronic media outlets must also work closely with other department heads—particularly the program director. Nowhere must this cooperation be closer than in determining the amount of *barter* programming (see chapter 4) that will be aired. If the program director accepts too many barter contracts, the sales department's revenue-generating prospects will be seriously squeezed. This, in turn, means not only less income for the company but also fewer commissions for its salespersons. As in most commercial businesses, time sellers supplement their basic salary (if they even have one) with substantial commissions—often in the area of 10% to 18% —based on the dollars they generate. An avail filled by a barter spot means no revenue for the station—and no commission for the salesperson. If the salesperson is employed on a "straight commission" basis (with no base salary), a high volume of barter can be especially contentious.

The sales manager must also come to agreements with programming and promotion heads as to the acceptability and volume of *trade-outs*. In a trade-out, an outlet exchanges its airtime for goods or services provided by the advertiser. A long-standing example of this was *TV Guide*'s providing space to station promotion departments in exchange for free airtime to pitch *TV Guide* purchases by consumers. The sales department itself may create a barter deal such as giving a restaurant no-cost spots in exchange for that eatery's free catering of a station client party. Trade-outs may have some tax and resource-providing advantages, but as in the case of barter, they also consume valuable avails. A sales manager who allows too many trade-out and barter deals to flourish may discover that there is no longer enough revenue potential in remaining airtime to generate a profit!

To complicate the sales manager's life even more, an increasing number of advertisers are seeking *value-added services* from the media with which they do business. As an incentive for buying an extensive spot schedule from the station, an automobile dealer, for example, may want the outlet's popular morning drive personality to make a personal appearance at the dealer's tent sale. Or the dealer may demand that the morning show be broadcast live from under that tent! In such circumstances, the sales manager and the program director must make a joint assessment as to whether or not either one or both of these activities will do the station image or air sound more harm than the sales contract with the car dealer is worth. If he or she has the clout, a sales manager may sometimes decide to provide such a client service over the program director's objections—but such heavy-handed tactics will only make an enemy of the person who produces the only product that the sales department gets to sell.

Still, for executives able to deal with all of these factors, the future seems brighter than for those on the programming side. As Allen Shaw, vice chairman of the Beasley Broadcast Group, points out, "Although consolidation may continue to centralize programming functions, both content and on-air talent, the large groups that inherit all the stations will still have to have local operations in sales. Sales is the largest opportunity for young people seeking to enter the industry. Good financially oriented Market Managers who can efficiently organize local operations, direct sales, and manage expenses will always be in great demand."[14]

General Managers

If a disagreement, such as the one just mentioned, between a program director and a sales manager persists, the general manager (GM) will probably have to settle the issue—the GM is the boss to whom both of these unit heads report. Settling turf wars is, however, a very unproductive use of a general manager's time and can result in no more than a temporary truce. As the top executive at the outlet, the effective general manager knows that it is much more important to engage in team building than in arbitration activities. In trying to construct a winning, profit-making operation, the person in the head office must marshal everyone's efforts toward the competitive arena that exists outside. When conflicts inside are so acute that they distract everyone's attention from building success, the facility is headed for financial disaster.

Management expert Peter Drucker has often stated that management is the task of making ordinary people perform in an extraordinary manner. The astute general manager knows that you can never achieve this extraordinary performance with people who are preoccupied with mundane grudge matches.

Unfortunately, today's general managers often have less time than did their predecessors to achieve extraordinary results. After a decade of mergers, buyouts, corporate takeovers, and defense against such takeovers, a number of facilities faced significant *debt servicing* by the early 1990s. In acquiring the property—or fending off the attempts of other companies to acquire it—owners had taken on high interest payments to lender banks or other financiers. Thus, general managers were required to generate increased revenue to cover this expense. In radio, this debt-servicing situation—along with the fact that the FCC had allowed too many new stations to come on the air—made leased marketing agreements (in which one station is operated in conjunction with a separately owned outlet) and duopolies (common ownership of multiple stations in the same market) necessities. Many general managers thus found themselves overseeing several radio stations in the constant struggle to realize a profit from the cluster as a whole.

This trend accelerated markedly after the 1996 Telecom Act's liberalization of ownership caps. With up to eight radio stations in a market owned by a single company and multiple television combinations also permitted, the general manager's task has grown much more complex. "A new breed of broadcaster is emerging," states Infinity (radio) Broadcasting's senior vice president David Pearlman, "with a new multitasking, entrepreneurial skill set unprecedented in the industry. Future local management will be asked not only to supervise more people in more jobs but also to oversee operations that were once the domain only of group operators of the preregulation past. Not long ago, a major group executive would be asked to manage six stations in six markets in what was then considered a big job. Today there are multitudes of GMs who manage that many in one market on one floor."[15] A similar dynamic now rules the visual media. "We all realize that the world in which we operate single TV stations in individual markets is going to end," KUSA-TV's general manager Roger Ogden observes, "whether the other lines of distribution are duopolies, robust Internet sites with video, smart uses of the digital spectrum or all of the above. Now is the time to act. We all need to operate multiple platforms in local communities to be survivors over the next decade."[16]

No matter how large the job gets, or what he or she may be able to delegate to subordinates, the GM can never pass on responsibility for P & L (profit and loss). A substantial debt load and/or stockholder expectations for higher dividends only deepen this responsibility. Under such pressures, rating/share performance and the advertising rates tied to it become concerns of almost overwhelming importance. The traditional saying that a program director is allowed two (ratings) books in which to succeed—and a general manager two program directors—is no exaggeration in many situations.

Eduardo Fernandez
Vice President and General Manager, Telemundo TV-44, Chicago

Being a general manager of a media entity such as a television station has been characterized in many ways. Some have compared it to being the general of an army where a station must be prepared to battle a host of enemies who are constantly bombarding your position. Others have made comparisons to being the ringleader of a circus where an unusual array of activities and talents make keeping track of it all very mesmerizing. Both characterizations are probably right. The consistent challenge of rallying a uniquely diverse group of departments (newspeople, producers, salespeople, engineers, accountants) along with delving into one of the most fascinating industries around is what makes being a GM of a television station or media company a rewarding experience.

I have always viewed my role as consisting of a number of facets. One is to ensure that what we do as a station serves the community in which we live. Whether it is providing news, weather, or public service initiatives, we always must remember that our license grants us the responsibility to educate and inform our viewers. The next key role is to run a profitable business and ensure that we are gathering and collecting valuable assets (whether it be programs or viable sales opportunities) that entice viewers to want to watch and clients want to purchase. The current broadcast TV business model only has one revenue stream. That is from advertisers who hopefully see the value in what we provide and are willing to pay a price for it. Overall, the general manager must keep a keen eye on expenses, which can spiral upward at any moment. Cash flows (revenue minus expenses) for TV stations still are fairly healthy compared to most industries. But as rising technical costs, personnel costs, and new media alternatives threaten to siphon dollars from TV's ad pie, the expertise and necessary skill sets for a GM have changed dramatically.

The challenge moving forward for future general managers lies in the ability to adapt to the constant changes that daily impact the industry. How do we utilize new media to our benefit? How will PVRs (personal video recorders) change the way people use our product? How do we remain viable in this century when the business model was created in the mid-1900s? While the answers are still unknown, the truth is that the business has continually evolved and has always met the challenges that have been presented. It will continue to do so in the future. Part of the answer lies in companies aggregating audiences (through duopolies, cross-media partnerships, purchase, or creation of cable outlets) in order to survive. Another answer lies in vertical integration, whereby organizations control all aspects of the business (from creation of programming to distribution to syndication) in order to maximize their competitive position. All these issues play a significant part in the decisions made by today's general manager.

Despite the current unsettled nature of the industry, running a television station is still a fascinating and rewarding role. The opportunity to work with talented, creative, and committed people provides a great deal of satisfaction every single day.

We often lose sight of the impact one station can have upon individual audience members or even an entire community. True broadcasters appreciate and do not take for granted the power of the medium. The personal challenge we face and question we ask daily is: "What can we as a station do today that will change how our community lives, works, and plays?" Very few industries enjoy that position along with a dynamic work environment that continually challenges us to be better.

The person who can cope with such pressures while still energizing and encouraging the rest of the staff is much in demand in our industry. If you aspire to such a position, it is essential that you identify and gradually acquire experience relative to the four main task areas in which a GM functions. As inventoried by professors William McCavitt and Peter Pringle, these tasks include

(1) planning, or the determination of the station's objectives and the plans or strategies to accomplish them; (2) organizing personnel into a formal structure, usually departments, and assigning specialized duties to persons and units; (3) influencing or directing—that is, stimulating employees to carry out their responsibilities enthusiastically and effectively; (4) controlling, or developing criteria to measure the performance of individuals, departments, and the station, and taking corrective action when necessary.[17]

To keep track of the financial information necessary to deal with many of these aspects, the general manager often is aided by a *comptroller* or *financial officer*. This staffer monitors and keeps the ledger on the outlet's cash flow and all the related business transactions. In larger operations the comptroller, in turn, is assisted by a *credit manager*. The credit manager works closely with the sales department in ensuring that invoices for airtime are mailed out and that payment is received. With comptrollers and credit managers policing the daily debits and credits, the general manager can concentrate on overall business and organizational goals. Still, the GM must always be prepared to intercede when one of these executives isolates a potential or recurring financial problem that threatens revenue stability.

The comptroller, for example, may point out that the promotions department is committing to too many trade-out deals for contest prizes, or the credit manager may sound a warning that too many slow-pay clients are still being allowed to obtain large numbers of prime avails. In such cases, the general manager must step in to enforce the financial discipline necessary for the maintenance of a healthy bottom line.

Even though actual *profit* is not in the picture, the general manager at a noncommercial facility deals with the same fundamental responsibilities as exercised by commercial colleagues. Revenues must be balanced with expenses. However, noncommercial income is most likely to derive from grants, corporate donations, and audience pledge drives. There is no sales department as such, but there probably is a grant or donor development office that coordinates the station's fund-raising and program-underwriting activities. Money must still be raised to fire up the transmitter, secure the programs, and pay the staff. The noncommercial GM can no more delegate this responsibility than the commercial manager can shuffle away profit-and-loss concerns.

Cable facility general managers (usually called *system managers*) are not that different in their overall perspective from GMs in other electronic media facilities and face the same increased responsibilities that flow from increased consolidation. As trade publication *Broadcasting & Cable* points out, "In the old days . . . cable systems were fairly localized. A system might be one franchise area—Chicago, for example—or might cover several adjacent suburban towns. It would have its own customer-service operation, its own technicians. Key decisions were made by the local general manager. Today, clustering has changed the nature of cable systems. With the wide deployment of fiber optics, a single cable headend might serve a large region. . . . Even in supposedly decentralized companies, a lot of operating power that once lay with local general managers now resides with regional and super-regional managers."[18] Cable general managers do differ from their broadcast counterparts in that they deal with multiple revenue streams: subscriber fees for basic and pay programming, local cable advertising income, and payments received from shopping services or other leased-channel users. In addition, the cable manager must monitor expenses related to such things as the franchise fee tendered to the local governmental body, copyright payments for the importation of distant broadcast signals, and charges from cable networks and satellite suppliers for retransmission of their programming. As a *marketer,* the cable GM also must watch the system's *churn rate* (subscribers disconnecting compared with those adding cable service) and attempt to control the technical and administrative costs associated with such customer turnover.

Marketing is a key to fiscal success because it drives and enlarges demand for a product. As just mentioned, cable GMs are marketers—as must be the general managers of outlets everywhere else in our industry. In accomplishing this marketing task, GMs must oversee the development of marketing plans that serve as the blueprint for system—or station or network—business survival and growth. According to marketing consultant William Foster, the components of this marketing plan should include "customer and prospect targets; an understanding of their needs and wants (market and customer research); our products and inventory; our prices, our selling structure, and sales force recognition and rewards; how we will inform our targets of our products and how we will persuade them we can meet their needs (selling materials, advertising, promotion, publicity, etc.), and how we will service our customers once on the air."[19]

General managers must rely on their staffs to suggest and operationalize the individual parts of this plan, of course, but they are ultimately responsible for giving the plan overall shape and a firm stamp of approval. Such a partnership between employees and top executives can be sustained only by managers who recognize that their job is as instructive as it is directive. Successful managers, says radio group owner Larry Patrick, understand that "management decisions present challenges and opportunities for teaching employees and demonstrating leadership."[20]

News Producers and Directors

General managers oversee a media enterprise that Edward R. Murrow once called "an incompatible combination of show business, advertising and news." The GM nonetheless must forge these three disparate activities into a coordinated whole by working through the subordinate executives responsible for each. We have already looked at two of these three elements in our discussion of program directors (the show business aspect) and sales managers (the advertising dimension). To complete the picture, we now examine the executive responsible for the news function—an executive appropriately labeled the *news director.*

Depending on the station's organization chart, the news director may either be of equal rank with the show business and advertising counterparts or be a subordinate who reports to the program director/manager. In music radio, particularly, news is seen today as no more than one element of the format and so must be subject to the program director's decision about how that format should best be developed. This obviously puts the news director (if the station even has one) in a precarious position. "Our biggest challenge in many situations is survival," admits radio news executive John Rodman. "The next is listener acceptance (credibility in news parlance), followed by questions of quality, productivity, and job satisfaction."[21] Although audio newspeople are not happy about this hierarchy, the fact remains that, since the deregulation of radio over the last several years, there are no requirements that a facility air any given amount of news—or even any news at all!

Generally speaking, this has meant that music-oriented stations have de-emphasized the role of news (and therefore the role of the news director, who now may constitute no more than a one-person department). On the upside, however, stations choosing an information-based format have expanded their news staffs and enlarged the news director's role. Some all-news outlets, for example, have no need for a program director because their total programming comes under the news director's span of control.

As you know by this point in your reading, television stations remain more broadly targeted than do radio facilities and so most of them still try to reflect a significant news presence. One reason for this, of course, is that a local news show is about the only program matter that can be completely unique to the facility that airs it. TV stations thought to be number one in their markets are thus normally also expected to have the number one news team. Nevertheless, because news occupies a comparatively small proportion of the broadcast day, it is a rare station on which the news director carries as much clout as does the head of programming.

This should not suggest that the job of television news director is a minor one. Compared with radio, where a news director may be the only journalist at the station, even an average-sized TV facility employs at least twenty news staffers. The larger the station, the more likely that the news director will concentrate on such administrative matters as budgets and personnel while delegating day-to-day news judgments to other people. Sometimes an *assistant news director* post is created to carry some of the burden of either administration or news content oversight—depending on where the news director most needs assistance.

Below this level, an *executive producer* is likely to exercise responsibility for the actual content and quality of the newscasts. Sometimes, this person serves as the producer of record for the early-evening newscast (which is the most important because it can generate the largest audiences). Other *producers* then serve similar functions for late-night, early-morning, and (if programmed) noon newscasts.

The producer job is a critical one in determining what gets covered in a typical newscast. As executive producer turned college professor Chris Tuohey describes it, "Producers are traditionally the foremen, or forepersons if you will, of their newscasts, with considerable editorial and supervisory clout. They choose what stories will get into a newscast, in what order they will appear and how much time they will get. They normally have the option to edit or rewrite stories written by reporters and anchors."[22] Yet, in today's television world, the producer role is often filled by the youngest and least experienced people in the newsroom. Because so many newcomers seek to enter the field as on-air reporters, they often ignore those behind-the-scenes aspects of broadcast journalism that are central to newscast construction. Consequently, there is a continual producer shortage. People enter (or fall into) the role immediately out of school. The good ones advance rapidly from lightly viewed time periods to profit-center evening newscasts and from smaller to larger markets, creating constant turnover. Some producers have been known to move from the smallest cities to the top 30 markets within two or three years after graduation.

As the competition for the lucrative news audience has heated up, local stations have increased the amount of news programming they air, thereby creating even more producer openings. In fact, opportunities for entry into the electronic media are probably greater for news producers than for almost any other job category, and the chances for career advancement are equally good. "Ask any group of news directors what their biggest problem is," comments Post-Newsweek News vice president Mark Effron, "and they'll tell you it's attracting young, smart producers with at least a passing knowledge of what happened the day before yesterday, a smattering of news judgment and good writing skills."[23] Functioning as a news producer requires what Chris Tuohey calls "a unique combination of production, management and organizational skills."[24] For young television professionals who can develop these attributes, and learn to collaborate with anchors who are sometimes twice their age, career success can come quickly. This may even lead to producing at the television or information radio *network* level, where producers are expected to be much more seasoned and the size of their potential audiences is much greater.

Mimi Levich
Editor and Executive Producer,
VOA English Programs Division

If you tell most people in the United States you work for the Voice of America, a common response is "Oh." And then after a pause, "What is the Voice of America?" It is not that what we do is top secret, controversial, or trivial. But by law, VOA is not allowed to publicize itself or its broadcasts in the United States. However, if you tell people who live outside the U.S. that you work for VOA, most of them not only know about VOA but can tell you the time programs are broadcast to their area, the names of everyone they hear on the air, and on what AM/FM or shortwave frequency they tune to hear VOA.

For people in developing nations, our broadcasts may be the only source of unbiased and credible news available. Over the years, governments such as China, Russia, and Cuba have so desired to prevent our news from reaching their citizens that our broadcasts have been jammed, our website made difficult to access, and our journalists told to leave the country.

Every week, approximately 94 million people tune in to hear VOA broadcasts in fifty-three languages. The languages have changed over the years and today range from Amharic to Urdu to Mandarin—and, of course, English.

A few years ago, I was on assignment in Nigeria and happened to be in the Leki Beach Market outside Lagos at the time our Hausa Service went on the air for one of its afternoon broadcasts. All around me, radios went on and people began listening to the latest news. On that same trip to Nigeria, I was in the office of Nigerian president Olusegun Obasanjo for an interview. He too said he was a VOA listener and began to question me about a news story he thought he had heard us broadcast.

VOA broadcasts have also aided people facing difficult situations around the world. Thomas Sutherland, an American who was held hostage in Lebanon for more than six years, came to VOA after his release to thank us for our English broadcasts. He said that on occasions during his captivity, he and his fellow hostages had access to a shortwave radio. They would listen to VOA, and through our broadcasts they learned that efforts were under way to secure their release. Burmese dissident Aung Sang Su Kyi has told our correspondents that our broadcasts were one of the few ways she had of staying in touch with the world during the many years she was held under house arrest.

Journalistic standards are high at VOA, just as they are at many commercial news organizations. We broadcast accurate, objective, and comprehensive news. In fact, our congressional charter requires it. VOA employees are all on call 24/7 and unlike most other federal agencies, VOA stays open every day of the year. Even when most government workers were sent home in 1995 because Congress and the White House were unable to agree on a federal budget, VOA was given an exemption to the furlough and told to continue broadcasting.

The first VOA broadcast in 1942 included a very important statement: "The news may be good or bad but we shall tell you the truth." At the time, the announcer was referring to the news of World War II. But that promise continues to guide all of us working at VOA more than sixty years later.

As an editor, fulfilling the "VOA promise" is sometimes a daunting responsibility. As an international news organization, "credibility" is *all* we have to offer. We cannot afford to be wrong, come across as biased or unfair in any way. There are a number of first-rate writers and reporters at the Voice of America, all of whom strive for excellence in reporting accurate and comprehensive news. As the final person responsible for what goes out over the air in English, I am constantly reminded of how I must always be at my best. Because like any news director, the reputation of my colleagues and my organization ultimately rests with me.

At larger television stations, the producer works closely with an *assignment editor* in determining and gathering the stories to be featured. The assignment editor matches available resources to breaking stories by selecting and dispatching crews as events warrant. Because a great deal of news is, by its nature, unpredictable, a skillful assignment editor always makes multiple contingency plans so that reporters and videographers can be available quickly when a big story breaks. If an assignment editor is a poor planner or is slow to react to events, a producer's efforts to improve newscast quality will be largely in vain.

The smaller the station, the more these jobs are compressed. In such instances, a news director might serve as producer and assignment editor for the early-evening newscast with an assistant or two performing similar combination functions at other times of the day and on weekends. As in most corporate settings, the larger the operation becomes, the more its top executive must be pulled from actual production activities to concentrate on long-range administrative and planning tasks.

However large or small the market, every news director strives to make his or her unit the best news-gathering operation in town as a matter of both personal and professional pride. In pursuing success, veteran news director turned general manager Tom Kirby maintains that there are four essential components to any TV news operation: (1) anchorpersons; (2) managers/producers; (3) reporters; and (4) videographers (like the award-winning cameraperson in figure 5-3). Kirby found, through his experience, that any station able to be the best against competing stations in any *two* of these elements will ensure stability in the audience its newscasts draw. Any operation dominating *three* of these areas is almost certain to improve its ratings. Kirby further points out that, while anchors are the most critical element and get the most attention from news directors, few of these executives concentrate anywhere near as much on the quality of their videographers. This factor is thus the easiest one to win if a news director works to develop it.[25]

As in most aspects of our highly competitive industry, a successful news operation can stand out from its rivals by establishing a point of comparative superiority. Thus, news director veterans like KVBC's (Las Vegas) Rolla Cleaver always seek to establish what he refers to as "a franchise in something you can own." Being the best in business news or local entertainment features or city government coverage will move a station well beyond the perception that it is just "giving the news" like everybody else.[26] News directors must also work closely with their promotion departments to make certain that this franchise is vividly communicated to the viewership at large.

Owning a news franchise is important for many radio stations, too, of course. But securing that franchise is much more difficult when the news director comprises the entire department. Under

Figure 5-3. Skilled videographers are a vital but often underestimated ingredient in a TV news department's success. *Courtesy of Pattie Wayne-Brinkman.*

such circumstances, it is usually best to program a limited number of newscasts and stack them with the information in which this particular station's listenership is most interested. This information might be agricultural news, business/financial developments, weather forecasts, street/traffic conditions, or school events. Because radio formats are targeted to more and more specific lifestyle audiences, news that conforms to those lifestyle preferences makes good sense for both business and listener service.

Cable systems are also promoting new local news services. In the largest markets, these might be self-standing operations of the cable system itself. In other instances, the local cable news channel is a cooperative effort with a radio station or newspaper. In a different arrangement, stemming from *retransmission consent* discussions (through which a popular television station gives a cable system the right to carry it for free), the station secures all or part of a second cable channel on which it can repeat its broadcast news programming and/or create additional newscasts from footage there was no time to air on the station itself. Now that television duopolies are permitted, a similar procedure can be followed in which the stronger station produces or reairs newscasts on its weaker sister outlet. Technology entrepreneur and former National Association of Broadcasters executive John Abel points out that the typical TV station news department "only uses 5–25% of the material it produces each day. Imagine how ludicrous this would be if it happened in other manufacturing industries."[27] Exploiting an additional channel to market more of this news "product" thus makes perfect sense. As stations fully roll out their digital capabilities, new or reaired newscasts also can be broadcast as one of the several program streams that digital technology allows each station simultaneously to transmit.

Such second-channel efforts not only make fuller use of news product but also, of course, produce additional revenue for the originating station. In today's rigorous economic climate, anything a news director can do to make his or her unit a profit center must be seriously considered. "Media companies expect local news operations to operate at 50 percent profit margins in big markets and 40 percent in smaller ones," reveals Tom Rosenstiel, director of the Project for Excellence in

Journalism. "And the easiest way to improve your profit is to manage costs. The hardest way is to improve your product."[28] It is just such financial concerns that have multiplied the use of those brief "newsbreaks" that are now sprinkled across the television schedule. Newsbreaks do provide the audience with additional kernels of information. But the main reason for their existence, reveals rep executive John von Soosten, is that "they provide additional news adjacencies to sell. And news adjacencies command higher rates than other times!"[29]

While there is only one news director at the station level, most broadcast and cable news networks employ legions of news executives. Even though their functions are primarily the same as those of their in-station counterparts, top managers in network units carry titles such as "president of the news division" or "vice president for news programming" rather than news director. Producers and assignment editors are just as common as at the local level, however, and the pressures to be first with the story and first with the target audience are just as great. Whether they function at a national or local desk, today's electronic news managers must be aware that "the new world is going to be one in which news organizations provide news to people how and when they want it over a wide variety of different media," advises David Westin, the president of ABC News. "Our challenge is making sure that we do everything we can to make the news available for all of those different forms of media. So the focus is content, because a good story is a good story whether it ends up in print, radio, TV or streaming video."[30]

News directors who prove they can handle all of these responsibilities and anticipate these new trends and challenges are becoming good candidates for general manager positions. "News directors probably make more decisions in a day than another department head may make in a week," asserts Jim Keelor, himself a former news director who became president of Cosmos Broadcasting. "They become action-oriented people. And if they have the good sense to understand the importance of sales and marketing, they have a very good chance of becoming managers."[31]

Entertainment Producers

Having examined the role of news producers as well as that of the other prime executives who work within radio and television outlets, we can now look at *entertainment* producers who function outside the stations and networks but in closely associated enterprises.

Executives overseeing entertainment programming typically work for the large studio companies or for independent entities that frequently join with the studios to create or distribute on-air product. Some of these people, like Aaron Sorkin, John Wells, and David E. Kelley, are *writer/producers*. They combine directive and creative functions to place a distinctive stamp on every phase of a show's development. With more and more options now available to the viewing public, top writers who have proven that their scripts can attract an audience are in great demand. They have acquired the clout to insist they be true partners in the projects on which they work. The designation *writer/producer* is often the outcome. "Writer/producers are now in control of the series television business," asserts Richard Katz, a business affairs executive for GTG Entertainment. "They are the dominant force," agrees CBS vice president for business affairs Bill Klein. "Actors and directors are important. But there is no place to start without the writer/producer."[32]

A top-notch writer/producer may be able to negotiate a share of up to 50% of a show's net profits and so is compensated very well for fulfilling two separate functions. In addition, the exerciser of a dual role wields a good deal more creative independence than can a producer working directly for a studio. Marcy Carsey, who, with partner Tom Werner, has developed such series as *The Cosby Show, Roseanne,* and *3rd Rock from the Sun,* describes such freedom in these terms:

We're placing the bets we want to place in the way we want to place them. With a studio, there's a whole other voice and set of reasons for making business and creative decisions, some of which have nothing to do with the quality of the show. . . . Then there are the larger issues, like when a network is unhappy with your concept or with a cast member or with a key creative talent. The studio's more likely to go with the network's feeling for the sake of the relationship. We would hold the ground for the sake of the show.[33]

Such money and freedom do not come without a price, of course. The pressures of wearing both conceptual and directive hats have caused more than one writer/producer to suffer heart attacks, marital breakups, and stress-intensified exhaustion. Garry Marshall, whose developmental credits include *The Odd Couple, Happy Days,* and *Laverne and Shirley,* noticed this occupational hazard several years ago when he observed: "It seems like there's an old guard of writer/producers, who have become over-worked by the networks. They are pressed by the networks to do too many shows. When they do too much, they become tired. They start to move out, like Larry Gelbart [*M*A*S*H*], [Norman] Lear, and myself. During the interim, businessmen come in to run it for a while until a new guard of writer/producers rise up and kind of take over for a while, then they get tired and—"[34]

Why do networks continue to go back to the same writer/producers until these creators are burned out or retire? Because "network executives are risk averse," points out former media development executive David Antil. "So they pick producers they've known for a long time—regardless of recent track record for success or failure."[35] With all the major networks now owning their own studios, this risk aversion has intensified because a failed project can adversely impact two different arms of the corporation.

While these comparatively few "star" writer/producers attract the bulk of trade and popular press attention, a far larger number of traditional producers are employed in the industry, working almost anonymously to integrate the various productional and creative elements required to mount a show. Even though a producer's actual duties vary depending on whether other producer-titled positions are assigned to the project as superiors or subordinates, the essence of the job is always to make certain that the project comes together on time, within budget, and in line with the agreed-on concept. The producer, in other words, is the person with the broad enough view of the show that he or she can coordinate the production process from beginning to end without becoming buried by creative or technical details.

These executives thus direct both *above-the-line* and *below-the-line* operations. Above-the-line elements are generally considered to be the creative aspects of the project: actors, writers, directors, and the producers themselves. Below-the-line items are what we discuss in chapter 3 as production functions: camerapeople, designers, editors, and other craftspersons who technically execute the concept. Producers must bring integrative and cost-control skills to bear on both sides of the line if the show is to have any hope of achieving its business and content goals.

Beyond these generalities, the producer's role can vary widely from company to company. In some series situations, different producers are assigned to monitor different episodes, but with all of them reporting to the same *executive producer*. The executive producer has overall responsibility for the series and its direction and is more likely to have been involved in hiring the talent and technical expertise on which the series draws. A one-shot project, like a movie of the week (MOW) or a miniseries, may use the executive-producer title in a similar way, with individual producers assigned to individual segments of the project. The project may even have *supervising producers* on the scene as intermediaries between the unit or field producers and the executive producer's office.

Nancy Meyer
Television Program Producer

There are two main categories of television entertainment producers: *long-form* (mini series and movies of the week, or MOWs), and series. The *long-form* producer originates concepts, works with writers on original ideas, options books, magazine articles, and true stories, sells the stories to broadcast/cable networks, works with the networks to find a suitable scriptwriter, collaborates with the writer designing a dramatic, narrative story, and oversees multiple drafts of the teleplay.

A television producer's skill set includes storytelling ability, research sense, recognition of saleable stories, knowledge of the television marketplace, a collaborative nature, and maintenance of a large address book (electronic or otherwise) of agents, managers, and key network personnel. You're only as good as the last produced project and always selling the next six projects. I don't know any successful producer who doesn't have an infomercial pitchperson in them.

In *series*, the producer is usually a writer-producer who has worked up through the ranks of the writing staffs on various series, executive producer being the highest rank. As a nonwriting producer, I once developed a drama series based on mystery novels about a detective in Alaska. I knew that I needed a network-approved showrunner (writer/producer who had written on a number of produced series) to sell the project. So I set the series up at a studio that had showrunners under contract. One of the showrunners responded to the material and shared my vision for the series. We were able to pitch the project at all the networks with the winning combination of the book, the studio, and the showrunner.

When the network green lights a project for production, the producer hires a line producer who creates a budget and shooting schedule which will help in negotiating a license fee. My busiest time is in pre-production, when consultation with the director takes place, when casting takes place, when hiring key department heads takes place, and when final notes for the shooting draft of the script are given. The usual two-hour MOW filming schedule is four-week preparation, three and a half weeks filming, and six weeks to complete postproduction. This period entails fourteen to sixteen hours a day in the office and on set, six days out of the week. When filming is completed, I'm responsible for delivering all the postproduction elements to the network, to domestic and foreign distributors, and fulfilling any other delivery requirements. I also promote the project in whatever way I can as airtime nears.

On a series the schedule is brutal because you're shooting the twenty-two episodes back-to-back in a full schedule. Story and script development are ongoing; while one episode is in preparation for production, a second is being filmed and a third is in postproduction.

To produce *True Stories*, a half-hour dramatic anthology series, I worked on a pool of twelve fact-based stories. In at least three of the stories I had to track down the court reporter, secure copies of court transcripts, and read transcripts that were thousands of pages long, marking pertinent passages for the writer and later annotating the scripts for legal and Errors & Omissions insurance (protection against litigation). The stories based on individual accounts required detective work, tracking down the key participants (including an ex-convict conman who didn't want to be found but wanted to be found), and negotiating an option on their stories.

A producer has to see both the big picture and the small details. She has to be mindful of obligations to the network/cable buyer, artistic wishes of the writer, director and actors, and her own vision of and expectations for the project. It doesn't hurt to have lots of energy, the expertise of a therapist, the perseverance of a drill sergeant, a resilient ego, and a sense of humor.

Especially in corporate video and commercial projects, the title of *line producer* is utilized. According to production company head David Leathers, the line producer "is practically defined as the person responsible for the implementation and operational management of a video project. The line producer collects and develops information from the production company, executive producer, director, clients or others and makes arrangements with key staff, vendors and others as schedules evolve and work gets underway. On small productions, the line-producer's role may be handled by the producer, director or an assistant director."[36]

However the workload is divided, and whether corporate or entertainment program material is developed, "the producer is usually the center of the creative process," states *All in the Family* and *One Day at a Time* mastermind Norman Lear. "From working with the writers in the development of the treatment, to writing the teleplay, to casting, to rehearsal, to rewrite, to taping, to editing, to promoting the show, to the airing . . . a producer, in my opinion, is someone who touches and affects the whole."[37]

At the top of the producer pyramid is the head of the production company or studio. In a very small organization, this executive may still function as a line producer, overseeing the actual assembly of individual projects. A corporate video unit or radio program packager is a likely environment for such a short chain of command. At the other end of the spectrum is the large studio with several series and other projects being developed simultaneously for both network and first-run syndication markets. Here, the organization's president or head is truly a chief executive officer (CEO), with several executive producers as immediate subordinates. Though this top executive may be called on to deal with major production crises (such as the threatened walkout of a series star), for the most part, these CEOs concern themselves with setting top policy, planning goals, and determining priorities. Lower-level producers translate these decisions into completed projects.

To a significant degree, network presidents and/or their programming vice presidents are mirror images of the studio heads. And with the consolidation of networks and studios, these executives are now corporate colleagues. Nevertheless, network executives remain the customers whose business the studio producers must aggressively pitch. Network executives will be loath to buy a project they see as weak even if it is produced by their parent company's studio. And studio executives are driven by the profit motive to sell to a network that competes with their own sister network if the deal is sweeter. Broadcast and cable networks have the capacity to give a show vast exposure. But a continuous tug-and-pull relationship remains between studio and network executives about the creative and scheduling decisions that determine how much *positive* exposure is actually achieved. In first-run syndication, program directors at key station groups replace network executives in this give-and-take process, although the dynamics of the process itself remain much the same. Networks and stations are all in the business of manufacturing salable audiences, and it is the production executives who must develop and deliver the programmatic bait that lures these audiences into the appropriate media tent.

A producer's job can be all-consuming. To people taking on the task, Chris Carter (executive producer of such shows as *X-Files* and *Millennium*) advises: "Get your flu shots and don't plan a social life."[38] *Law & Order*'s Dick Wolf adds that much of his success as an executive producer comes from his knack for hiring obsessive people. "Production schedules require months of 18-hour workdays. . . . It takes an obsessive person to give up any outside life to focus completely on the show."[39] Adding to the burden is the fact that the chances of failure are always much greater than the chances for success. Long work hours and scrupulous attention to detail never guarantee a success—but the absence of these elements does guarantee a failure. "You have to love the process," says Emmy Award–winning producer Suzanne DePasse, "and if someone says you did a great job at the end, that's just cherry on the cake. Very often you learn more from the ones that don't work than the ones that do. Every project is a baby miracle."[40]

Creative Directors

Even though the messages whose birth they oversee are much shorter, advertising agency creative directors are to commercials what producers are to programs. It is to the creative director that copywriters and art directors bring their scripts, layouts, and storyboards. Like any frontline entertainment producer, the creative director must then work with these conceptualizers to refine their message products so that they conform to the needs of the client. Here, the client is an advertiser with a certain market strategy. Commercials are expected to conform to this strategy while they artfully and persuasively coax their target audience to watch or listen. The creative director thus must encourage art directors and copywriters to be innovative in message design while still adhering to the gameplan that account executives (see chapter 6) have worked out with the client. The job, observes trade reporter Mary Connors, is "equal parts quality control, referee, salesman, shrink and shaman."[41] *Adweek*'s Ann Cooper adds that "it isn't enough to win glittery statues or ego-enhancing pencils anymore; now, more than ever, success is dependent upon clients' results as well. . . . In addition to den mother, psychologist, cheerleader, arbiter of taste, basketball coach, team player, dictator, historian, jack-of-all-trades, showman, social convener, architect, designer and Renaissance man, today's more evolved specimen is also required to be strategist, businessman, planner, financier and new product developer."[42]

Creative directors almost always have been promoted to that directive post from the ranks of the copy or art departments. They thus have a certain affinity for the conceptual properties of the commercial. Nonetheless, it is their responsibility to ensure that a message adheres to the business dictates of the client's marketing plan. Writers and artists occasionally may let an innovative idea meander off target, but a creative director is paid to nudge it back *on* target—even if that prized innovation must be reined in. Advertising executive Paul Goldsmith sketches the scene:

> When a creative person sits down to conceive an advertisement, he will often find it tempting to wander away from the agreed-upon strategy. Clearly, a desired quality is the discipline to keep the work within the confines of the product positioning. A scintillating piece of copy for a new car that highlights the fact that you need never carry a spare tire, when the strategy talks about better mileage, is a flop no matter how cleverly or brilliantly executed. "That's brilliant!" "That's funny!" "Beautifully done!" All are wasted words when the client says, "This is not the strategy we agreed upon."[43]

Creative directors, then, are selected not only for their past conceptual brilliance but also for their demonstrated discipline in matching message execution to assignment goals. The strategy behind the commercial in figure 5-4, for example, is to show homemakers that Reynolds Oven Cooking Bags are the easy alternative to "steamy" cooking pan cleanup. The creative director has caused the spot to be focused in such a way that even fifteen seconds is enough to establish the product benefit in an entertaining—but still highly relevant—manner. What is more, the visual alone tells the essence of the benefit story, an important attribute for successful television communication.

Like a program producer, a creative director must also keep the work flow moving so that projects are completed on schedule—even if a number of rewrites have had to be ordered. As one major agency's guidebook for its copywriter employees explains it, the creative director

> is caught in a Quantity-Quality Squeeze. He's got to meet the quantitative demand, while satisfying the qualitative standards. So he sits in one of the hottest spots in the agency—if not *the* hottest.

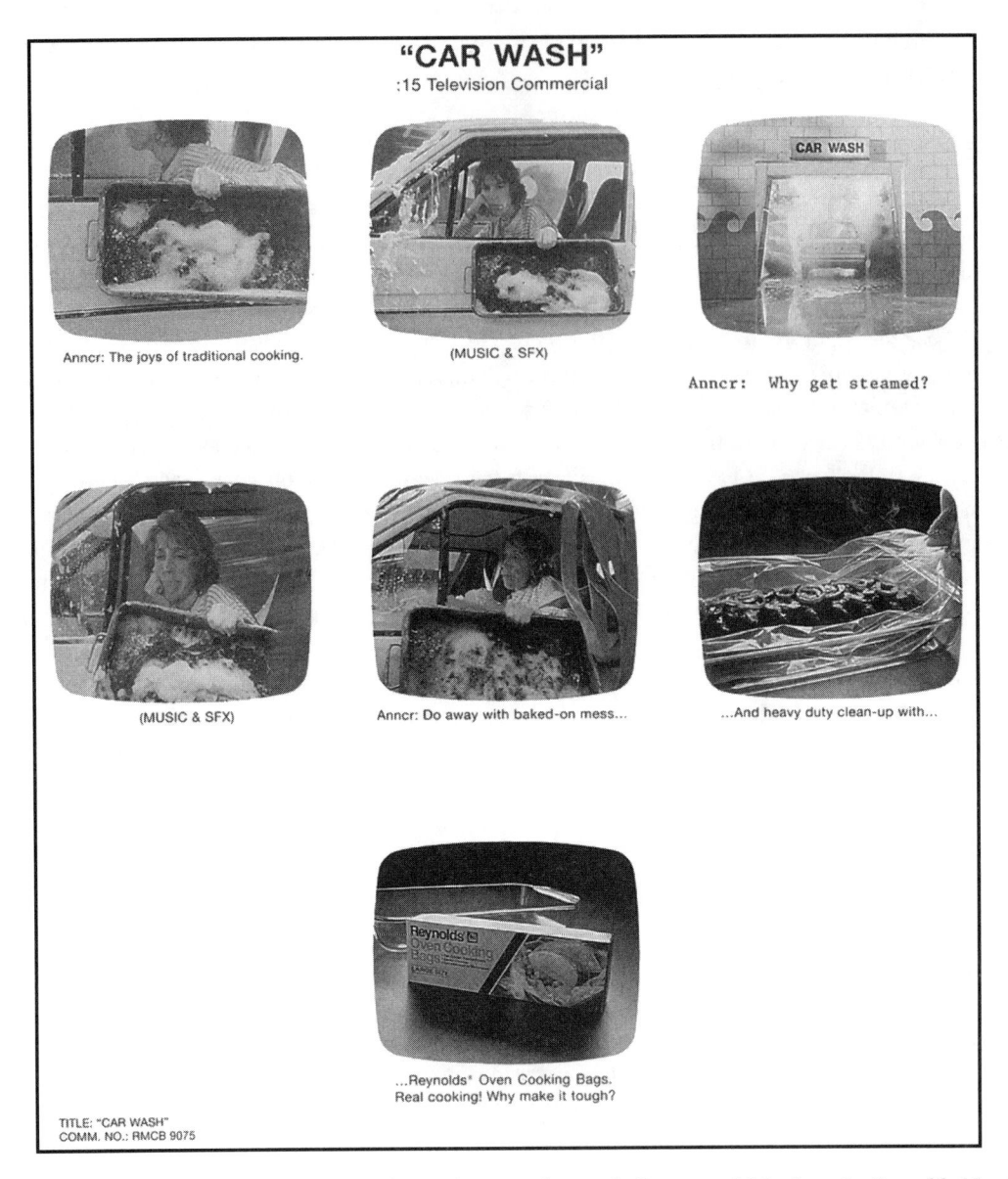

Figure 5-4. Matching message execution to the campaign goal. *Courtesy of John Lowrie, Reynolds Metals Company.*

He's paid to assume responsibility for copy on as many as half a dozen accounts. He's paid to supervise a number of talented people—often the hardest kind to handle—and to fit them all into the working day. He's paid to issue assignments, articulating the objectives as clearly as he can, in each and every case. He's paid to *inspire* his people to outdo themselves, and to see that they outdo themselves by the appointed hour.[44]

Pressure, as a result, is a big part of any creative director's job and the reason some commercial writers and artists do not aspire to the position. "At every turn," points out veteran creative director Roger Bodo, "it is your job to receive, translate, and transmit clear and focused communications.

This will be *the* greatest challenge to your creativity because you are the person in the middle trying to keep everybody happy—including yourself. . . . The parking lot is dark and empty. Your dinner is 30 degrees beyond cold. But you created—and it's good."[45]

Sometimes, as in most aspects of our profession, it is not possible for the creative director to keep everybody happy if the task at hand is to be fully accomplished. "The desire of the creative director is to be liked by people," executive creative director Lee Garfinkel observes. "But you can't let that get in the way of work. I work hard. I work long hours, and I expect the same from the people who work for me."[46] Agency founder Hal Riney echoes Garfinkel's comment: "A creative director can't make everybody happy, so you have to make yourself happy. You must have a strong point of view, and believe in your intuition. And you must be able to not just lead people, but to inspire them to follow you."[47]

As with program producers, the creative director function tends to be exercised at more than one level as the size of the organization increases. A *group head,* for example, may be responsible for the work of several creative directors who all service the same large advertiser. A *creative vice president* or *executive creative director* is the boss to whom all the group heads report. In a midsized agency, the group head post may be absent, with creative directors operating directly under the vice president/executive director. Whatever the precise organizational chart, these managers must somehow maintain their directive perspective without losing a feel for their creative and conceptual roots. Some former creatives find this balancing act impossible to perform, especially at the larger agencies. "There's more and more bull the higher you go," laments former agency creative director and president Malcolm MacDougall. "And as agencies get bigger and bigger, you can go pretty high. You can even wind up sitting in boardrooms talking about the disaster that will befall the company should the yen fall."[48]

Gerald Downey
Vice President and Creative Director, Alan Frank & Associates

Creative director is about as high as you can go on the creative (or fun) side of the advertising business. The next step would be agency ownership, retirement, or becoming a sharecropper!

Typically, a creative director begins life as either a copywriter or an art director. It has been my experience that the art director–turned–creative director often has a bit of an edge over the copywriter–turned–creative director, in that skills as an illustrator enable a person to sell a raw concept more easily to the creative team, account executives, or clients. So my first bit of advice is this: Writers, learn how to draw!

As a good advertising agency creative director, you must know a little about every aspect of creative—and a lot about one specific area. For example, you must be at least as good a copywriter as anyone in the creative department and yet also be able to communicate intelligently with art directors and the production staff about every aspect of their jobs. As creative director, you must distinguish good from bad artwork, recognize a good layout, and know when a storyboard is on target.

And because, at many ad agencies, the creative director is also the producer of that agency's radio commercials and video productions, you must know your way around radio studios, sound stages, and editing suites. A creative director should be conversant with broadcast production terminology, the titles and roles of various crew members, and also with videotape and film formats and the advantages and disadvantages of each.

You must be prepared to critique without criticizing, because creative temperament is fragile. And you must be quick to congratulate your staff for creative victories, because creative egos must be fed.

Some of the best creative directors are people with fat Rolodexes. The creative department of an ad agency is equipped to take most projects only to the client-approval stage. From that point, the creative director must know where to find the best people and facilities for such specialized work as finished art, illustration, animation, photography, performing talent, radio production, video production, or whatever else is needed to complete the job.

The best creative directors often have experience in client contact. This enables them to sell their concepts first to their own account executives and later, if necessary, to the client as well. The creative director should be able to provide solid reasons for the approaches taken and copy used in every commercial, video, and print ad.

Being a creative director for an advertising agency isn't exactly a day at the beach. It has more than its share of headaches caused by unreasonable deadlines, insensitive clients, nervous account executives whose allegiances often seem to lie more with the client than with their own creative department, and a creative staff filled with sensitive egos and, shall we say, unique personalities.

On the positive side, the job exposes you to a variety of people, places, and things that few other endeavors can match. Knowing that your work is often seen or heard by thousands of people, affecting the spending decisions and lives of many of them, is a tremendously satisfying experience. And it beats the heck out of working on the assembly line.

Keep your ambition under control early in your career, solicit and accept advice, and recognize your weak points. Inevitably, later in your career, you'll review some of your earlier efforts and recognize that your work matures even as you do.

Somehow, astute creative directors manage to maintain their equilibrium. They can distance themselves from their staff's commercials far enough to remain conceptually objective, but not so far that they can't appreciate a breakthrough execution or suggest how to attain such an execution. They can also sense "how high is too high" in terms of their own career ladders and what they most enjoy doing. Rejecting promotions to top management may be a good decision for people who want to stay involved in the idea side of the business—but there is a power price to pay. "Being a creative person at some of these places is like being a woman at an English dinner party," cautions creative chief Ralph Rydholm. "You get talked to, adored, even fawned over. But then, when the real business takes place, the men go off with their cigars and brandy."[49]

Very small *boutique* agencies have been formed by disillusioned creative directors who want to own or manage their own shops while still being closely involved in the conceptual arena. Some of these boutiques have been notable successes. Others have gone bankrupt because their executives spread themselves too thin. Still others have found that success only brought them full circle to the same quandary. As Grey Advertising's chairman emeritus Ed Meyer has noticed, "As a hot-shop boutique grows up and seeks to organize itself and its systems, as it attempts for the first time to reach the mass market instead of local markets or elite groups . . . larger clients begin to rein in the agency's high-flying impulses."[50]

Like entertainment writer/producers, creative directors in any setting must always play a split-personality role that encompasses both directive and conceptual responsibilities. Not everyone can perform such a juggling act or keep all the balls in the air for an extended period of time. Every allegedly completed assignment that a staff member brings to your office requires a quick decision—and the higher up the creative director is on the corporate ladder, the higher are the stakes. Do you

Directive Functions

play it safe on this project or push your copywriters and art directors for something more? "It's knowing which ones to roll the dice on," says executive creative director Nina DeSesa. "And being right more times than you're wrong."[51] Advertising legend Roy Grace put the matter even more simply: "The essential job of a creative director is taking out the garbage."[52]

Owners

Speaking of future prospects, the people who seem most able to chart their own fates are the people who own outlets, agencies, and other entities associated with the electronic media. Owners, it would seem, can exercise the ultimate in directive functions and determine the entire shape of our profession.

In fact, however, most owners are just as susceptible to continuing restrictions and unanticipated events as are the rest of us. Competitive business pressures, changing governmental regulations, economic fluctuations, key staff turnover, and technical malfunctions/limitations are but a few of the forces that impinge on and limit an owner's options.

The discretionary power of ownership is also impacted, of course, by how the company is organized. At one end of the spectrum is the *sole owner,* who holds full title to the enterprise and is the direct beneficiary or victim of any profit or loss. Many stations, cable systems, advertising agencies, and even networks were created by individual entrepreneurs who built them from the ground up. Such closely held companies were a family business, and that family's fortunes rose or fell with how the business itself fared. In most instances, however, there is a limit to how far a single-owner entity can grow before it must diversify control in order to stay fiscally healthy. Ironically, the more successful a media company becomes, the more it becomes necessary to bring in outside money and partners to sustain growth and maintain its competitive position. In the case of stations, a single independently owned outlet will have difficulty competing with same-market rivals that are part of a multicity group. This difficulty will be compounded when group owners operate LMAs (one station managing a separately owned outlet under a "leased marketing agreement") or duopolies (two or more co-owned stations) in the independent's own backyard. Because it controls several outlets, a group can negotiate first and hardest for programs since syndicators thereby can secure carriage in several markets with a single deal. A group also has a better chance of securing lucrative regional and national advertising because its size makes it a more appealing client for the larger rep firms which, in turn, have access to the larger agencies and their media budgets. Even the day-to-day operations of the group station can be run more economically because the group can secure volume purchase discounts on everything from paper clips to transmission line. And multistation clusters within the same market are able to assemble much more attractive advertising packages than can most single outlets. The situation, in short, is not much different from the corner grocery store trying to compete with several chain supermarkets. It is possible to survive, but only if the independent's owner is extremely responsive to customer needs and takes special pains to deliver unique local services at reasonable prices.

Station groups may be the possession of a sole owner, of course, but the number of individuals with the financial resources to build and maintain such an enterprise is limited. Furthermore, the progressive liberalization of ownership rules has allowed powerful media companies and conglomerates with significant media interests to enlarge their holdings at the expense of smaller, less financially flexible groups. Smaller groups thus have been swallowed up by a larger entity or broken up in piecemeal sales to several companies that are seeking to establish footholds in new markets or to convert existing properties into duopolies.

The larger the group becomes, the more likely it will broaden its ownership base from a few individuals and their companies to a larger participation by banks and investment entities outside the broadcast industry. Ultimately, the corporation might move to publicly traded status, with stocks being sold on the open market and boards of directors concerned with moving as much money to the bottom line—and stockholder dividend checks—as possible. Many people now lament that the broadcasting business is no longer run by the media professionals who founded and developed it but rather by outside financiers who see broadcasting as serving no higher calling than any other goods-and-services enterprise.

Either to increase their own ownership interests or to fight off takeover attempts by outsiders, many broadcast entities over the last two decades were forced to borrow large sums of money from banks or through limited partnerships, stock sales, or other instruments. The monthly interest due on these borrowed funds thus increased station/network operating expenses and required leaner and meaner running in order to divert more income to *debt service* (interest and principal payments). This phenomenon occurred in both radio and television, of course, but the dollar amounts were much larger in television, where both incomes and operating expenses are substantially higher. Over the last ten years, most of the ultimate directive functions in our industry have shifted from general managers' and group executives' offices to Wall Street and outside boards of directors. Figure 5-5 presents the top ten media companies for a recent year and evidences the prominence of the publicly held conglomerates on today's media landscape.

As the list of companies in figure 5-5 illustrates, media concentration is a fact of life in cable and satellite as well as broadcast communication. Individual cable systems have been combined into *MSOs* (multiple-system operators). MSOs themselves have been swallowed up by giant companies like Time Warner and Comcast that seek greater regional concentration that makes for economies of scale in plant operation and signal distribution. Broadcast, cable, telephone and DBS (satellite) ownership is increasingly interlocked and complemented with extensive Internet involvement. The corporate goal of media titans like News Corporation's Rupert Murdoch (see figure 5-6) is to build multichannel delivery systems and secure the in-house production capability necessary to supply much of their content—and some of the content for other people's program systems as well.

Electronic media ownership patterns are thus much more complex than before, especially in terms of those major corporations that are increasingly coming to dominate the industry. A vast infusion of new capital and new business interests into the electronic media has brought a wealth of new professional opportunities, but it has also made the concerns of ownership much more like those of any other commercial enterprise. The old concept of the self-contained broadcast or cable company that was thought to have a higher public service calling than other business ventures is probably gone forever.

Advertising agencies have been even more subject to consolidations, takeovers, and buyouts. In several instances, ultimate ownership passed to overseas interests in Great Britain, France, Japan, or Australia. (This might have happened with broadcasting, were it not for the Communications Act of 1934's prohibition on station foreign ownership.) As with electronic media companies, advertising agencies are now driven by profit-and-loss concerns that transcend what is happening in the individual unit itself. The fortunes of the giant parent and its partners, investors, or stockholders have pushed final directive authority into higher, more global, and more remote boardrooms. (The same condition pertains to some of the largest U.S. movie studios, which also have been purchased by overseas interests.)

Key advertising executives, who would formerly have spent most of their careers within a single shop, are now likely to jump from one agency chain to another, thereby blurring the creative dis-

TOP TEN MEDIA COMPANIES IN THE U.S.
August 2003 - - - based on 2002 U.S. Revenue

Time Warner	$28,629
Viacom	16,326
Comcast Corp.	16,043
Walt Disney Co.	9,763
NBC-TV*	7,390
Cox Enterprises	7,394
News Corp.	6,645
DirecTV	6,445
Clear Channel	5,851
Gannett Co.	5,617

(figures are in millions of dollars)

*subsidiary of General Electric

Figure 5-5. Figures are in millions of dollars and are based on 2002 data from U.S. revenue. *Compilation by* Advertising Age *extracted from several industry reports.*

tinctiveness that used to set each agency apart. As advertising industry observer Richard Morgan sees it, there is

> nothing wrong with these moves, aside from their being inconceivable a few years ago, except they do transform the industry into a machine of interchangeable parts. They do move agencies closer to the category of commodities—as classically defined. (An ominous aside: commodities, as classically defined, compete only on price.)
>
> This trend is in keeping with the rest of industry. Charles Fombrun, a professor of organizational strategy at New York University, says global and technological forces have imposed "major realignments" on many industries. When these realignments lead to takeovers, he adds, you can count on 50 percent of the team taken over being gone within three years. So much for corporate culture in the era of takeovers.[53]

If you aspire to owning your own media enterprise, certain opportunities are always present, especially in Web-based endeavors. Most of these opportunities are inherently small and (therefore)

Figure 5-6. Rupert Murdoch, architect of the Fox network and aggregator of multiple newspaper, satellite, and cable interests. *Courtesy of Rubenstein Associates, Inc.*

risky ventures. Those that do succeed must usually grow in order to sustain that success, which means raising outside capital. Beyond a certain point, capital needs can be met only by sharing ownership rather than by borrowing money. The ultimate directive function of possession must therefore be shared as well.

This is why, in today's media world, many top professionals prefer to be managers with perhaps a small piece of ownership. Such an arrangement motivates them to do their best on behalf of the company while preserving their personal flexibility to move on to other ventures. As we stress throughout this chapter, all directive roles in our profession encompass weighty responsibilities. The roles entailing ownership involve the additional factor of significant personal financial risk. At some point in your career, you may have to decide how much of this risk you wish to assume.

Chapter Flashback

Directive functions are performed by executives whose decision making determines how electronic media personnel and material resources are used in serving the requirements of clients and the preferences of audiences.

Program directors shape and schedule content progression in a manner calculated to attract intended listeners or viewers. In radio, this means maintaining and refining a particular format that is most likely to appeal to an audience sliver attainable in that particular market and of interest to an identifiable group of advertisers. Executives exercising the PD function in television may be known by a variety of titles. Like their radio counterparts, they must make decisions in light of what competing stations are airing. However, the TV program director focuses on individual and often divergent shows rather than on a single uniformly targeted format. Independent station programmers have a much greater task than do their network-affiliate counterparts because they must fill the entire schedule themselves. Cable system programmers pick and arrange complete program services to use available *shelf space* in a way that will garner the most revenue.

Directive Functions

Sales managers direct the efforts of the people selling airtime on their outlet. They must be skilled forecasters and motivators. Together with the facility's program director and *general manager*, they must decide when *barter* and *trade-out* deals can replace cash sales. The general manager (or *system/regional manager* in cable) has the ultimate responsibility for how the facility is run—primarily in overseeing profit and loss. *Comptrollers* and *credit managers* may help the GM keep track of cash flow and threats to it. *News directors* must make certain that money spent in news gathering can increase the outlet's visibility and demographic salability. In radio, the news director may be the only full-time journalist, but at a news-active television station, staffs of twenty or more are common. Though often young and relatively new to the business, producers are the key determiners of TV newscast content. Some cable systems are also moving into local news coverage, often in cooperation with a newspaper, radio, or television station.

Entertainment producers work at studios and similar enterprises to fashion program product that will be of interest to media outlets. *Writer/producers* are especially prominent in creating and packaging scripts. Their power accrues from their past conceptual successes, which enhance the prospects for their future projects. A variety of other producer titles are also used in the business to describe varying degrees of responsibility for program series and episode development. Within advertising agencies, *creative directors* are to commercials what producers are to programs. Creative directors must lead copywriters and art directors in turning out on-strategy commercials within client-mandated time frames.

The ultimate decision makers in our enterprise, as in any other, are the *owners*. Their power may be immense or diluted depending on the fiscal organization of the company. Government regulations also limit owner prerogatives, as do business conditions in general. Consequently, many top professionals prefer to retain career flexibility by remaining as full-time managers with perhaps a minority ownership share.

Review Probes

1. What is *waste circulation?* How might it undermine a program director's seeming success at generating substantial audience numbers?
2. Define *hot clock, strip, checkerboard,* and *vertical programming* as well as *tentpoling* and *hammocking*.
3. What is the difference between delivering *share-of-market* and delivering *budget?* Who determines what standard is to be used and to which employees that standard is applied?
4. What did Edward R. Murrow call "an incompatible combination of show business, advertising and news"? What is the title of the executive responsible for each of these functions?
5. What is a *boutique* shop and why is it difficult for such an operation to remain small?
6. List the forces that have contributed to media consolidation.

Suggested Background Explorations

Bergmann, Ted, and Ira Skutch. *The DuMont Television Network: What Happened?* Lanham, MD: Scarecrow Press, 2002.

Block, Peter, William Houseley, and Ron Southwell. *Managing in the Media.* Woburn, MA: Focal Press, 2001.

Compaine, Benjamin, and Douglas Gomery. *Who Owns the Media? Competition and Concentration in the Mass Media Industry.* 3rd ed. Mahwah, NJ: Lawrence Erlbaum, 2000.

Croteau, David, and William Hoynes. *The Business of Media: Corporate Media and the Public Interest.* Thousand Oaks, CA: Sage, 2001.

Doyle, Gillian. *Understanding Media Economics.* Thousand Oaks, CA: Sage, 2002.

Eastman, Susan, and Douglas Ferguson. *Broadcast/Cable/Web Programming: Strategies and Practices.* 6th ed. Belmont, CA: Wadsworth, 2002.

Gage, Linda, Lawrie Douglas, and Marie Kinsey. *A Guide to Commercial Radio Journalism.* 2nd ed. Woburn, MA: Focal Press, 1998.

Gates, Richard. *Production Management for Film and Video.* 3rd ed. Woburn, MA: Focal Press, 1999.

Kawamoto, Kevin, ed. *Digital Journalism: Emerging Media and the Changing Horizons of Journalism.* Lanham, MD: Rowman & Littlefield, 2003.

Keirstead, Phillip. *Producing Television News: Computers and Automation in the Digital Age.* Mahwah, NJ: Lawrence Erlbaum, 2004.

Kelleghan, Kevin. *Supervisory Skills for Editors, News Directors, and Producers.* Ames: Iowa State University Press, 2000.

Lee, John. *The Producer's Business Handbook.* Woburn, MA: Focal Press, 2000.

Miller, Phillip. *Media Law for Producers.* 3rd ed. Woburn, MA: Focal Press, 1998.

Orlik, Peter. *Broadcast/Cable Copywriting.* 7th ed. Boston: Allyn & Bacon, 2004.

Picard, Robert. *Media Firms: Structures, Operations, and Performance.* Mahwah, NJ: Lawrence Erlbaum, 2002.

Shane, Ed. *Selling Electronic Media.* Woburn, MA: Focal Press, 1999.

Thompson, Robert, and Cindy Malone. *The Broadcast Journalism Handbook: A Television News Survival Guide.* Lanham, MD: Rowman & Littlefield, 2003.

Warner, Charles, and Joseph Buchman. *Media Selling: Broadcast, Interactive, and Print.* 3rd ed. Ames: Iowa State Press, 2003.

Wicks, Jan LeBlanc, et al. *Media Management: A Casebook Approach.* 3rd ed. Mahwah, NJ: Lawrence Erlbaum, 2003.

CHAPTER 6

Facilitative Functions

In this chapter we examine the roles of the professionals who provide services that the directors discussed in the previous chapter need to effectively run their operations. Some of these facilitative occupations generate key data to help inform management choices. Some assist with client servicing and coordination. Some provide transmission vehicles or help in these vehicles' purchase. All of them conduct behind-the-scenes functions of which many consumers are completely unaware. But despite their low profile, these professionals' roles are essential to the fiscal health and growth of our industry. In the next few pages, you may discover an interesting but heretofore unrecognized career option you'll want to explore further.

Account Executives

In fulfilling their facilitative duties, account executives (AEs) perform a delicate and continual balancing act. As the liaison between the client and the agency charged with creating that client's advertising, the account executive must try to maintain two corporate loyalties with equal allegiance. On the one hand, the AE must attempt to sell the agency's creative work to the advertiser as an appropriate and effective response to that client's marketing needs. On the other hand, an account executive must convey client concerns and objections back to the creative team, even when these objections may result in scuttling advertisements of which that team is inordinately proud. "The account person, in a way, has the toughest job in the agency because he's had to interact with client, creatives, media, research," says Chiat/Day president Dick Costello. "With really terrific account people, they have to be good at twelve different things, unlike other good advertising people, who have to be good at two or three things."[1] "When I go into a meeting with a crazy idea, my role is not to be the radical one," confides Fallon account executive Kathy Spraitz. "I have to be the link of trust and confidence because clients look to [account people] for faith. We buoy the creative process. We're the 'sane ones' in the room."[2]

This delicate task is a constant test of an account executive's diplomatic and persuasive skills. The AE who slavishly panders to every client comment will lose the respect of agency creatives. These creatives consequently succumb to doing less than their best because their work is never supported at the account level. Conversely, the account executive who presents creative work to a client on an uncompromising take-it-or-leave-it basis is likely to find the client and its business walking out the door in search of a more responsive agency.

The AE must strive to project a sense of objectivity toward both parties while keeping them both reasonably content and on the road toward a common goal. This cajoling must be done by individuals who are usually outranked by the people with whom they must communicate on both the client side and the agency side. Some senior AEs may acquire more clout, but in most cases this liaison executive must deal "upwards" with agency creative directors and with client marketing vice presidents. Each of these forces can bring considerable muscle to bear on the less powerful account executive who is supposed to be servicing them. In a sense, the AE is like a United Nations peacekeeper. The two sometimes distrustful sides have all the weaponry, and it is the AE's task to bring them together without getting shot in the cross fire.

Lisa Albyn Drummond
Senior Account Executive and Media Supervisor, Caponigro Marketing Group

Advertising is a dynamic, fast-moving, ever-changing part of our lives. Just look at the attention (and revenue) the commercials that run during the Super Bowl garner. Creative gets most of the attention, but in order to make those impactful commercials and print ads, there is an important process behind the scenes.

The account services team controls that process. And at the very heart of the advertising account team is the account executive. This pivotal position is the gatekeeper to each project the account team tackles. As an AE, you will need to be familiar with both the big and the small pictures. You serve as the liaison between the client and the agency. Sometimes you will represent the client viewpoint, and sometimes the agency point of view. This can put you into the proverbial space "between a rock and a hard place." However, because you are aware of both sides of the issue, you are also in a unique position to offer up suggestions for compromise, which both the client and the agency can agree.

You will be responsible for cultivating a working relationship with the client. The more you understand the client's business and needs, the more valuable you become to the agency. That means making the most of those much maligned "people skills." You will need to practice active listening, patience, simple courtesy, and diplomacy. Clients may come to entrust you with proprietary information. In return, they will expect your respect, professionalism, discretion, and assistance at all times. As the steward and champion of the client's brand, it will fall to you to ensure that all communications, no matter the media vehicle employed, reflect the strategy of the brand.

The account executive is responsible for getting the job done correctly, on time, and within the established budget. This means shepherding every detail of each project. At any given time, you will be expected to know where each project stands and what are the appropriate next steps. It is important to become familiar with each department in the agency and each phase of the process. This access exposes you to all aspects of the agency, from traffic to legal to finance, media, creative, production, and shipping. And it provides you with an extraordinary opportunity to learn.

Details become paramount. Your knowledge of project management is tested each step of the way. Internally, you become "the motivator"—adept at cajoling, scolding, reminding, motivating, praising, nudging, begging, guiding and, if all else fails, playing the "enforcer."

The first person held accountable if something goes wrong is the account executive. You will make it a point to never go to the client with a problem without having a solution or alternate plan ready. Murphy's Law is reality, and with some practice, you will learn to be prepared.

Time lines, back schedules, status reports, and conference reports become invaluable tools. You document, document, document because you are the keeper of the information, and it is your responsibility to disseminate it to the appropriate members of the team.

Finance people hold you accountable for the bottom line, which is why you must stay within budget. Your relationship with the creative department is love/hate—depending on the status of the project and the phases of the moon. Traffic people prove to be essential as they move materials from one step of the process to the next. As they discover new media venues to use to showcase the client's products, media planners and buyers continually change the landscape. You learn to consult your legal department, if you are lucky enough to have one. And when material deadlines get pushed to their limits, production people morph into lifesavers.

You may be asked to present creative concepts, budgets, media plans, and strategy. Being an articulate, persuasive business professional will help make your case. Polish your presentation skills.

Over the course of your career you will give out more kudos than you receive, because that is part of being a good motivator. However, you will have the satisfaction of knowing that when a job is done, it is done well, and you were key to the outcome.

My best advice? Remember that you are part of a larger group, with each department integral and interdependent on each other. Protect your sense of self and your sense of humor. And, lastly, learn to juggle. It will help with the stress, and the heightened dexterity will come in handy for multitasking.

In short, an account executive must perform a facilitative function that subtly "directs the directors." Each side must perceive this executive as its loyal employee even though this employee is, in fact, responsible to both in the clarifying and execution of advertising strategies. Along the way, the AE must acquire extensive knowledge about the client's business in order to analyze its long- and short-range marketing issues. He or she must then be able to convey these problems clearly and accurately to the agency personnel whose creative output must address these issues. In dealings with the client, meanwhile, the account executive must be able to explain how the submitted advertisements meet client needs and mesh with client marketing plans. The executive must also be totally familiar with the agency's entire range of services so they can be marshalled as needed to help achieve client objectives. Account executives "have to do more things now than in the past," observes TBWA president Tom Carroll, "like coordination of all different media and acting like an old-fashioned Hollywood producer who had to produce a hit movie. We're trying to produce a hit campaign."[3]

Above all, the AE must be a *planner* so that agency output is sufficient for client marketing needs and is available at the agreed-on deadlines. An account that is characterized by frantic last-minute responses to late-developing marketing plans is usually indicative of an account executive who is out of touch with the people or processes involved. Occasional problems may be someone else's fault, but they are still the AE's responsibility. As Jerry & Ketchum president Jack Taylor puts it, "The best account people accept the responsibility (not just the accountability) for managing (not just running) the account, with wisdom (not just knowledge). This includes being creative: having ideas, knowing a good idea, helping to sell it. It includes knowing the business well enough to stay ahead of the client and his requests. It includes an entrepreneurial spirit which looks for ways to make the budget work harder than its size. It means being the one person the client thinks of as being involved with his whole business."[4]

In recognition of the complexity and importance of this job, some agencies upgrade the position of account executive on larger assignments to *account manager*. This upgrade carries with it increased clout so that the manager is more a peer of counterpart creative directors and marketing vice presidents. Junior account executives may also handle pieces of the business and report to this manager. As with the various ways the entertainment producer function can be shared (see chapter 5), such a division of authority does not change the fundamental nature of the job, but it does better distribute the workload.

Recently, a number of stations and cable systems have appropriated the title of *account executive* for their own salespersons. In so doing, they are signaling a shift in emphasis from a purely sales function to a facilitative one. Rather than just selling a client airtime, an outlet account executive assists that advertiser with the production of commercials and the refinement of the marketing strategies that motivate those spots. These professionals thus strive to provide the client with a total package of assistance that will solidify the business relationship between the advertiser and the AE's station/system. The thrust of these account executives' duties is the same as that of their advertising agency namesakes, but the end goal is to capture business for a particular media outlet.

Whether they are employed at an agency or outlet, account executives can benefit from asking themselves the following key questions posed by J. Walter Thompson creative director J. J. Jordan:

> Was the process better for having gone through you? Is the work better for your involvement? Did you help solve a problem and make a great idea actually achievable? Did you quell the misgivings of a vocal naysayer? Did your insight into the client's business contribute in any way to the advertising and ultimately the brand? Did you help a nervous client gain confidence in something that could actually help their business? Did you help enhance your agency's [or outlet's] reputation? Are you personally proud of the work?[5]

As you may well have detected by this point in your reading, that last question can—and should—be asked continually no matter which electronic media career you are pursuing!

Media Services Executives

The placement of client advertising in various outlets is the job of media services executives. Working with the account executive and the appropriate client representatives, the media person determines which channels of communication to use and the specific vehicles to be selected within each channel. Outdoor (billboard), transit (displays on public transportation vehicles), direct mail (correspondence to individual homes/businesses), and newspaper and magazine layouts may all be used in addition to (or instead of) electronic media options. Any client, even the largest, has only so many dollars to spend on advertising placement. Thus, the media service executive's goal is to maximize the consumer impact that these dollars can buy. Saving just 2% on ad placements for a client like Burger King, for example, can mean that millions of dollars annually stay in BK's coffers.

Media-buying services may exist as independent companies or as units within (or self-standing subsidiaries of) full-service advertising agencies. In independent companies, the client must derive and provide his or her own commercials and layouts. The buying service concentrates on selecting media and then secures time and space from the outlets chosen. The buying service traditionally receives a commission back from the individual media vehicles it patronizes. A full-service agency,

Facilitative Functions

on the other hand, handles creative and other duties in exchange for the right to receive commissions from placing the client's advertising dollars. For many creative or consulting needs, the agency often charges the client a fee in addition to the media-derived commissions, but this fee is negotiated as part of a total service package. In fact, the trend in the industry is away from commissions in favor of flat fees for all kinds of work.

Over the past few years, media-buying units have become much more powerful and independent as clients seek greater advertising efficiencies and accountability. "In a sexy business, media departments were always the business, not the sex," recalls trade reporter Craig Reiss. "They were the low-paid numbers crunchers, working long hours in back rooms to put other people's brilliant ideas on the air."[6] Things are far different now. Media-buying executive Michael Drexler chronicles the evolution this way:

> When "media" was under one roof with other ad functions, most "full-service" agencies were never willing to devote substantially more of their financial resources to their media departments. To many, media departments were pure overhead—and a drain on profits. The big bucks went first to the creative department and then to account management. Independent media organizations, meanwhile, developed more resources and better skills; they won more business; and they consequently forced more of the big full-service agencies to "unbundle"—setting up their own media departments as independents to pursue a la carte business.... Today, almost all large agencies or agency holding companies have created media entities. They've taken on a different name, a new positioning and most important, a focus devoted to enhancing media performance and accountability through greater investment in staff and resources.[7]

"Media buyers are no longer the invisible men and women of the ad industry," adds *Adweek*'s Cristina Merrill. "In fact, they're now the center of attention and being courted like stars."[8]

But they are stars who must juggle a great deal of complex data. Media buying—and the essential planning function that goes with it—are sophisticated processes that rely on computer analysis and retrieval of a mass of cost and demographic information. The main principles of media selection would more than fill this book and so cannot be discussed in depth here. Fundamentally, however, any media plan must begin with decisions as to *reach, frequency,* and *cost* goals.

The concept of cost-per-thousand (CPM) was introduced in chapter 4. An advertiser wishes to reach as many people as possible for money spent, but the types of people reached are an important consideration. That part of an outlet's audience not made up of potential prospects for the client's business is *waste circulation*. Thus, a station offering a low cost-per-thousand may not be a wise media buy if there is a lot of waste in its audience (such as over-thirty-five listeners on a station used by a pimple cream manufacturer). Once these unwanted consumers are subtracted, the cost-per-thousand figure for the remaining *qualified prospects* might be unacceptably high. Media executives must decide if an outlet's audience composition merits a buy in terms of this specific client's *target universe* (desired audience).

Then there is the trade-off between *reach* and *frequency*. *Reach* is how many qualified prospects will be exposed to the commercial via a certain media placement. *Frequency* is the number of times each of these prospects is likely to be exposed. Because almost all advertising depends on repetition in order to be remembered, any media buy must be calculated to deliver multiple *impressions* (contacts) with the listeners, viewers, or readers whom the advertiser is targeting.

The media services professional must manipulate these and many other specific variables to obtain the most cost-effective message circulation for any client. And with new media options and more sophisticated plans for segmenting the market continually appearing, the media-buying job

Pat Wallwork
*Media Director and Partner,
McKee Wallwork Henderson*

Media planning and buying is both an art and a science. For more than twenty-five years, I have spent my career planning and negotiating media schedules on behalf of clients. From McDonald's to Pampers, it is important to always create a win-win situation between the media and client.

In the spring of 2003, a new baseball team—the Albuquerque Isotopes—and stadium debuted. I knew the baseball games would be well attended and strategically wanted my client, Presbyterian Healthcare Services, to be a part of the new season. I looked into purchasing a Trivision scoreboard ad at the stadium. The asking price was more than the client's budget. I proposed adapting our existing outdoor billboard messages with Presbyterian's logo to read "Watching The Isotopes Can Make You Feel Better." Not only did we give the new baseball team "freeway" presence, but they also received an endorsement from the state's largest healthcare provider. In return, Presbyterian paid half price for the stadium's Trivision scoreboard ad.

To be a good negotiator, you need to uncover the media's real objections to your buy. For a not-for-profit client, I was trying to obtain a free matching schedule for the New Mexico Department of Health. (Most media outlets offer a matching schedule to non-profit organizations that parallels the time or space they pay for.) The sales manager told me he would get in trouble with "corporate" if they saw an order without any revenue. To complete the deal, I suggested that we buy the entire schedule for half price.

Media buying at an advertising agency has many facets. I feel my success has come from big-picture planning, while understanding the specific implications for my clients. In the media business, the "devil is in the detail." It's important to develop the media strategy based on specific research of each media outlet and its potential impact in the marketplace.

Multitasking is the best way to describe my day-to-day routine. Deadlines and disappearing advertising inventory guide my day. At any given moment, I could be working on ten different clients/projects at various stages of media research, planning and buying, and proof of performance. This involves a lot of telephone conversations, meetings, and computer deskwork. In addition, part of my time is devoted to overseeing and training the media department. Media buyers need to be good communicators and motivators while possessing analytical skills. Many of my employees have college majors or minors in psychology.

I put a lot of emphasis on client relationships. Throughout the years, I have learned that clients value my contribution in a variety of ways including my expertise, organizational skills, time management, and education. They also appreciate a bit of entertainment. After all, this is advertising and most clients think of it as "Hollywood"—even in the buying of airtime.

As you embark on a media career, it is important to know that this business changes frequently. From increased media fragmentation to strategic planning to negotiating media schedules, you are assured of an exciting profession.

Facilitative Functions 153

becomes more complex each day. "When I started in the advertising business and you wanted a packaged-goods item to reach a housewife with three kids it was easy," recalls veteran media executive Larry Spiegel. "You used television and a few magazines. Now it's not that simple. . . . Basically, media should be as creative as creative is creative."[9]

An evolving tool to help make these plans and manage this complexity is called an *optimizer*. Fundamentally, an optimizer is a sophisticated computer program that helps determine the best overall mix and value for a media buy rather than focusing solely on raw CPM, reach, and frequency numbers. "Equal ratings points within and across dayparts do not have equal impacts for a variety of reasons," warns media research director John McSherry, "including the amount of time a viewer spends watching a show and viewer loyalty."[10] Well-designed optimizers can build in show loyalty, commercial environment, and a host of other factors to move well beyond the old practice of simply dividing up a media buy by daypart and CPM. As an example, what is the difference between watching a popular sitcom on network prime time and watching a different (and older) episode of that same series as a local station's syndicated offering earlier that evening? For one thing, the audience may be younger and/or have their attention more distracted around the dinner hour. For another, commercial loads are heavier in syndication than in network prime time, meaning more competition for your spot, a functionally shorter show, and therefore less viewer fixation to the screen. A state-of-the-art optimizer can factor in such distinctions and move beyond mere cost and reach/frequency computations to determine the true placement value of the buy.

Optimizers are one way through which media planning and buying becomes, in the words of media consultant Erwin Ephron, "an ongoing process, not an event" and one that moves far beyond the tyranny of a single evaluative standard. To illustrate, Ephron cites this story: "The Planning Ministry in the old U.S.S.R. had production goals for every factory. One in particular, a brass foundry, had an annual target of 350,000 pounds of finished brass lamps. Each year it beat the target easily by shipping lamps with an average weight of 170 pounds. In Russia, these were called 'elephant lights.' The moral is a single number system is easily fooled. Performance measures that concentrate on cost-per-thousand are no exception."[11]

In trying to avoid "elephant light" oversimplifications while still keeping up with their vast planning, data accumulation, and data analysis tasks, most media service units consist of multiple levels. *Assistant buyers* are entry-level persons whose main task is to execute and track the specific decisions made by their superiors. Above them, *buyers* exercise more latitude in determining which vehicles will be bought within a given medium. They may, for example, be given a radio budget of $150,000 with which to purchase a certain number of *gross rating points* (*GRPs*) emphasizing certain demographic groups. The buyer chooses the stations, either by working with outlet salespeople themselves or with station reps. *Associate buying directors* and *buying directors* have greater responsibility and may be involved in discussions with the client about how to apportion the advertising budget among the various types of media (radio, television, Web, outdoor, etc.) and help set campaign reach/frequency/CPM and increasingly, *value* goals via the use of optimizers and other analytical tools.

At the top of the pyramid, the *head of media services* (or a similarly titled executive) oversees the media-budgeting and media-buying activities for all of the agency's clients. This job may include combining time or space purchase orders for several separate clients or products in order to obtain a better volume discount (a cheaper rate) for all concerned. This executive may be involved in the production of barter program series created specifically to carry client advertising. Proctor and Gamble's mounting of a soap opera in conjunction with its advertising agency is one longstanding example of such activity.

As with studio producers, the smaller the shop, the more compressed the media department's hierarchy and the greater the combining of responsibilities. The essential ingredients of the media service executive's job are the same across a buying unit of any size, however. The job includes (1) helping the client set audience delivery goals; (2) determining the media apportionment best calculated to reach these goals; (3) selecting the specific media vehicles for the buy; (4) negotiating the best possible rate for time or space; (5) tracking the buy to make certain advertisements run as scheduled; and perhaps (6) cooperating in follow-up research to ascertain the effectiveness of the overall media plan. In many agencies, an associated *traffic department* (see the following section) may coordinate the actual distribution of message material to the media outlets and assume the auditing function entailed in task 5, tracking the buy.

Media services is neither the most glamorous part of our profession nor the most visible. But it has always been the part that makes the financial wheel go 'round and increasingly is becoming a force in total campaign development. According to Wieden + Kennedy media director Lawrence Teherani-Ami, people in his line of work now should possess "the ability to visualize a concept, understand the lifestyle and media consumption of the target, and extend ideas in innovative ways the creative team haven't considered. . . . A smart media person residing inside the development arena can unify all the communication tools and consider at the outset the context in which the communication lives."[12]

Nevertheless, a media director's focus cannot be solely on the client advertiser. To build the solid relationships on which so much beneficial deal making is based, the media outlet's interests must be taken into consideration as well. "What I try to do," concludes nationally respected media buyer Jon Mandel, "is to understand the marketer's market problem and the seller's selling problem and find the best deal. You have to motivate the seller to go the extra mile for you; you have to understand the seller's problem."[13]

Traffic Coordinators and Directors

As we have just mentioned, "tracking the buy"—making certain that paid advertising runs as scheduled—is a vital part of campaign auditing and execution. The people who perform this function within media services units are known as traffic coordinators. If they deal in only one media category, they are more narrowly identified by such titles as *broadcast* or *print* coordinators.

Traffic coordinators execute the media game plan that media services executives and account people have derived. In our industry, it is these coordinators' job to see that the right electronic outlets get the correct spots and run them as scheduled. This delivery may be by surface mail/express, high-capacity landlines, or satellite. If there is a breakdown in the preferred delivery system, traffic coordinators must have an alternate avenue at the ready. Spots cannot air if they are not at the station, and delay can mean an entire campaign is thrown hopelessly out of sync. Many commercials are constructed with alternate middles (called *donuts*) or endings (*tags*). So the traffic coordinator must check to be certain not only that the right spot is airing, but that it is the right *version* of the right spot. Last-minute changes by the client or account team can be particularly worrisome in this regard and can force traffic coordinators into unexpected overtime or weekend work.

In some cases, traffic staffers are also in charge of collecting affidavits from broadcasters or other media agencies that spot flights were run as scheduled. This is necessary to validate client payment to the agency for the media buy. Clients pay their media service unit (whether it is self-standing or part of a "full-service" agency), which then pays the outlets for running the advertising. The traffic function must work to ensure that there is adequate "proof of performance" that spots paid for were spots aired—and at the times specified in the contracts with the various outlets. There are thus three parties involved in the transaction: the client, the media service unit, and the media outlet. A breakdown in the documentation among these three cannot only mean substantial revenue loss to one or more of them but payment disputes that can result in client losses for the agency and revenue losses for everyone. In the worst cases, poor validation work by traffic staff results in civil legal action for "failure to perform" or even criminal charges if fraud is alleged.

Gary Blackwell
*Broadcast Traffic Coordinator,
Doner Advertising*

At first glance, the responsibilities of a broadcast traffic coordinator may appear boring and monotonous. However, constant flight revisions and unexpected last-minute pushes on tight deadlines sometimes give this admittedly tedious job an exciting edge.

As a broadcast traffic coordinator, I am the main intermediary between my agency's account teams and television and radio station traffic managers. Account supervisors outline the specifications of upcoming commercial flights (made up of commercials running during a specific set of air dates). It is my job to convert that basic information into formatted traffic instructions that tell stations what spots should be on the air and what each should be airing. After creating these traffic instructions, I turn to the second major part of my job: having the commercials dubbed and shipped to each station before the flight material deadline.

While the concept of the job is fairly simple, the process is complex. Deadlines are often the most difficult hurdle, since clients are known to make last-minute decisions on whether a particular commercial is acceptable to air. These revisions wreak havoc on a broadcast traffic coordinator, as multiple e-mails, urgent fax memos, and phone calls must be made to ensure that stations will miss the fewest number of scheduled commercial spots possible. In the case of running revisions (which occur when a revised commercial is sent out to take the place of one already on the air), the same measures must be taken to ensure that the revised commercial will be put on air as soon as possible to maximize its exposure.

Outside of revisions and deadlines, I deal with a variety of other concerns on a day-to-day basis. I spend a considerable amount of time on the phone, answering questions about commercial flights, clarifying traffic instructions, and ensuring that each flight I send out will run as smoothly as possible. I search online databases of the shipping companies that send out my commercials to check if a commercial has already been trafficked, to get shipping information on a particular station or shipping address, and to see that materials for a commercial flight were indeed sent out and delivery confirmed. In this busy environment, it is not uncommon for shipping companies to misdirect materials or lose traffic instructions. It is my job to correct these mishaps and get the materials delivered as close to deadlines as possible.

There are two skills that every responsible broadcast traffic coordinator eventually learns: multitasking and effective time management. Being a broadcast traffic coordinator can be a trying job: working on multiple accounts after hours, trying to massage deadline requirements, convincing station traffic representatives to give me more time. However, there are other periods where I may be working on just one or two small commercial flights a day. Obviously, the size of the agency affects a broadcast traffic coordinator's workload, although the biggest impact on workload is the advertising "season." Back-to-school and holiday seasons are extremely busy, especially for agencies that work with department stores and other resellers.

I am fortunate enough to work for a fast-growing agency, and there's very little downtime or slow traffic stretches. As the agency has grown, I've come to realize exactly how important my job really is. I am the last person to handle the commercial flight information for my accounts. As the last line of defense for the agency, it is entirely up to me to ensure that nothing goes wrong during the exchange of commercial flight materials between the agency and broadcast stations. This position has been great for developing effective time management and multitasking abilities, and the close interaction with clients develops important interaction skills. Plus, the amount of responsibility which this position carries looks good on a résumé. In short, working as a broadcast traffic coordinator is a great place to prepare for a career as an advertising media representative or an advertising account team member.

The traffic coordinator's job is thus a highly responsible one. Surprisingly, however, most agencies consider such positions to be entry level. A traffic job is the way many people break into the advertising side of our profession. It is in this high-pressure, media services "boiler room" that one has the opportunity to really learn the interrelationships of the business—an important training and proving ground for later advancement to media buying, account work, or the conceptual side. When copywriters are late delivering custom donuts or tags for new spot versions, account persons fail to get client authorization for a new spot of flights, or media services executives mispackage or misdirect a "buy," it is the traffic coordinator who is left to pick up the pieces. Little wonder, then, that traffic experience makes for better copywriters, account people, and media buyers. Cleaning up after other people's mistakes is a good way to learn how to anticipate and avoid those mistakes when you rise to their positions.

At larger media outlets, a traffic *director* position is the traffic coordinator's counterpart. Station traffic directors facilitate the scheduling for the programming day, with special emphasis on when commercials, public service announcements (PSAs), promos (outlet self-promotions), and other continuity are to air. A key part of their job is to interact with agency traffic coordinators in getting the right spots to air at the right time and validating that this has, in fact, been accomplished. Similar functions are performed at cable systems and Web entities under a variety of job titles. As in the case of the traffic coordinator position, the traffic director slot is generally entry level. Effective performance here can lead to higher outlet positions in sales, programming, and management. Therefore, whether you seek a career at the outlet or agency branch of our profession, a traffic job may be the best way to get your foot in the door and pave the way for solid career advancement.

Audience Measurement Executives

As we discussed in chapter 4, the estimation of audience delivery as benchmarked by ratings, shares, and other quantitative numbers is a central part of the pricing structure that drives most elements of our industry. Professionals working in audience measurement are relied upon to compile the data that justify the expenditure of billions of advertising and program-development dollars. Through a variety of sampling techniques, these facilitators and the companies for which they work seek to determine the size and composition of station/system/network/website audiences so that advertisers can determine the most efficient places to "rent" ears and eyeballs.

For print media, audience measurement is a comparatively concrete and straightforward operation. Since 1914, the Audit Bureau of Circulation (ABC) has monitored and documented the distribution of most magazines and daily newspapers in the United States and Canada. Because a copy of a publication is a tangible commodity, the dissemination of these commodities is relatively easy to track and use in estimating probable readership figures. The only significant variable is the "pass along" rate, via which an educated determination must be made as to how many people read each printed copy. An electronic signal, on the other hand, does not bundle itself into countable copies. Instead, its audience is determined by how many people choose to tune to its impermanent transmissions. Broadcast measurement services arose as attempts to estimate how many people within range of the signal were actually exposed to it.

Beginning in the 1930s with the Arch Crossley's Cooperative Analyses of Broadcasting, followed by the Hooperatings, broadcasters and advertisers initiated a variety of sampling techniques to ascertain who is attending to what. Hooperatings (from the C. E. Hooper Company) applied the *telephone coincidental* technique, whereby a statistically selected sample of people were called in a given quarter hour and asked questions about their radio listening at that moment. From their responses, estimates were derived as to the actual size of the listening audience during that quarter hour and the percentage of those listeners who were tuned to a given station or network. Ratings methodologies and technologies subsequently have become much more sophisticated, but the basic task of population sampling to estimate audience size and apportionment remains the core of modern audience measurement activity.

Ed Cohen
Director, Domestic Radio Research, Arbitron

What do media research people actually do for a living? When it comes right down to it, no matter what company we work for or what our role may be, we support the growth of the companies we work for. In plain English, we help our companies make more money (and even if you're working for a non-profit, you will still be involved in trying to increase revenue).

Research people go at it in any number of ways. But probably the best way to divide the field is by primary and secondary research. I'll start with the latter, because there are far more people working on that side, whether they wish to admit it or not.

Secondary research involves the use and manipulation of data generated by others to support your employer's goals (generally, that entails revenue maximization, whether directly or indirectly). For many people in the field, this means sales research. Sales researchers weave together quantitative data from the ratings companies, companies supplying qualitative data, and other sources of information (government agencies, trade associations, etc.) into a coherent and cogent "story" that helps sell your medium and, more directly, the entity you represent. The recitation of data can be both confusing and off-putting to many buyers. A good sales researcher makes the data accessible and uses it in a clear and concise fashion to help in a successful "pitch" to potential advertisers.

Secondary analysts also use data to help others in their operation understand their outlet's competitive position in the marketplace. The role of research director is often that of interpreter, the person that others go to for a better understanding of data and what it may mean. As an example, the traditional "book breakout" after a new ratings report is received is usually led by the research director. The research director highlights the positives and negatives of the station's (or in the case of many radio and television operations, multiple stations') position in the market.

Primary researchers, on the other hand, create the data. Whether you work as a primary researcher for a ratings company (as I do) or for another data supplier, you should be well versed in the survey process. Today, that generally means an advanced degree and experience in working on surveys, whether media related or not.

I'd like to believe that primary research is the more interesting work and just plain more fun. Still, when you work for a ratings company, your work is seen as selecting the winners and losers in the marketplace, complete with the resulting financial implications. And a ratings company supplies the "currency," the data that are used to buy and sell advertising time. The value of currency means that we generally charge more money for the data—and that means clients who aren't always happy to see you or talk to you. (But it's rarely personal!)

The ratings business faces numerous challenges: respondent cooperation issues, new methods of surveying people, how to properly measure different subgroups, technology, you name it. That's why it can be a lot of fun. For my job, it takes a number of skills together with the ability to work on multiple issues and problems at once and to sound intelligent about all of them. It involves explaining what we do and why we do it in front of clients who do not have the same level of research sophistication. But they sign the checks and need to understand why we do what we do.

To put it simply, if you like intellectual stimulation in a business environment, consider the possibilities in media research.

Today, two companies provide the bulk of ratings data about U.S. broadcasters and cablecasters. Television ratings are compiled by Nielsen Media Research, while radio measurement is dominated by the Arbitron Company. In the 1940s, Nielsen pioneered use of the audimeter, an electronic attachment to a receiver that, via continually improving methods, kept track of every signal to which that set was tuned. Like their more advanced descendents, these early audimeters chronicled when a set was turned on and to what station. But they were unable to determine how many members of the household were listening/viewing at the time. In fact, they could not even detect if anyone at all was in the room!

In 1987, therefore, Nielsen converted its national television sample from audimeters to *people meters*. These devices pinpoint audience composition by requiring members of the household to punch in their personal code numbers when they start and cease viewing. Unfortunately, the greater involvement on the part of consumers may also inject greater error into the system. Children's viewing patterns in particular seem subject to distortion because young people either neglect to punch in or use incorrect coding. (A similar problem impacts measurement of elderly viewers.) In measuring *local* television markets, Nielsen relied exclusively on audimeters and paper diaries mailed in by viewers until 2002, when it introduced *local people meters* (LPMs) in Boston. As the rollout of LPMs continues, the aim is to be able to measure local audiences more quickly and accurately. LPMs are also seen as a major boon to cable systems because they make possible the measurement of local cable advertising. On the downside for local cable, however, LPM data can track DBS viewing as well. As some local viewing of cable networks occurs via satellite, not cable, the local cable system cannot claim that the spots it inserts into a CNN or History Channel feed are automatically viewed by people watching those nets.

Arbitron radio research continues to be primarily diary based, via which selected listeners compile seven-day entries to generate quarter-hour ratings (see figure 6-1). The books are then mailed back to Arbitron. As with the placement of meters or phone calls, diaries are distributed so that the active sample at a given time mirrors the composition of the total population being measured. "Although the diary will continue to be a vital mainstay in audience research, emerging technology

Facilitative Functions

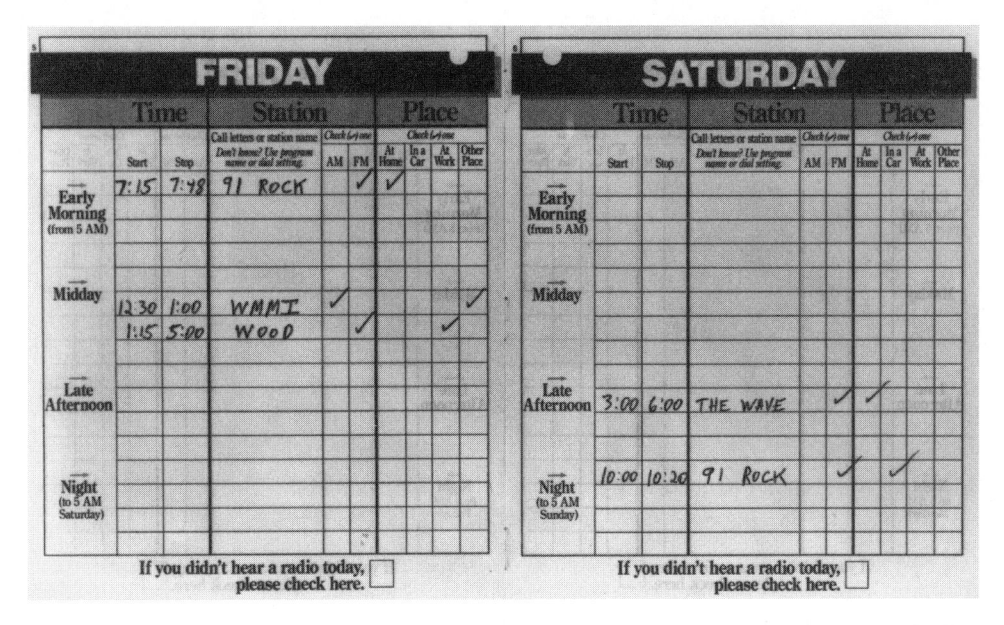

Figure 6-1. Completed pages from an Arbitron diary used by a respondent to report radio listening. *Courtesy of the Arbitron Company.*

is presenting us with the opportunity to add an entirely new approach to the ways in which we measure and report radio audiences in this country," writes Arbitron president of U.S. media services Owen Charlebois. "Embracing this technology and adding it to our research repertoire will provide enormous benefits to the radio industry. It will also create new research challenges and opportunities, for both practitioners and academics."[14]

One new technology that Arbitron has been testing is the *personal people meter* (PPM). Unlike the conventional people meters that sit on the set and ask each audience member to identify themselves, the PPM is a personally worn device that automatically records what a user is listening to or viewing. The devices weigh only $2^1/_2$ ounces and can be carried or clipped on. Trade reporter Linda Moss describes their use:

> As participants go to work or school, shop or stop at a restaurant or bar and then return home, the PPM quietly goes about its job, automatically picking up inaudible codes in radio and TV programming and storing the data. At bedtime, the participant takes off the PPM and places it in a docking station, where the data for the day is uploaded to Arbitron. Even when it is "docked," the PPM can still pick up media exposures. . . . The PPMs have built-in motion detectors to ensure that people don't just leave it on a table in the family room, instead of wearing it. A PPM must register as being in motion at least eight hours a day for its results to be considered valid, or "in-tab," for that day.[15]

Like any audience measurement system, however, the sophistication of data gathering must always be balanced with its expense—and the PPM is a more costly system than any currently available alternative.

Audience measurement services make their money from substantial subscriber fees paid by advertising agencies, networks, individual stations, and similar interests. Because billions of dollars (and thousands of careers) hinge on the results and accuracy of "ratings" data, these interests are constantly looking for ways to make these data as reliable and current as possible. Broadcast researchers Herbert Howard and Michael Kievman long ago pointed out that whatever the measurement company or system, "modern audience research essentially relies on the concept of statistical inference, which permits the estimation of characteristics of a population (such as all TV households in an area) from data obtained by sampling a cross-section of the population being measured. When a sample is carefully selected, following valid statistical procedures, the information gained may be projected confidently to the total population."[16]

Mass communication is most often a one-way enterprise. But trustworthy audience measurement can provide an after-the-fact feedback that enables electronic media professionals to capitalize on discovered success and to remedy detected failures. These professionals must be adept at research techniques while still factoring in the twin concerns of (1) research cost and (2) the impact of methodological changes on our industry's financial health and corporate interrelationships.

Although Nielsen and Arbitron are the most prominent measurement firms, there are scores of other companies that provide more segmented and specific market analyses for media clients. In addition, advertising agencies, local stations and/or their groups, networks, and major clients themselves employ measurement specialists to evaluate Nielsen, Arbitron, and other externally provided research as well as to generate proprietary (private) research themselves. Audience researchers may be derided as mere "number crunchers," but how deftly they do their crunching can mean the gain or loss of jobs for all the rest of us.

Audience measurement professionals must also anticipate new technologies and the new research techniques these technologies might make necessary. A case in point is the conversion to digital television. "For the first 50 years," trade reporter Steve McClellan recalls, "television has been an analog-based medium where Nielsen has measured channels on different frequencies. It worked because, at any one time, only one program was transmitted per channel. But that all changes in the digital world where signal compression allows for the simultaneous transmission of multiple programs in a single channel. As a result, digital meters must measure individual programs; channels are no longer relevant for ratings purposes."[17] Encoders must therefore attach separate and unique audio and digital signatures to individual programs at the point of transmission. As no two code sets are identical, individual shows can be identified regardless of what station, system, or network distributed them. This will make for a great deal more research material to "crunch"—and has the potential to fundamentally alter the way advertising time is purchased.

Audience measurement challenges are even greater for the Web because there remains disagreement on what exactly should be measured. "While Media Metrix and Nielsen NetRatings [the two major Web measurement firms] have built their businesses by hyping things like overall traffic, time spent on a given site and clickthrough rates, these things may say very little about the user's lifestyle or their psychographics," observes trade reporter Hassan Fattah. "In short, they say nothing about the people advertisers want to speak to."[18] Measurement professionals both inside and outside the Web sector thus have the exciting task of moving beyond "counting" to actually defining their business. As with the older electronic media, this task ultimately will be achieved. "When TV first came out," states Media Metrix president Doug MacFarland, "they measured it according to the number of homes. It took 30 years to get people to figure out it's about the people watching at any given time. Data always begs for more data, and then it begs to be understood."[19]

If you enjoy the pursuit of such numerical gathering and analysis activities, audience measurement work may be as appealing as it is powerful in determining electronic media winners and losers.

Brokers

One benchmark for computing media winners and losers is to see who is buying media properties, who is selling them, and at what profit or loss. The professionals in our industry who facilitate these core transactions are known as media brokers—and the very nature of their role makes it highly competitive. "While it is probably true that winning isn't everything, I've never found anything redemptive in the alternative," reveals media broker Gary Stevens. "In some ways the beauty of the broadcasting business is the incredible exhilaration in the win. Even the depths of losing is something worthwhile, however, for it only makes the next win that much more poignant."[20]

Whether they specialize in broadcasting, cable, or more fringe delivery systems, media brokers are transactional experts who locate buyers and sellers for electronic distribution properties. In one sense, they are like commercial or residential real estate specialists. But their responsibilities go beyond the listing and selling of physical property. A station's physical plant is worthless without an FCC license, and a cable system cannot operate without a local franchise. Thus, a broker must also calculate the dollar value of these intangibles in helping a facility seller set an asking price.

One index used is past and anticipated *cash flow*, defined by business journalist Geoffrey Foisie as "revenue left after necessary operating expenses, but before taxes, overhead, interest and excluding non-cash expenses."[21] Other benchmarks include such things as projected market advertising growth/decline, local media competition, the calculated predictability of license or franchise renewal and, for cable, homes passed as compared with homes wired as well as DBS penetration and whether or not the local telephone company may launch a competing *overbuild*. The broker combines both tangible and intangible factors like these in arriving at an appraisal. Then the tasks of listing (publicly or confidentially) and selling can begin.

W. Lawrence Patrick
President, Patrick Communications, LLC

Station brokers represent buyers and sellers of broadcast properties. Much like realtors, brokers are asked to help owners sell their properties, acquire additional stations, determine the value of stations, and handle the negotiations between parties. Most brokers are experienced broadcasters who now focus on the transaction side of the broadcasting business. Some brokers are station owners themselves.

Brokers range from one-person companies and small firms with three to five brokers all the way to full-service, Wall Street investment houses. There are currently about 150 full-time media brokers in the United States. No formal training program exists for brokers, but most brokers have some financial experience in a station or group, an M.B.A. degree, or considerable years of experience operating and buying or selling stations for their companies. Brokers play a crucial role in determining the value of a radio or television station and in making sure that the negotiations go smoothly for a station or group.

In a typical transaction, a broker will sign an engagement agreement to represent the owner of a station who wants to sell a property. This agreement will outline the duties of the broker, the compensation to be paid to the broker if a deal is consummated, and any conditions on the sale. The agreement will also have a set term, generally six months to one year, and is often an exclusive contract. This means that only that brokerage company has the right to represent the seller.

In terms of compensation, brokers generally are paid from 1 to 5% of the price of the stations sold. The larger the price, the smaller the commission percentage. Most brokers have a standard 5% commission on the first 3 million of value and 2 or even 1% commission against all monies paid over this 3 million level. On a 5 and 2 commission structure and a deal worth $10 million, a broker would earn a commission of $290,000 on the sale of the stations.

There is one area where media brokers vary from traditional real estate brokers. There is no "multilist service" where realtors share listings in media brokerage. Station sales are kept very confidential, and many sellers do not want many people to know that they are attempting to sell their stations. They are concerned with staff defections, inside information concerning their stations becoming available to their competitors, and the general unease that often surrounds station sales. Brokers also want to be the exclusive listing agent so that they can control the sale process.

Once a station is listed for sale and an engagement agreement is signed, the broker will begin discreetly to contact companies that might be likely buyers. Brokers also will prepare a detailed presentation book on the station to be distributed to these prospective buyers. These books include information on the market itself, including population trends, retail sales figures, the major employers in the market, and economic or demographic trends. The book will also include detailed information on the financial performance of the station, a list of equipment included in the sale, real estate details, and contracts and leases to be assumed by a buyer. There will also be a section on the station's ratings, the other media competitors in the market, plus anything special about the station.

Prospective buyers will normally be required to sign a confidentiality agreement before being given the book. By signing this document, they agree not to discuss the confidential information in the book with others and to keep this information private. After a limited time to review the book, ranging from only a few days to a few weeks, buyers will ask questions and talk with the broker about the station. The buyers will be asked to submit written offers.

At this point, the broker negotiates with each buyer, attempting to both maximize the price for the station and to determine who is the best buyer for the property. One buyer may be offering all cash at a lower price, while another may be offering a higher price but needs the seller to allow him to pay part of the price over time in a note. Other issues that come up may include whether one buyer will keep the station's staff and the format. And in some cases, when there are two equal offers from different buyers, it simply comes down to who the seller likes better.

Once a verbal deal is reached, a buyer will write a letter of intent outlining his or her offer for the station. All of the basic terms of the offer are contained in this letter. The broker takes this letter to the seller, suggests changes, and negotiates a final version of this document. Once the document is signed, the buyer is allowed to inspect the station in more detail, while the attorneys for the parties step in and begin to prepare a full contract for the purchase and sale of the station.

Once the inspection period is over and the contract is signed, the deal is filed with the Federal Communications Commission for review and approval. The seller has to run on-air and newspaper announcements alerting the public to the sale. The public has the right to express any opinions about the seller and buyer to the FCC. After a period for public comment and the FCC's internal review process, the deal is normally approved, and the parties schedule a closing. At closing, the parties exchange many papers and the payment for the station. At that point, the buyer assumes control and owns the station.

The broker enjoys closings because that is when he or she is paid for their services. As soon as the buyer pays the seller, the seller, in turn, pays the commission due the broker. The broker than normally places some advertisements in trade publications announcing the closing of the sale. This helps generate publicity for the brokerage company and hopefully generates more business contacts.

Most brokers rely on word of mouth, trade press advertising, direct mail, and especially personal contact to generate sales leads. Good brokers may sell several dozen stations a year and earn significant incomes. Smaller brokers may sell only a few stations per year but still earn a very good living.

In addition to station sales, some brokers also conduct appraisals or valuations of stations. Brokers will be asked by companies to tell them, in written detail, how much their stations are worth and why. These appraisals range from off-the-cuff opinions to very sophisticated economic analysis of the value of the stations. For instance, family station owners often use appraisals when they are giving stock to younger family members as they have to determine the value of that gift for tax purposes. When broadcast station owners die, brokers often have to appraise the stations for estate tax purposes. Partnership buyouts, divorce, and even some court cases often require brokers to render opinions as to the value of broadcast stations.

One final element of some brokerage firms' expertise is investment banking. This is an area where some brokers help raise either bank financing (debt) or investment capital (equity) to help a broadcaster acquire stations. Not all brokers are capable of performing this task, but it is a critically important one for many station owners. Brokers can not only help a broadcaster to locate and purchase new stations to build his or her group but can also help arrange the financing to support this growth.

Brokers are an often unnoticed but important part of the broadcast industry. With generally 1,000 to 1,500 station sales a year, brokers help owners maximize value through sales and help buyers grow their broadcast companies into larger groups. A broker with the right combination of industry contacts, financial understanding, and good salesmanship can be a very successful entrepreneur involved in the trading of hundreds of millions of dollars worth of stations over his or her career.

In the 1980s, the expansion of the cable business and the associated acquisitional appetites of MSOs together with the Reagan administration's deregulatory initiatives spurred both system and station brokerage activities to unprecedented levels of intensity. This intensity declined markedly in 1990, however, when a recession found many outlets struggling to meet the *debt service* they had taken on in more optimistic times. Debt service is not only interest payments but also includes amortization (gradual extinguishment) of the principal borrowed. And because most loans to media companies are not fixed, level payments but increase each year, the debt service increases each year as well. To meet it, owners need constantly higher cash flows or they must refinance or resell the outlet.

Fortunately, with the improving economy and further loosening of ownership limits in the mid-1990s, the market again heated up. It then dropped drastically after the turn of the century as many groups had reached their FCC-allowed limits, and a souring economy constricted both cash and expansive decision making. All of this is reflected in figure 6-2. This chart documents in real-dollar detail the upswing in the business of station trading in the mid-1980s, followed by the negative impact of the 1990 recession and the 1993–94 recovery. Especially in radio and cable, the pressure for mid-1990s consolidation increased as investors realized that only very large companies would be in a position to profit from economies of scale and amass the clout to compete. The 1996 Telecommunication Act's liberalization of ownership caps power-boosted this trend and made brokers' lives especially hectic (and profitable).

All of these consolidations have drawn new investors into the industry. Because of this, brokers have had to learn the more sophisticated valuation procedures that outside financiers with no background in electronic media understand and are used to dealing with. In earlier times, when broadcasting and cable were still in the hands of their founding families, a broker could rely on insider relationships and common understandings to bring buyer and seller together. But as veteran media broker Dick Blackburn pointed out in 1989, with the mid-1980s boom, brokerage turned into a "science. . . . I think that anyone in the last five years who has relied on the 'old boy network' type

Fifteen Benchmark Years in Station Sales
(Total purchase price for U.S. radio and TV outlets sold)

Year	Amount
1954	$60,334,130
1959	$123,496,581
1964	$205,756736
1969	$231,697,570
1974	$307,781,474
1979	$1,116,648,000
1985	$5,668,261,073
1987	$7,509,154,473
1990	$1,976,626,100
1992	$1,045,373,000
1994	$4,970,400,000
1996	$25,362,820,000
2000	$33,700,000,000
2001	$8,700,000,000
2002	$8,123,180,000

Figure 6-2. Station trading volume from 1954 to 2002. *Source:* Broadcasting & Cable Yearbook, *2003–2004.*

thing has found themselves out in left field. . . . If you're up to meeting the demands you'll do fine, you have a place. If you're going to cruise, people will blow by you."[22]

Like conventional real estate agents, media brokers are usually paid through commissions based on property selling prices. Because of the vast sums often associated with a media outlet, however, the percentage of this commission usually decreases as the sale price gets larger. Thus, a common model (known as the "Lehman Formula" in honor of the Lehman Brothers investment firm that first devised it) is the "5-4-3-2-1" commission, through which the broker receives 5% of the first million dollars, 4% on the second, and so on. Another standard template is for the broker to get 5% of the first 3 million dollars and 2% over that. For deals approaching or exceeding $100 million, the broker's fee may decline to 1% plus a variety of incentives.

An additional variable, and one that makes a broker's life more complicated, is the perceived value of the on-site management at the facility. How much have these executives contributed to the success or failure of the property? If they have been successful, are they likely to remain in place after the sale if the buyer so wishes? More than one outlet's profits have plummeted when a locally well-connected general manager or sales manager has left in the wake of an ownership change. As we have observed throughout this book, talented *people* remain our profession's most important resource despite the preoccupation with hardware and technological developments.

To make many deals work, a media broker must also be able to locate sources of capital and financing for the prospective buyer. A good residential real estate agent has the same responsibility, of course, but few home sales are multimillion-dollar deals. Knowing which banks, other investment houses, and individual companies are interested in media propositions and the financial instruments they are willing to use can be the crucial ingredients in a broker's ability to consummate the sale. It is the broker's responsibility to help create the "capital structure" of the company buying the outlet. This might involve investment bank financing, seller financing, consulting agreements with the former owners, and a host of other considerations. Every effort also must be made to minimize the seller's tax obligations.

As a sidebar to all of these financial considerations, "understanding the parties' emotional involvement is critical, and the seller and buyers should consistently be refocused on the contemplated transaction," advises longtime broker James Gammon. "Nevertheless, in the final analysis, station acquisitions are rarely emotional buys. The parties must become aware of the financial feasibility factors, which are paramount."[23]

Perhaps the most important attribute that media brokers must possess in carrying out all of these responsibilities is an unwavering respect for confidentiality. Most sellers do not want their staff members to know about a sale until it is announced for fear of losing key personnel in the interim, thereby disrupting the outlet's stability.

Satellite Services Executives

While media brokers facilitate deals on Earth, it might be said that satellite services executives facilitate deals in space. Though their work is largely transparent to the general public, it is these professionals who provide the avenues through which the stations and networks serving that public receive much of their programming.

Fundamentally, executives working in this branch of our industry lease communication satellite carriage time to entities seeking to transmit data via these orbiting "birds." The business was born in 1962 with the launch of AT&T's *Telstar I*. An active repeater bird (rather than just an inert sphere off of which to bounce signals), *Telstar* possessed the capability to receive microwave signals from Earth, amplify them 10 billion times, and then retransmit them to the ground. On July 10, as the satellite came in line of sight of the Andover, Maine, tracking station, the three U.S. television networks presented a twelve-minute news program that interrelayed TV pictures between the United States, France, and England.[24] Even though broadcast television was not *Telstar*'s primary purpose, this breakthrough event paved the way for today's commonplace use of satellite relaying of both international and domestic program content. Five years later, *Early Bird* was put into space as the first satellite intended primarily for television transmission, and our electronic media entered the space age for good.

Initially, satellite communication was a governmental/public corporation affair. In 1964, eleven nations joined together to establish Intelsat, the International Telecommunications Satellite Consortium. Intelsat's objective was to create a global commercial satellite system that would assist all nations, on a nondiscriminatory basis, to expand their telecommunications services. The United States joined Intelsat through the Communications Satellite Corporation (Comsat), which was established by Congress in 1962 to own and operate this country's initial satellite communications system. Ultimately, once the infrastructure was fully established, Comsat's monopoly in the United States was allowed to sunset, and a number of private satellite companies came to the fore as self-standing, profit-seeking businesses.

Tim Jackson
Senior Director, North American Video Services, PanAmSat

When one hears of "sales" in the electronic media industry, the first thought that comes to mind is advertising sales, which is typically the subject of many sales courses in college communications departments. There are, however, many other goods and services that need to be sold in our industry, including distribution deals for film, syndicated television programming, and cable television networks; broadcast and film equipment; and many more ancillary services that include satellite contribution and distribution capacity.

The satellite communications industry is a relatively new one. Deregulated commercial satellite communications serving the radio and television industries only became available in the mid-1980s. Since that time, several domestic, regional, and international satellite communications providers have evolved, including PanAmSat, Loral Skynet, SES Global, Intelsat, Eutelsat, and NewSkies Satellites, among others. Virtually every radio and television network and cable television programmer worldwide uses one of more of these (and/or other) providers to deliver their programming to their affiliates. Large television programmers, like CNN, use over twenty different satellites to insure that their programming is made available to the greatest number of viewers around the world.

The goods we are selling in the satellite communications field include full television transponders, which a programmer or network will typically use to deliver ten to fifteen digitally compressed standard definition television channels (or two to four high-definition television channels); individual digital channels, which smaller programmers use to deliver a single compressed (approximately 4 mbit/s) channel to their affiliates; and occasional use (or ad hoc) services, which are typically used for news or sports contribution and are made available on a per-hour basis. The capacity is leased to our customer for periods ranging from under an hour (for occasional use capacity) to fifteen years (for a full transponder on a new satellite).

The decision makers who procure satellite capacity are senior executives with the world's largest media companies. All of the effort in creating, producing, and originating product is for naught if the distribution pathway is inadequate or inferior. Most of these customers have a very strong technical background with over fifteen years' experience in the industry. Their companies entrust these individuals with responsibility for the integrity of their entire networks' distribution, and they expect us, as the vendor, to have the same depth of understanding. For that reason, most of the people working in my position have over ten years of experience in a wide variety of electronic media positions, and nearly all of them have very strong technical backgrounds. Understanding signal flow, digital compression and multiplexing, advanced technologies (such as video-over-Internet, and HDTV), and the needs of broadcast and cable affiliates are some of the requirements for successfully engaging these decision makers. The industry is very small, and most of the people that participate in the business of distribution have known (or know of) each other for many years. In many cases, one's reputation is a major benefit toward moving the business forward.

As in most sales jobs, the potential for growth is great, and the challenges and results are pretty clear with each quarter's sales report. Unlike retail sales, we do have a fairly finite client list, and keeping in touch with them regularly is a must. With customers in all four continental United States time zones, Asia, and Europe, my working hours are varied and unpredictable. When we are finalizing major contracts, I'm expected to be available on nights, weekends, and holidays. Last year, we finished three major agreements on Christmas Eve and New Year's Eve! Regular travel is also expected, both to stay in touch with existing customers and to visit the new ones.

Entry-level jobs in the satellite communications industry are typically available to people with a sales or business interest and to those with a technical interest. The sales track will typically begin in the booking center or network operations center and move into a junior sales position. Senior sales positions result from years of varied and reputable work in the industry and a strong Rolodex.

The satellite communications industry offers people a dynamic and important role within the program distribution chain. The ability to work closely with major radio, broadcast, and cable programmers; help new networks get started; and stay on top of new technology makes my job a challenge, but most of all fun.

A communications satellite is launched into a *geostationary* orbit 22,300 miles above the Earth and may have a usable horizon (a *footprint*) that embraces up to 30% of the planet below. The term *geostationary,* or *geosynchronous*, means that the "bird" remains over the same point of the Earth's surface because both are orbiting at the same speed. This 23,000-mile-high arc in which geostationary satellites are parked is known as the Clarke Belt, in honor of science fiction author Arthur C. Clarke, who wrote *2001: A Space Odyssey.* Each satellite carries a number of transponders, and each of these transponders now can accommodate several different programs depending on how much digital compression is used to fit multiple content streams within a given bandwidth.

As more and more of these satellites were deployed, not only did they take over some of the relaying of broadcast network signals previously done by landlines, they also created whole new enterprises. In fact, the establishment of cable as a competitive industry came in 1975 when Home Box Office (HBO), owned by Time, Inc., married satellite technology and cable's selective distribution capacity to create a national premium TV service delivering uncut movies and exclusive sporting events directly into subscribers' homes. Before HBO, each cable system wishing to originate material had either to obtain and show the movies themselves or to enter into expensive landline or microwave interconnections with other systems in the region. With HBO and a satellite dish, however, pay programming literally dropped out of the sky at the cable system's head end for easy and immediate sale to local households.

HBO paved the way for scores of other pay and basic (advertising-supported) cable networks. These networks greatly expanded viewer choice, both through original material and recycled programming previously aired by broadcasters, and forced these broadcasters to cope with a wide new range of competition. Satellites, like the one pictured in figure 6-3, converted cable from a mere repeater service for broadcasters to a powerful originator in its own right. Ironically, while on the one hand the satellite built the cable business, on the other it afflicted it with a new rival in the form of DBS (direct broadcast satellite). On December 17, 1993, Hughes Communications launched the first of its high-power satellites to carry its own Direct-TV DBS service along with that of Hubbard Broadcasting's USSB, with which it subsequently merged. DBS brought multiple channels of programming directly into subscribers' homes through dishes rather than a cable system's wires. First concentrating on rural areas uneconomical for cable to wire, the DBS industry subsequently has expanded to fiercely compete with cable in more concentrated population areas as well.

Figure 6-3. A communications satellite with solar panels fully deployed. *Courtesy of GE American Communications.*

The satellite has come to both assist and compete with the radio industry as well. On the one hand, satellites provide a way through which individual stations can be inexpensively programmed from central locations around the country. A single satellite feed may bring a *voice-tracking* disc jockey (see chapter 1) to stations in dozens of markets, and a single satellite programming company may offer dozens of feeds in a variety of formats. On the other hand, DARS (Digital Audio Radio Service) companies XM and Sirius began offering their programming direct from "bird" to listener in 2001 and 2002 respectively, bypassing the local station entirely. In-car (and in-truck) listening is a core part of their business plans, and each company has arrangements with several automotive manufacturers for compact factory-installed satellite-receive antennas. A DARS service can provide the listener with scores of music choices, most if not all of these commercial free, and offer a wide range of format choices no matter how remote the area through which a vehicle is traveling. Of course, it can offer these same services to the home or business as well, all for a monthly subscriber fee like that paid for cable or DBS television service or Internet access.

The satellite services executive is the person facilitating all of this activity. In some cases, where the bird is wholly owned by the company whose programming it carries, this executive functions in-house, working to ensure that transponder time and backup are sufficient to meet the needs of the content being delivered to customers. In most cases, however, electronic media organizations do not own their own satellites but lease transponder time from provider companies. This time can either be purchased on a *dedicated* basis (such as by a broadcast network that knows its needs a certain number of transponder hours each and every day) or an *occasional-use* basis for the relaying of special sports, concerts, news or similar events. For example, in the case of breaking news stories too far away from a station to use an ENG microwave relay, an SNG (satellite news gathering) truck like the one in figure 6-4 can uplink a signal to a satellite on which transponder time has been secured for downlinking to the station or network news department.

Working with regional, national, and many international clients, the experienced satellite services professional is expected to provide a clear and reliable pathway for electronic signals to get from their point of origin to their often multiple points of reception. Particularly in news, sudden demand can surface that exceeds the capacity a client has customarily booked, and alternative bird space must be located. Technical glitches and signal failures require backup plans and swift deci-

Facilitative Functions

Figure 6-4. An SNG truck on location. *Courtesy of Wendy Watson, San Francisco Satellite Center.*

sion making so that the transmissions on which so many media companies and audiences depend get through with minimal interruption or, ideally, no interruption at all. Though the satellite services business is a very small one in the numbers of professionals that it employs, its importance to the functioning of all electronic media—mass and nonmass—cannot be overstated. The facilitators who work in it must bring a seamless blend of selling, technical, and planning skills to bear every moment they are on the job.

Webmasters

Like satellite services executives, professional webmasters must be marketers as well as technical facilitators. They must build and maintain a programmatic vehicle that compares well with a myriad of other content-rich corporate sites while advancing the marketing plan of the company for which they work. Some webmasters in our industry facilitate self-standing sites that constitute a discrete brand and media product. Others are in charge of sites that are an extension of another media unit—such as those affiliated with stations, networks, groups, MSOs, and even individual program series. In either case, they try to capitalize on an intrinsic characteristic of the Internet. As advertising agency CEO Peter Arnell describes this quality: "The Web allows for a more intimate and more seductive and longer-lasting relationship with our consumers. So we use the Web to manage relationships, as well as portray the brand and product."[25]

This managing of relationships can be a particularly powerful dynamic with younger consumers—the very demographic that is growing up with the Web and is the key to its expansion and acceptance as a central fixture on the media landscape. More than any other medium, the Internet encourages its users to fashion their own program schedules and contribute to the content of those schedules in a very interactive manner. In this way, the Web is following up on a capability pioneered by the jukebox industry in the 1930s. With the advent of the jukebox, young consumers (like those pictured in figure 6-5) were freed from the tyranny of what the radio station wanted to feature. The jukebox allowed them to select and arrange the music to fit their own preferences. They were also given an economic alternative to having to purchase their favorite recordings—they could sim-

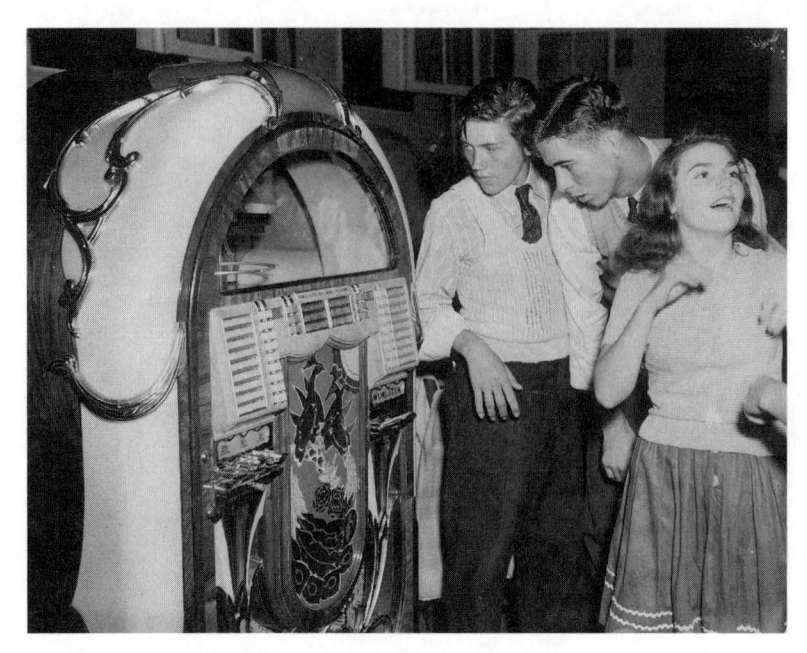

Figure 6-5. West Virginia teens groove to swing music from a Wurlitzer jukebox in 1942. *Farm Security Administration photo by John Collin, courtesy of the Library of Congress.*

ply "rent" them for a nickel a play. Older consumers soon followed the teenagers' lead. Today, instead of physically gathering together around glowing neon crates, young people (and the older consumers who emulate them) are gathering separately at glowing computer screens where program progressions and participations can be determined with the click of a mouse. They can "rent" this content for the monthly price of an Internet connection, perhaps supplemented by a download fee. In a sense, the Web is today's much more flexible jukebox industry and each website its own jukebox—with its own content inventory from which to choose.

The webmaster attempts to make this inventory something that is appealing to target consumers while making the execution of their choices user friendly. No matter how many content developers (see chapter 2) may be contributing to a site, the webmaster must make certain that the look, "feel," and personality of that site remains consistent and strengthens those earlier-referenced relationships. There should be nothing accidental about website design and maintenance, nothing inadvertent in the way content is arranged, even though there may be a great deal of it. "We've all heard that a million monkeys banging on a million typewriters will eventually produce a masterpiece," observes commentator Eyler Coates. "Now, thanks to the Internet, we know this is not true."[26] No webmaster wants to have a site that seems to confirm Coates's quip.

Most professional sites contain material *repurposed* from other sources. This is particularly true when the site is an extension of another electronic media outlet. It is the webmaster's job to see that both the format and content of this material contributes to the image and purpose of the site. It must not just be the blurred replication of "stuff" from the originating channel. "Instead of just making the site a dumping ground for leftover media or archived bits that didn't make it into the televised program, we try to emphasize storytelling in our feature choices," says KQED San Francisco's webmaster Angela Morgenstern, "publish original narrative and perspectives, and provide tools to help visitors learning about a particular story understand the world around them."[27]

Facilitative Functions

To accomplish such enrichment, webmasters must administer templates that make it easy for staff members at sister outlets to contribute content without requiring a lot of technical intervention on the webmaster's part. Nevertheless, template design must be sophisticated enough that consumers are not bored with what ends up on their monitors and speakers. Today's media-wise consumers expect the same professional sheen from the sites they visit that they are used to receiving from the television channels to which they tune. Once a medium moves beyond its trial-and-error infancy (as radio did in the 1920s and TV did in the 1940s), audiences become highly intolerant of amateur presentation.

Brian Demay
*Webmaster, WBQB/WFVA,
Fredericksburg, Virginia*

My career has included webmaster duties for almost a decade now, and it's been fascinating to watch as the Internet (and how people use it) has transitioned so dramatically from a novelty to a necessity. My first website consisted of maybe a dozen pages and was not much more than text, photos, and a radio station log: cutting edge at the time, but paltry by today's standards.

Today, it would be impossible to handle any modern website without a firm grasp of several different operating systems, Adobe Photoshop, Dreamweaver (or comparable layout program), and a working knowledge of ASP, DHTML, and FLASH—not to mention CGI and JAVA scripting.

I am operations manager and webmaster for the successful Mid-Atlantic network chain of stations in Fredericksburg, Virginia. When we relaunched our websites in 2001, the goal was a more interactive Web experience; we needed something that would allow each department to handle its own content, freeing up my time for more "big-picture" duties.

The answer was a Web template based on ASP, or Active Server Pages. Now, every page on the website has the same basic "shell" (masthead, navigation bar, ad column), but the mail content area is changeable. By setting up a hierarchical system, allowing staff members varying degrees of access to specific Web pages, I am able to keep the content of the site fresh with daily updates, while not sacrificing the basic Web design. There is a large database of images and text that are generated dynamically by the Web server for each Web page, eliminating browser compatibility issues between Netscape, Internet Explorer, or the different versions of each.

With the template system, as webmaster I no longer have to single-handedly edit each page individually whenever a change in content needs to be made. Each department can upload text, graphics, and HTML directly to the site:

- My morning team recaps the day's show, teases the next morning's events, and posts pictures and links daily.
- My music director updates our "Artist of the Week" and concert listings.
- The news department publishes its daily newscasts.
- Our daily requests are tallied and posted by my evening personality.
- The PSA director updates the community calendar.
- Secretaries here type in school and business closings!

The site design is my own, graphics are generated in-house, but each department controls the content for its pages. I am more of a watchdog now, making sure the morning material in particular is family friendly, and copyediting where necessary. Overall, it is a nicely integrated system.

For me, technological advances have made the daily maintenance of complex sites like mine much less labor intensive. For someone looking at a career in this aspect of communication, I would suggest learning the basics of art and design first. The design of a site is what catches the eye. But content keeps them there. If your Web design is clean and cohesive, that's half the battle. A Web-design course can give the student the software tools needed to further his/her webmaster education, followed by a Microsoft Networking certification to fill in the technical gaps. The rest will be trial and error and on-the-job training.

The key to getting started is to *get started*. Don't wait for a class to develop a personal website of your own! Many online service providers give you the tools to publish a rudimentary site at no cost. Let that be your stepping-stone, and expand your skills and education from there. As the world increasingly relies on the Internet for information, entertainment, and e-commerce, choosing a career as a webmaster would be a wise choice indeed.

This does not mean that visual pretense should be allowed to overshadow ease of audience access. Web design guru Jakob Nielsen argues that fancy, extraneous website effects "exist for plain ego. That's the website's way of saying, 'We are so important, we can impose on you this extra step.' That is contrary to the nature of the Web, which is user oriented: 'I offer something, you click on a link and go get it.'"[28] Like other effective facilitators, good webmasters ultimately work to promote valued transactions among communicators, clients, and consumers—not make such transactions more difficult. So their websites project a sense of responsive service instead of remote self-centeredness.

Business Development Specialists

Websites often are launched not only to enhance a company's relationships with its current clients and audiences but to initiate new relationships. When it is clients being courted, a webmaster can be seen as fulfilling the role of business development (or "new business") specialist. Virtually every electronic media sector is seriously concerned with such development activities. Growth is important in any enterprise—but it is absolutely essential in the publicly owned entities that now dominate our profession. In most cases, if a business is to grow faster than the market as a whole, then new clients must be found to fuel that growth.

Advertising agencies pay special attention to such prospecting or "rainmaking" work, with many shops employing a high-level executive as a vice president or senior vice president for business development. This person takes the pulse of potential clients and evaluates the chances of converting them to *actual* clients. Because agencies service only one company in a given product category to avoid client conflict, the development specialist isolates areas in which his or her shop has, as yet, no substantial business and then trolls for clients within the most promising of those areas.

Despite continual attempts to quantify it, advertising remains an intensely subjective and image-oriented business. Thus, new business professionals must move beyond media budget figures and expenditure patterns and attempt to assess whether the personality of the agency will match the particular character of the company being studied. Many client/agency affiliations that looked wonderful from a purely financial perspective have quickly degenerated when their corporate cultures failed to mesh. The development specialist is paid to be sensitive to such matters before hasty marriages result in equally hasty and financially damaging divorces. The signing of a big new account can mean a rapid expansion in billings and the addition of new staff. It might even result in the opening of a new branch office to better service the client in its headquarters city. The quick loss of

Facilitative Functions

that same client, however, will mean terminations, career dislocations, and professional embarrassment that more than offsets any temporary increase in the agency's profits.

After the facilitative spadework has been done, and if the opportunity to pitch that "prospect business" can be secured, it frequently falls to the development specialist to be one of the lead presenters in such a pitch. At this point, presentational and persuasive skills come to the fore. Professionals in this field must be multitalented if they are to meet the consecutive challenges that pursuit of a new account entails. People who are only investigative number crunchers or hospitable glad-handers need not apply, because they do not have the breadth of skills that rainmaking requires. "In the old days," confesses agency new-business veteran Charles Decker, "it was pretty much an old-boy network. You could go out and golf [a client] to death, jam a quart of scotch down his throat and, after midnight, ask him for the business." In today's increasingly consolidated and cost-conscious world, conversely, new-business pursuit is a difficult, expensive endeavor that must be led by sober individuals who know how to isolate and capture an objective. "It's analogous to detective work," Decker states. "Ninety percent is drudgery and shoe leather."[29]

Besides agencies, business development personnel are also found in larger stations or station clusters, though with different job titles. Usually located in the sales department, they may be known as *co-op* or *local sales* specialists. Co-op staffers study reams of market data to find out which manufacturers or distributors offer programs that split the cost of outlet advertising with local dealers. Using this information, the co-op executive isolates relevant retailers in the market and facilitates their obtaining of co-op funds. These dollars are available only if the retailer uses them to purchase airtime, of course, so the station has thus hooked a new advertiser.

Local sales developers may also use co-op in their efforts, but they are more broadly concerned with stimulating new business from any retail or service enterprise regardless of co-op availability or of whether these enterprises have previously advertised over the electronic media. Sometimes, unfortunately, this task is foisted off on the station's newest salesperson, who must build his or her commissioned account list from scratch. At more entrepreneurial stations, however, a more experienced staffer is paid a firm salary to conduct these facilitative explorations—thereby upgrading the activity and increasing its chance for success. Meanwhile, many cable system sales staff members are, by definition, at least partly new-business developers because local cable sales activity is still in its infancy as compared to the broadcast competition. The same condition applies to Web marketing as well.

Finally, development officers are also active at public broadcasting outlets and systems. In such noncommercial settings, their duties are twofold. First, they devise campaign strategies to encourage contributions from individual viewers and listeners. These strategies may take the form of auctions, pledge drives, and the extension of membership opportunities to people who donate. Second, public broadcasting developers structure underwriting plans through which businesses can support the cost of prestige programming in exchange for on-air credit (*donor spots*) and carefully detailed tax deductions.

Agency or station; broadcast, cable, or Web; commercial or noncommercial; development is a grueling, unending process that is usually segmented from the glamour portions of our industry. It also requires mental toughness. For, as longtime agency rainmaker Don Peppers once revealed, "[I]t is not possible to win all the pitches. In fact, it is unusual to win more than 10 to 20 percent of them. Obviously this means we end up losing a lot. So one of the most important personal traits that any good development manager must have is resilience."[30] Despite the high rate of failure, it is this ongoing task that unearths the money that pays for the sparkle and sheen audiences have come to expect in their electronic media commercials and programs.

Chapter Flashback

Professionals who perform *facilitative* functions play essential behind-the-scenes roles that provide the services, data, and vehicles that media managers need to direct their enterprises.

Agency *account executives* serve a crucial liaison function between clients and creative directors and their staffs, working as guardians of each party's best interests. They help each side collectively to fashion and deliver advertising messages that are audience attracting and on strategy. Some media outlets now refer to their sales staffs as account executives to signal greater focus on assisting clients with their total marketing plans. *Media services executives* cooperate with account people in selecting the outlets best suited to reach target consumers in the most cost-effective manner. Among the benchmarks they use are *reach, frequency,* and *CPM*. Agency *traffic coordinators* execute and track the outlet placements chosen by the media services executives to ensure that media schedules are met as planned. On the outlet side, *traffic directors* make certain that paid advertisements are aired correctly and also handle the scheduling of other continuity such as public service announcements and on-air promotions.

Audience measurement executives are relied upon to compile the data that justify the expenditure of billions of advertising and program-development dollars. They work for companies that compile "the ratings" to measure audience size and composition. While a number of firms are active in various aspects of audience measurement, the two dominant companies are Nielsen Media Research in television and the Arbitron Company in radio. People meters are now bringing more precision to audience measurement—but also higher costs. *Brokers* use their specialized knowledge of media properties, financing mechanisms, and governmental regulations to facilitate the sale of electronic outlets. In addition, they must not neglect buyer and seller emotional factors in consummating a deal and must protect the confidentiality of all the parties involved.

Satellite services executives provide the orbiting relay capabilities on which so much of our industry depends to distribute consumer programming and business data casting. They lease time on geostationary birds to a variety of clients via either dedicated (long-term) or occasional-use arrangements. Their labors have been key to creating whole new segments of our industry, such as HBO's paving of the way for a robust cable network business. *Webmasters* facilitate the newest elements of our profession by shaping self-standing websites as well as sites that are extensions of other media enterprises. Effective webmasters provide user-friendly sites that encourage visitors to navigate (program) the site to their own specifications. Finally, *business development specialists* seek to identify and attract new corporate customers to their firms. This task is especially important for advertising agencies. In this process, these "rainmakers" must consider both fiscal and corporate culture compatibility. Other business development personnel function at commercial outlets as co-op or local sales specialists and as corporate and consumer solicitation experts at noncommercial stations.

Review Probes

1. Why is the job of account executive sometimes referred to as "controlled schizophrenia?"
2. What is the relationship between the terms *qualified prospects* and *waste circulation*?
3. Why are traffic positions in our profession called "traffic"?
4. What advantage does the people meter have over previous audience research methodologies, and what advantage does the *personal* people meter have over previous such electronic devices?
5. In what ways is a media broker similar to and different from a residential real estate broker?

6. List the new sectors of our profession that have been made possible by the communications satellite and the Internet.

Suggested Background Explorations

Alexander, Janet, and Marsha Tate. *Web Wisdom: How to Evaluate and Create Information Quality on the Web.* Mahwah, NJ: Lawrence Erlbaum, 1999.

Einstein, Mara. *Media Diversity: Economics, Ownership, and the FCC.* Mahwah, NJ: Lawrence Erlbaum, 2003.

Elmer, Greg, ed. *Critical Perspectives on the Internet.* Lanham, MD: Rowman & Littlefield, 2002.

Govoni, Norman. *Dictionary of Marketing Communications.* Thousand Oaks, CA: Sage, 2003.

Gunter, Barry. *News and the Net.* Mahwah, NJ: Lawrence Erlbaum, 2004.

Howard, Philip, and Steve Jones. *Society Online: The Internet in Context.* Thousand Oaks, CA: Sage, 2003.

Iuppa, Nicholas. *Interactive Design for the World Wide Web.* Woburn, MA: Focal Press, 2001.

Katz, Helen. *The Media Handbook: A Complete Guide to Advertising Media Selection, Planning, Research, and Buying.* 2nd ed. Mahwah, NJ: Lawrence Erlbaum, 2003.

Plum, Shyrl. *Underwriting 101: Selling College Radio.* Mahwah, NJ: Lawrence Erlbaum, 2003.

Schroeder, Kim, et al. *Researching Audiences: A Practical Guide to Methods in Media Audience Analysis.* New York: Oxford University Press, 2003.

Shyles, Leonard. *Deciphering Cyberspace.* Thousand Oaks, CA: Sage, 2002.

Strauss, Roy, and Patrick Hogan. *Managing Web and New Media Projects.* Woburn, MA: Focal Press, 2001.

Tellis, Gerard. *Effective Advertising: Understanding When, How, and Why Advertising Works.* Thousand Oaks, CA: Sage, 2003.

Webster, James, Patricia Phalen, and Lawrence Lichty. *Ratings Analysis: The Theory and Practice of Audience Research.* 2nd ed. Mahwah, NJ: Lawrence Erlbaum, 2000.

Wicks, Robert. *Understanding Audiences: Learning to Use the Media Constructively.* Mahwah, NJ: Lawrence Erlbaum, 2001.

CHAPTER 7

Evaluative Functions

Ultimately, the success or failure of any enterprise must in some way be *measured*. In the electronic media, this evaluative function is both internal and external. It is accomplished by professionals within media companies as a way of making their own companies stronger and as a means of analyzing competing companies or those with which their enterprise is contemplating doing business. These inside professionals are assisted by consultants, researchers, lobbyists, and public relations experts who help the organization run more smartly and protect its reputation. Evaluation is also undertaken by outsiders. Regulators and attorneys ascertain whether a media organization is behaving legally. Media analysts evaluate whether it is well and profitably run. Professional critics advise the public on media programming, practices, and trends. And professors help future professionals make sense of the media business and their potential role in it. All of these evaluators are essential to our industry's well-being and further development. Taken together, they provide the program standards, governmental linkages, financial appraisals, information systems, critical commentary, and training progressions by which the electronic media's activities are shaped, sustained, and promoted.

Consultants and Program Researchers

Generally speaking, consultants are *outside evaluators* brought *inside* to advise a media business on what it should do. In most cases, their determinations are based on media research that they themselves might conduct or that they commission separate program researchers to conduct on their clients' behalf.

Consultants offer a wide range of expertise to their clients, expertise that would be too expensive or specialized to have on the payroll full-time. Some consultants work with the station's on-air product. Often referred to as *program* or *news doctors,* these professionals formulate suggestions about how an outlet can make its service more attractive to its target audiences. An old industry truism is that "much of what a consultant sells to a client is his own success story."[1] Thus, most programming consultants have proven track records in improving the performance of facilities for which they previously worked. Operating from their own reservoir of experience, these visitors offer program and news directors insights to success by articulating what has and has not worked in similar situations elsewhere. Rather than reinventing the wheel, in-station managers get a refined view of options from consultants much more quickly and efficiently than by their own trial-and-error experimentation.

Tim Moore
Managing Partner,
Audience Development Group

In the past thirty years, radio has evolved though the engineering era, to the programming era, then into a period of deregulation and its current state as an economics-driven industry. Wall Street has funded radio's largest groups, and a relatively small percentage of publicly traded national companies own a disproportionate number of radio stations in America's top 150 markets.

As strategic and tactical advisers to these companies, and those remaining independent ownerships who oppose them, our firm has necessarily needed to evolve along with today's concentrated clusters of multiple-format stations. The process began with our acknowledgement for the need to practice the willing suspension of disbelief: everything that went before needed to give way to the do-more-for-less model being imposed in station clusters virtually everywhere. New York or New Haven, the magnitude may be different, but the scope is the same. Companies are requiring their core teams to get better ratings, supporting an even more urgent need to raise the company's "top line."

My company, Audience Development Group, engages these clusters with a *Mayo Clinic* philosophy: highly capable multiformat advisers—assessing and advocating programming strategies and researching music—so that our client stations have the luxury of *one* adviser practicing multiple format guidance. Our success comes from strong evidence of efficiency and, of course, the industry's scoreboard. Consulting anyone in today's deregulatory environment requires an understanding that increasing amounts of time must necessarily be devoted to the "behavioral" characteristics of a station cluster, in addition to the X's and O's of programming strategy. Most multiple-station clusters have coalesced former rivals still nursing competitive grudges, still populated by people doing multiple jobs on and off air, under a modicum of pressure and disharmony. Understanding their role, assets, liabilities, and concerns is often a major key for cluster managers to find ways of reducing *relationship tension* (the bad kind) so that *task tension* (the good kind) can rise.

Where pure programming competency was once a strategic advantage, I'm today committed to the proposition that "skill" is merely the price of admission: a base from which to convince radio talent to practice something they *may* not want to do, so that they can become something they've always wanted to be.

Sales consultants perform a similar function by providing suggestions about how an outlet can improve its billings. Some of these tips may be technical in nature, such as redesigning the rate card or advocating a different system for dividing account lists and responsibilities among the existing sales staff. Other advice might be more motivational, such as offering techniques and philosophies to boost sales success by restructuring how the account executives make their client pitches or changing the method and type of system that management uses to reward the selling staff.

Other consultants may be called in to assist a media business with engineering, accounting, data management, or bill collection difficulties. "Typically, a media problem is not just a vertical problem," consultant Russell Mouritsen advises. "It tends to cross over into several departments. Therefore, a media consultant must have access to many resources in order to do the job."[2]

Whatever the subject, a skilled consultant will first analyze the existing situation and then lay out options. He or she will never come to town with a predetermined answer or attempt to jam any single solution down the local manager's throat. In our industry, as in most others, prized consultants are careful analyzers and evaluators rather than off-the-cuff dictators. What award-winning radio consultant Donna Halper says of her branch of the business should be equally true of other types of electronic media consultants: "Everything we do is based, at least in part, on research findings. We are anything but rash in our judgments. Suggestions and recommendations are invariably inspired by some type of on-the-scene or nationally conducted research. Instincts are important in this business, and I pride myself on having some of the best, but I am also long enough on experience to know the value of careful, methodical study."[3]

As was mentioned earlier, some of this research is provided by separate specialist firms that, through a variety of proprietary tools, gather data that the consultants can use in the formulating their recommendations. One such firm is Marketing Evaluations, Inc., developer of the Q Ratings discussed in chapter 1. Through application of Q-based benchmarks, consultants can advise stations on which syndicated fare will work best in which situation, help advertising agencies identify potential product spokespersons, and assist producers in the casting of new on- and off-air projects. The researchers working for Marketing Evaluations and other such companies are skilled in data collection, analysis, and packaging. Their data-driven insights provide other media enterprises—and the consultants who advise them—with helpful guidance on a wide range of consumer attitudes and behaviors. These data go beyond the quantitative audience measurement numbers from Nielsen and Arbitron (see chapter 6) to attempt to provide *qualitative* descriptions of the media marketplace.

Francine R. Purcell
Vice President, Q Score Services,
Marketing Evaluations, Inc.

Marketing Evaluations, Inc. is a full-service market research company established in 1963. We handle a wide variety of custom research projects, including concept and product testing, advertising awareness and impact studies, promotion tests, tracking and diary studies, and spokesperson evaluations.

What we are most famous for, of course, are our syndicated Q Score studies, which measure both the familiarity and likability of everything from television programs (both broadcast and cable) to athletes to cartoon characters to actors/actresses to newscasters and sports reporters. And, in our Performers of the Past study (informally known as "Dead Q"), we even assess the current popularity of the deceased!

The basis for all of our Q Score studies is a historical framework of research, which has established the significance of "favorites" as an indicator of greater consumer involvement. With television programs, for example, previous studies have found that viewers who consider a given program to be one of their "favorites" are more likely to view the program more frequently, view more of each episode, view the program more attentively, and recall more program content. And, in turn, "favorite" viewers tend to be exposed to more commercials more often, view commercials more attentively, and recall more commercial content—which can be pretty important when it comes to getting the most bang for your advertising dollar!

And when it comes to celebrities, the "favorites" concept can be a key factor in the selection of just the right personality to star in your next movie or to serve as the spokesperson for your product or brand.

Our Performer Q study is perhaps our best-known research service. Twice a year, in the winter and summer, Performer Q measures the familiarity and likability of personalities throughout the media. Each wave of the study includes the names of more that 1,700 actors, actresses, comedians, musical performers, news and sportscasters, models, and athletes. The study is based on nationally representative samples of respondents (as are all of our surveys) drawn from our own People Panel, which is our nationwide consumer panel consisting of more than 55,000+ cooperating households in the contiguous forty-eight states. The celebrity names are divided into four questionnaire versions, which are administered through the mail. The final in-tab sample for each questionnaire version is 1,800 respondents aged 6+. (The in-tab number denotes those members of the sample who are actually included in the tabulation of results.)

Respondents are asked to evaluate each personality on the following scale: one of my favorites, very good, good, fair, poor—or someone they have never seen or heard of before. The sum of the scores for "one of my favorites" through "poor" is the "familiar" score, that is, the percent of respondents who know who the personality is. The Q Score is defined as the percent of those familiar with a personality who rate that personality as "one of my favorites."

Clients receive a complete demographic profile (broken out by respondent age, sex, education, occupation, geographic region, etc.) of the scores for each personality in the survey, as well as a number of target-audience rankings (in which all the personalities measured are ranked by Q Score against key demographics). The personalities are also grouped into categories (prime-time TV actresses/actors, male/female musicians, athletes, etc.), and normative data is provided for each performer type. This allows clients to determine whether a given personality's scores are above or below average relative to others in the same category.

As vice president of Q Score Services, my job is to shepherd all of these Q Score studies through the field from beginning to end—from questionnaire design to the dissemination of the findings to all of our clients. In the course of a given year, that consists of eight waves of TvQ, four waves of Cable Q, two waves of Cartoon Q, two waves of Performer Q, and one wave of Sports Q—a total of seventeen surveys, which often overlap on their way into and out of the field.

The job requires rigorous attention to detail, the ability to juggle several projects (at different stages of completion) at once and meet some stringent deadlines—all to ensure the highest quality in all of our research products.

Of course, there are frustrations from time to time. Producing such a large volume of research material every year requires the participation and cooperation of a lot of different people, including outside suppliers—printing services, data processing companies, the Post Office—not all of which are under my direct control. And there's no denying that clients can (very occasionally!) be difficult and demanding. So it certainly helps to be flexible and to have a sense of humor when things don't go exactly as planned.

But there's also tremendous satisfaction in meeting our clients' needs for timely, accurate, and actionable research results. Our role is to provide support and guidance for the media's decision makers: a role our company has filled with distinction for forty years.

Such media research activities do have their drudgeries. But they also comprise exciting detective work that can lead to programming innovation and breakthroughs in the understanding of consumers and their preferences. Researcher findings and consultant application of these findings to real-world challenges can fundamentally impact the way an electronic media company performs. Though research cannot compensate for a lack of experienced professional instinct—it does help give that instinct an edge. As Paul Lenburg of media research company ASI puts it, "There is no

Evaluative Functions

substitute for creativity, experience and just plain old guts. What we do is add a little objective information that connects the creative people, the distribution people and the studios with the consumer. Our business is connectivity."[4]

Media Analysts

One special kind of consultant is the media analyst. The sale, purchase, and operation of large media companies requires a tremendous amount of capital. Securities brokerage firms and investment banking houses employ analysts specializing in media businesses to make the evaluative judgments as to whether or not this capital will be made available. When they focus only on publicly traded companies, these professionals are often referred to as media *stock analysts*.

The pronouncements of these evaluators are tremendously influential to the viable functioning and growth of our entire electronic media system. Big electronic enterprises often must raise capital to expand or sustain their competitive position in a constantly changing marketplace. In fact, for the major players, expansion is no longer an option but a necessity. As industry analyst Debra Goldman explains:

> The rationale for getting bigger seems simple and necessary. The more the media splinter and take new shapes, the more the big boys need to own all the pieces, from studios to networks to theme parks to magazines to T-shirts. . . . Once they bulk up, the media megaliths becomes halls of mirrors in which each of their entertainment products can be endlessly reflected. . . . For the basic fact of life in a fractured media environment is that the more pieces there are vying for attention, the more the impact of each piece will depend on its relationship to the other pieces. The World Wide Web is nothing more than that principle brought to digital life.[5]

A good financial report card from key media analysts can affect a company's ability to secure expansion funding and also impact the advantageous or disadvantageous way its *debt servicing* (see chapter 6) will be structured. Consequently, the publicly held media companies especially make frequent and carefully packaged presentations to the analysts who head the media investment divisions at major banks and other investment firms. These analysts take this information, combine it with their own data, and then issue evaluative guidelines to their own bosses and investor-clients. *Broadcasting & Cable*'s deputy editor John Higgins portrays these analysts' clout this way: "Armed with billions of dollars managed on behalf of small mutual fund investors or huge pension funds and insurance companies, a core group of portfolio managers and buy-side analysts either lavishly reward or torture media executives."[6]

Each media analyst's determination will, of course, be shaped by the overall business strategies and philosophies of the institution for which he or she works. Some investment houses are more conservative than others. Some stress short-term dividend potential more than long-term growth. But taken together, these financial evaluators' pronouncements constitute a decisive diagnosis of the fiscal health or infirmity of the significant players in our industry. Former group CEO Trygve Myhren argues that this diagnosis should ultimately come down to "looking at what I call 'the four Ps': product, packaging, pricing and promotion. And you ought to think real hard about how well you're doing each of those. Because if you don't do them well, you're not going to be in that business very long."[7] As we discussed in chapter 6, media brokers make these same determinations in order to work with the financial analysts in structuring and facilitating outlet sales.

James Boyle
*Media Analyst/Broadcasting,
Wachovia Capital Markets, LLC*

The role of a media analyst on the "sell side" of Wall Street has notably evolved over the last ten years. Initially, one should be aware there are two sides to the banking and investment community of Wall Street—the "sell side" and the "buy side." The "buy side" is largely composed of the institutional mutual funds and the rapidly expanding hedge funds. Hedge funds, or "hedgies," are less regulated than the mutual funds and their charters are much more flexible in the type of investing they do. The "sell side" is the dozens of pure-play brokerages or integrated banks and financial companies that essentially sell financial products and services—thus their "sell-side" name.

Equity research is one of several research departments at a larger buy-side or sell-side shop. In addition, there is fixed-income research (investment-grade bonds or high-yield bonds). There is convertible bond research (convertible paper is effectively a hybrid, a bond that pays interest and may potentially convert into equity stock at a certain strike price.) There are commercial loan research departments that judge the probability of the timely payback of senior or subordinated loans. At a smaller shop, you might typically encounter only equity research, more often than not staffed with generalists covering multiple industries. At larger shops, one would more likely see industry specialists, like me, who focus on a specific business such as broadcasting.

The experience that is preferred for an equity research analyst varies. A basic grounding in economics and finance is generally expected, affirmed by an undergraduate degree and frequently a graduate degree. However, someone who has direct experience in an industry is increasingly attractive to shops that believe they can more easily teach the nuances of Wall Street to an industry person than they can teach an industry like broadcasting to a young, recent graduate. Either way, the new employee is often hired to start at the entry-level position of assistant or associate, where they can learn more about the industry and its specific companies while helping the analyst team generate reports, formulate views and, in general, service institutional clients and salespeople.

The various constituencies that an analyst deals with range from internal to external. Internally, equity research helps the institutional salespeople and the traders reach a better understanding of the industry, its companies, the long-term trends, the near-term influences, and the various valuation benchmarks that may drive a stock price down or up. In the recent past, the buy side has substantially beefed up their in-house research capabilities so as to have an edge over their peers. In other words, the more you know about broadcasting and the electronic media, the better advice you can give your clients as to investment opportunities in this industry. Analysts often talk to the publicly traded media companies that they cover. Also, to stay abreast of data and trends, you stay in contact with private media companies and people who do business with them. Other external conversations might be with trade publication journalists, business publication reporters, and the various electronic media journalists and news anchors who serve the general public.

An analyst is constantly multitasking and being pulled in different directions. It is a very time-sensitive business, where an analyst is asked for immediate opinions in reaction to breaking news. The hours tend to be lengthy. The travel is fairly heavy as one visits companies and clients and attends conventions. The stress level is relatively high. The competition is often fierce. Long-term trends can suddenly break down. Near-term influences crop up without warning. Sometimes a stock may move significantly in a short period, although the company's fundamentals, the macroeconomic information, and the industry data are all unchanged. Frequently, your opinion must be swiftly formed, with 50% or less of the data you'd like to have actually available before a decision must be reached.

There are multiple, interdependent, and conflicting reasons why stock prices move at any one time. Sometimes there is not a precise right answer. An electronic media analyst's primary job is to provide timely and value-added data flow, opinions, and reasonable explanations of our industry's past trends and prospective future changes, while being readily accessible to clients. It is an exciting job, but it is not for everyone. The successful broadcast or media analyst must be smart, quick, incisive, detail oriented, organized, and able to "sell" his or her well-reasoned view of an ever-changing world.

By whatever standard, if the networks, station groups, MSOs, studios, and Internet companies are judged by the analysts to be risky or poor-return ventures, such judgment can adversely impact the fund-raising ability of smaller companies throughout their respective branches of our business as well. In many ways, the one or two dozen media analysts at the top investment firms may be the most powerful people in our profession in charting its directions and anointing its victors.

Regulators

While media analysts work on the business side to evaluate the financial health of electronic enterprises, regulators function on the governmental side to evaluate whether these enterprises are in compliance with the federal rules and regulations that underpin much of our profession.

Even though there is a tendency to see them more as enforcers than as supporters, industry regulators serve a protective as well as a disciplinary function. Evenhanded rules that are impartially enforced create a stable and consistent operating environment for the players involved. Broadcasting could never have emerged as a serious business in the late 1920s if the federal government had not taken steps to bring order to the airwaves. A central outcome of the 1927 Radio Act and the Communications Act that followed seven years later was the establishment of an orderly allocation of a spectrum within which legitimate interests could conduct their business. No serious investors or advertisers would have bothered with a communication enterprise made up of frequency hoppers and unpredictable reception. Once the government had cleaned up the spectrum, defined its uses, and set the Federal Communications Commission in place as oversight agency, advertiser and audience services were free to prosper.

Subsequent regulation in other areas was not so welcomed by broadcasters, of course. *Ownership* restrictions on who could operate what and *content* strictures such as the equal-time law for political candidates (Section 315) inhibited broadcasters in ways unexperienced by their print media competitors. The fact that print and electronic media are not subject to uniform regulation has been the subject of continual debate for eight decades. There is no equivalent of the Federal Communications Commission that oversees how newspapers and magazines are run.

On the whole, however, the electronic media have profited more than lost from a federal regulatory system that has stabilized their operating environment and largely shielded them from potentially more quarrelsome state and local legislation. Contrary to what one would be led to believe from a few highly publicized FCC/industry clashes, the two forces have tended to work in more a consultative than a combative role. As communications law experts Erwin Krasnow and Lawrence Longley astutely point out:

At least to some degree the administrator [in this case, the FCC] can legitimately see his charge as including the preservation and encouragement of the regulated industry. The crux of this problem, then, is determining to what degree this goal should be subservient to other considerations, in particular to a larger conception of the public interest.... On a day-to-day basis, commissioners are forced to immerse themselves in the field they propose to regulate; however, the line between gaining a familiarity with an industry's problems and becoming biased thereby in favor of that industry is perilously thin. It is difficult for commissioners and their staff to operate closely with an industry without coming to see its problems in industry terms.[8]

Given this dynamic, it is not surprising that broadcasting was helped much more than hurt by FCC oversight from the 1930s into the 1980s. This was especially true in terms of the commission's protection of free TV against competing technologies. For example, over-the-air subscription television (STV) was delayed for some thirty years and cable was tightly controlled for more than two decades in order to preserve "free" television's turf.

Many broadcasters came to appreciate this protection only after they had lost it. In the early 1980s, President Reagan's FCC chairman, Mark Fowler, announced his intention to initiate sweeping deregulation of the industry. At first welcomed by the radio/television establishment, these policies were soon shown to be a double-edged sword. On the one hand, broadcasters were now much more free to make decisions as to program content and day-to-day operational matters. But on the other hand, they were no longer to be artificially insulated against new competition or against unfair practices by existing competitors. Fowler aides began to speak of a social Darwinism in which the marketplace should be able to decide which stations and which delivery systems would survive. "It is not the federal government's job to protect an unpopular station from bankruptcy" was a theme stressed over and over. Television, according to Fowler, was not much different than a "popcorn popper" or a "toaster with pictures"—the marketplace would determine which brands would succeed and which would fail without any interference from the FCC.

Suddenly, broadcasters faced unchained rivals like cable, STV, LPTV (low-power television), DBS, and satellite radio. At the same time, with quantitative guidelines removed for both nonentertainment program requirements and the number of commercials that could be aired per hour, every station was now free to go its own way. Under the old system, a broadcaster knew that the competitor down the street had to operate under the same commercial load restrictions as did his or her outlet. And everyone had to devote a certain amount of airtime to those public affairs discussions that large audiences avoided. But in a deregulated environment, this comforting predictability was gone; and competition became much more volatile.

Even more unsettling was the fact that the FCC would no longer protect a station from clearly unfair practices by its rivals. As communications attorney Arthur Goodkind prophesied in 1984:

Broadcast deregulation has two faces. On the one hand, deregulation means that broadcasters can spend less time and effort in complying with FCC rules and regulations. Yet at the same time, deregulation also means that the FCC will be far less available than in the past to protect responsible broadcasters from illegal or unethical practices on the part of their competitors. For that protection, station operators will in the future need to look increasingly to the courts and to other methods of self-help. . . . If, for example, a radio licensee operates with excessive power or fails to do what is necessary to keep its directional antenna in correct adjustment, other stations directly or competitively affected must now assume a major new burden in detecting, recording and reporting violations of the commission's rules.[9]

This same shift in regulatory philosophy resulted in FCC refusal to select a single standard for AM stereo, thus undercutting the viability of both that technology and the band it was intended to help.

Deregulation is never irreversible, however. In the case of cable, the Cable Communications Policy Act of 1984 and related FCC actions combined to lift a number of regulatory burdens from that industry. The functions of state and local cable oversight bodies were largely preempted and voided, and the remaining federal rules were few. Cable systems were able to charge virtually whatever they wished for whatever service they chose to provide. Abuses of this freedom piled up over the next several years and resulted in the 1992 Cable Television Consumer Protection and Competition Act that strengthened FCC supervision of the medium. In contrast, state and local regulatory clout was not restored.

Thus, in cable as in the other electronic media, the main—if not exclusive—policing agency is the FCC and its commissioners. When the number of commissioners was reduced from seven to five effective June 1983, each commission seat became that much more powerful in the decision-making process. A commissioner is nominated for a term of up to five years by the president, with confirmation by the Senate. No more than three commissioners may be from the same political party. The decisions reached by these commissioners are often difficult ones. Former FCC commissioner Rachelle Chong articulates three reasons for this difficulty:

> First, the FCC's scope of regulation covers many diverse industries, including television and radio broadcasting, cable television, telephony (both wired and wireless), international communications services, maritime radio, and amateur radio. The matters that come before the FCC for decision often require a thorough understanding of the relevant industry and sometimes can be quite technical and complex.
>
> Second, the matters can involve sensitive legal and constitutional issues, for example, the First Amendment rights of broadcasters. These matters can be very controversial and can involve weighing of the legal risk of a contemplated legal action.
>
> Third, politics can sometimes play a part. Congress created the FCC to be its expert independent agency to implement communications laws, so the relationship between the two bodies is close. Congress has oversight responsibilities over the FCC and approves our budget. Not surprisingly, the FCC works very closely with members of Congress and their staffs to properly implement the laws of Congress and to respond to their concerns.[10]

Because of this decision-making complexity, the background that nominees bring to the FCC has long been the subject of debate. Should they come from the communications industries in order to offer real expertise? Or should they be experienced regulators but without prior involvement in communications matters so that their past associations do not appear to influence their decisions? In reality, according to former FCC commissioner Glen Robinson, most members are picked based on much more politically pragmatic grounds than these questions imply. Robinson argues that the selection process

> is not a search for the "best candidates" if by that term is meant those qualified by experience, training or professional aptitude. . . . Appointees are not selected to represent political, social or economic interests. Nor do they represent any particular array of talents and skills that could be objectively identified as relevant to the job. . . . In a field of regulation such as communications, where there are not only one but several different industries and industry interests, the system of selecting candidates for appointments works strongly against the appointment of persons with specific views about regulatory issues because such views are likely to engender opposition from at least one major industry group

sufficient to offset any support from another. . . . These factors tend to eliminate from serious contention persons with established track records in the field of communication or regulation—unless the record is acceptable to all of the affected interests that exercise significant political influence.[11]

Dedicated, intelligent, and effective public servants like appointees Glen Robinson, Rachelle Chong, and Kathleen Abernathy do become FCC commissioners, fortunately, but perhaps more in spite of than because of the "least objectional option" mode that characterizes many appointment choices.

Kathleen Q. Abernathy
FCC Commissioner

I had just learned that the president would nominate me to serve as a commissioner at the Federal Communications Commission and I called my mother to relay the good news. "That's great, honey," she said. "Now what exactly will you be doing?" Good question, with no easy answer. My daughter, who is now seven, asks regularly what I do in the office other than talking on the phone and meeting with people. I tried to explain to her that in Washington, D.C., most people make their living by talking on the phone and meeting with people.

But back to the core question of what it is that I do. Fundamentally, my job is to represent the interests of the American people and craft communications policy that advances the public interest. This is often much more difficult than it sounds. The "interests" of the American people can vary dramatically depending on one's age, sex, political affiliation, state of residence, and level of income. Fortunately, I am not writing on a blank slate when it comes to determining how to advance the public interest. First, I am directed by specific statutory mandates provided by Congress and interpreted by the courts. In addition, I am able to draw on my twenty years of experience in the telecommunications industry to help inform my decisions.

But I nevertheless wanted a more coherent framework to help guide me during my term as an FCC commissioner—a framework that would be applicable across all industry segments and subject areas. I developed the following five-part regulatory philosophy to provide a foundation for decision making.

First, Congress sets the FCC's priorities in the Communications Act, and the agency should faithfully implement those priorities rather than pursuing an independent agenda. It is tempting for a regulatory official to follow her own policy preferences on any given issue, but where the statute dictates a particular course of action, it is my job as an appointed official to implement the will of the elected legislature.

Second, fully functioning markets invariably make better decisions, and better promote consumer welfare, than regulators. Therefore, unless structural factors prevent markets from being competitive, or Congress has established public policy objectives (such as universal service) that are not market-based, the FCC should be reluctant to intervene in the marketplace.

Third, where the FCC promulgates rules, it should ensure that they are clear and enforce them vigorously. Efficient markets depend on clear and predictable rules, and a failure to enforce rules undermines the agency's credibility and effectiveness.

Fourth, the FCC must be humble about its own abilities and reach out to consumer groups, industry, trade associations, and state regulators to maximize the information available in the decision-making process. We must recognize that it is impossible for a regulatory agency—particularly one with jurisdiction over a high-tech sector like communications—to duplicate the knowledge base of those it regulates.

Finally, the FCC, as a government agency in service of taxpayers, should strive to provide the same degree of responsiveness and effectiveness that is expected of an organization in the private sector. I know from my private sector experience that delay can be more debilitating than an adverse policy choice, because of the uncertainty it breeds. I therefore make it a priority to improve the timeliness of decision making at the FCC.

Even with this regulatory philosophy to guide me, I am continually challenged by complex legal and technical issues brought to the FCC. Sometimes my personal preferences may be at odds with the statute. Sometimes there is no obvious good answer to a regulatory dispute and I am forced to select the least-bad alternative. Sometimes the popular answer is not the right answer. At the end of the day, however, I must remain true to the law and to myself. Public service is a tremendous privilege, but that privilege has corresponding responsibilities. So, as I explain to my daughter, my job is about much more than talking on the phone and going to meetings.

The five commissioners are the most visible part of the FCC apparatus, but approximately two thousand other persons staff the agency and handle routine processing and decision-making duties as well as preparing advisory opinions and research for the commissioners. Actions involving individual media outlets seldom reach the commissioners themselves unless a precedent-setting or punitive issue is involved. Thus, the FCC staff, working under delegated authority from the commissioners, are the professionals who normally process the paperwork that keeps the regulatory wheels turning.

Many of these staffers are assigned to one of the FCC's six bureaus: Wireline Competition, Enforcement, Wireless Telecommunications, Media, Consumer & Governmental Affairs, and International. It is the Media Bureau that deals with most domestic mass media issues. In addition to the bureaus, the FCC contains no fewer than ten separate offices: Inspector General, Communications Business Opportunities, Workplace Diversity, Administrative Law Judges, General Counsel, Strategic Planning and Policy Analysis, Engineering and Technology, Legislative Affairs, Media Relations, and Managing Director. The commission also maintains eleven field offices, a research laboratory, and a call center.

Other than the heads of the bureaus and offices, their top assistants, and the commissioners themselves, most FCC staff are career civil servants rather than political appointees. The president (with Senate concurrence) appoints the commissioners as their staggered terms expire and can designate who will serve as chair. The chair selects the bureau chiefs and a few other top managers, who direct the activities of the dozens or hundreds of ongoing staff members in their units. Some staff may leave for the private sector or join the commission from that sector. This revolving door has been criticized as giving unfair advantage to firms for which ex-FCC personnel now work or those whose former employees are now at the commission. Various rules, however, prohibit current staff from handling matters dealing with recent employers and former staff from immediately representing parties directly affected by their previous FCC functions. At the same time, it is argued that people moving back and forth between the FCC and the private sector ensure better lines of communication and a sharing of expertise that government otherwise might not financially be able to afford.

The FCC is not the only regulatory agency in town. Being a creature of the legislative branch, the commission is policed by the Congress, which must approve its annual budget and creates the legal framework within which the FCC must function. As its sometimes pointless and repetitive hearings on such issues as television violence, broadcast indecency, and song-lyric labeling have shown, Congress has never been shy about intruding into FCC-delegated areas in response to constituent pressure. Not infrequently, this intrusion results in an FCC that is whipsawed between the interests of one party's Congress members and a president from the opposing party.

For electronic media advertising, the Federal Trade Commission (FTC) is involved in the regulatory game. Even though its activities were considerably narrowed during the Reagan years, the FTC's scrutiny is still a fact of life when it comes to commercial speech. The agency has been especially committed to its advertising *substantiation* program, whereby companies and their advertising agencies must be certain they have the proper evidence in hand before disseminating product performance claims. Meanwhile, as the FTC retreated from some types of oversight, the National Association of Attorneys General (NAAG) moved in to fill the vacuum. Drawing its membership from the chief law enforcers in the fifty state governments, this group has exercised its influence in such areas as airline ticket and car rental price advertising as well as tobacco-related health issues. Although it has no statutory legal standing, NAAG's membership uses its collective resources to approach advertisers and their media outlets with a united voice and, implicitly, with the threat of lawsuits if its recommendations are ignored.

Other federal bureaucracies also impact various aspects of electronic media operation. The Food and Drug Administration (FDA) may become involved in cases of pharmaceutical advertising and claims of medical endorsement. The Federal Aviation Administration (FAA) must be dealt with in terms of antenna tower placement, height, and lighting. The Bureau of Alcohol, Tobacco, and Firearms scrutinizes the methods by which companies market its namesake items. The Occupational Safety and Health Administration (OSHA) polices everything from radiation levels emitted by electronic equipment to control-room air quality. And a variety of units throughout the government monitor civil rights issues, including hiring and employment maintenance practices.

Lawsuits and similar actions also put the courts in the regulatory picture. Unlike the FCC, FTC, or NAAG, however, the courts become involved only when prompted by other parties in the form of cases and petitions. Because airwaves do not respect state boundaries, broadcasting and, by extension, cable and other electronic media have long been considered elements of interstate commerce, which thereby must be dealt with more or less exclusively at the federal level. Thus, the federal court system is the usual place where appeals from FCC or congressional actions are taken and where complicated copyright matters are fought out. Sometimes, the threat of court action is enough to provoke out-of-court compromises in order to avoid the massive legal bills that lengthy electronic media cases have often run up.

Communications Attorneys

The lawyers who litigate such cases (or fashion such compromises) are usually specialists in communications law. With such a large body of ever-changing regulation and policy pertaining to the electronic media, no general practitioner could hope to be conversant with even the main precedents and procedures in the field. So the "comm" lawyer is used by media clients, or their own in-house counsels, to represent their interests before the FCC, the courts, or other media businesses.

Over the years, the emphasis of communications law has changed depending on whether deregulatory or reregulatory pressures are in vogue. The attorney in this field must follow these shifts on behalf of clients in order to be the best possible preventer, protector, or advocate, as the situation demands. Indeed, many communication attorneys maintain that *preventing* problems through knowledgeable anticipation is the best service a legal counsel can provide. This prevention might include, for example, reminding client stations about the types of documents they should place in their public file or advising them on what actions to take to mesh with commission shifts in its EEO (Equal Employment Opportunity) expectations. It could also involve checking the station's contracts with a syndicator or network to ensure that they will conform to the client's best business and legal interests, or watching FCC dockets and notices of proposed rule making that might be relevant to a given network's, station's, or cable company's operations.

By closely studying both business and regulatory trends, the communications counsel may, in fact, advise actions that further influence these forces. As Krasnow and Longley point out:

> In the intricate and dynamic relationship between the FCC and the industry, the Washington communications lawyer plays a special role—not only in interpreting FCC policies for broadcast licensees but also in shaping the policy direction of the commission. In a study of Washington lawyers Joseph Goulden noted that while the lawyer's historic role has been to advise clients on how to comply with the law, the Washington lawyer's present role is to advise clients on how to make laws and to make the most of them. Goulden described how the Washington lawyer serves as the interface that holds together the economic partnership of business and government.[12]

Again we see that, in our industry, regulation may be as much a positive as a prohibitive force, depending on how it relates to the business and operational interests involved.

Alan C. Campbell
Founding Partner, Irwin, Campbell & Tannenwald, PC

The one critical factor that dramatically affects what communications attorneys do and how we do it is the evolution of telecommunications technology. While the substantive legal issues with which communications attorneys deal have remained somewhat constant, the technical method by which our advice is shared and for whom services are provided are constantly changing and will continue to evolve.

Much of the day-to-day legal work of a communications attorney focuses on interpreting rules enforced by the Federal Communications Commission. The vigor with which the FCC enforces its regulations varies, depending on which political party is in power; however, the FCC always must regulate communications in the public interest. For example, more conservative regulators historically have favored marketplace regulation, which means fewer federal mandates and more reliance on the marketplace to maintain order and influence communications companies to operate in the public interest. On the other hand, more regulatory-minded parties are of the view that detailed regulations are necessary to ensure that the public interest is served. Which view predominates mirrors a pendulum that shifts far toward the nonregulatory approach until abuses and violations occur (or at least are perceived to have occurred) and which then swings back to the regulatory mode. The role of the communications attorney varies with this regulatory ebb and flow. One of the attorney's main functions is to keep clients abreast of these changes and explain in understandable terms what these rule changes mean and how they impact operational issues.

Interestingly, as this is being written, the FCC is in a general deregulatory mode, with three Republican commissioners and two Democratic commissioners. (The Communications Act of 1934, as amended, mandates a five-commissioner setup, with a requirement that no more than three commissioners be from the same party [47 U.S.C. 154].) As noted, it is likely that in due course the pendulum will shift back to the adoption of more regulations.

Through all of these regulatory swings, communications technology is changing and evolving at light speed. Advances in technology affect the methods an attorney uses to deliver legal advice and the types of clients the communications attorney services. I would estimate that now about 80%–90% of an attorney's contact with clients, the FCC, and other counsel is through the Internet and e-mail. This new way of doing business is obviously not unique to telecommunications but is simply a fact of the modern business environment. Perhaps the most unusual factor in this evolution is that it has all happened so quickly in just a few years.

One final comment concerns an overused word in telecommunications—"convergence." Regardless of what it may mean in other aspects of the industry, to the communications attorney it can mean the types of businesses that are now involved with, for example, traditional broadcasting issues. In the past, for example, a communications attorney versed in broadcasting matters would often deal with carriage and nonduplication disputes between cable operators and television broadcast stations. Most recently, however, he/she would as likely be providing advice to a rural local telephone company that is offering broadband service to its customers, including the delivery of broadcast signals, and is in need of advice on the broadcast-cable carriage rules. Moreover, broadcast stations are now engaged in the same carriage disputes with satellite carriers that they had years ago with traditional cable companies.

The dominant factor by far in virtually every aspect of telecommunications is change in technology, which will occur at an ever-increasing rate and will continue to impact dramatically the communications attorney's role.

Through their constantly evolving familiarity with both the regulatory and business climate in our profession, skilled communications attorneys are often the essential spark plugs in driving a deal or motivating a compromise. Former FCC chairman and eminent attorney Richard Wiley once put it this way: "What lawyers bring to the party are a knowledge of the process and the players, and an appreciation of how best to structure arguments, based on past precedent and on what the commission is likely to do."[13] In an enterprise such as the electronic media—the purpose of which is the packaging and delivery of highly appealing *public* communication—it is nonetheless vital to recognize how essential *private* (behind-the-scenes) communication expertise becomes to the entire enterprise's well-being. The best communications lawyers practice their specialty with due regard for both its private and public contexts.

Broadcasting magazine once compiled the following inventory of what good legal counsel must do in the service of our profession. This inventory aptly summarizes the range of evaluative functions such attorneys perform:

> They shepherd applications for renewal or transfer of broadcast licenses through the labyrinths of the FCC. In the name of clients, they file tons of paper supporting or opposing (usually opposing) commission proposals. They defend clients and their interests (in the name of the public interest) before the commission and the courts, and lobby members of Congress and the commission. They are accused, sometimes with reason, of tying anchors to the ship of progress. They are credited, again sometimes with reason, with helping to pioneer new fields of telecommunications service. They are business consultants. They are, occasionally, a client's psychiatric social worker. And at times, they do public relations jobs. Probably no one lawyer provides all of those services, but some law firms do....
>
> Nor is that all. Given the nature of their practice before the FCC, an agency that is part judicial and part legislative, communications lawyers have an opportunity, which some have eagerly seized, to affect government policy in telecommunications.[14]

Standards and Practices Officials

Like communications attorneys, standards and practices officials also help keep media enterprises out of legal trouble. These in-house evaluators check content before it is aired to avoid altercations with audiences, affiliates, and advertisers. Disparagingly referred to as *censors*, standards officials have to apply their research and instincts to the difficult task of deciding what is and what is not appropriate for a given outlet's airwaves. Rather than offering a consultant's detached, objective, and outside view of what the station should be doing to improve performance in some area, the standards or continuity acceptance person brings an intimate and continual analysis to bear in a more preventive manner. As an in-house employee, the standards executive's job is to protect the operation from legal or public relations trouble through careful and tactful scrutiny of commercial and show content. Thus, this work has been dubbed *prophylactic programming*—taking action now that will prevent unfortunate consequences for the electronic media facility later. As NBC corporate executive Judy Smith once put it, "Any time there is an issue in our programming that viewers care about, that affects the bottom line."[15]

Standards divisions exist at most broadcast and cable networks and in the headquarters of the major broadcast group owners. Larger stations may also employ a standards executive, but at the local level the responsibilities of this position are usually fulfilled by the program director, with the advice of legal counsel sought as needed.

There is no question that the role and scope of standards activities have changed at both the network and local level over the years. Until 1982, the National Association of Broadcasters (NAB) Radio and Television Codes served as virtually industrywide guidelines for both programming and advertising content. When these were withdrawn in deference to antitrust concerns, each broadcast operation was left to its own devices. Some formulated or expanded the policies they already had in place. Others simply continued independently to follow the last version of the NAB documents under the reasonable assumption that the codes' utility had been proven over the years. Some others ignored the situation entirely—at least until they ran into trouble for something they aired.

At the same time, the fragmentation of the radio audience and the explosion of new cable services pushed back the boundaries of acceptable on-air program content—most noticeably for certain format radio outlets and limited-access pay-cable offerings. Meanwhile, the Reagan administration's narrowing of Federal Trade Commission activities left stations and systems to police the advertising that they aired without much guidance from the federal government.

The result of all these conflicting trends was a predictably mixed one. All three major television networks greatly reduced their standards offices—and then found they had to rescind these cuts in order to maintain credibility and consistency. Stations found themselves hurriedly authoring new policies to provide a rationale for turning down undesirable product advertisers who now sought radio/television exposure. And cable operations discovered that acceptance standards were critical if they were to project an image attractive to reputable, mainstream advertisers. Then, at the end of 1989, a Congress-pressured FCC began an ongoing crusade against aired indecency that raised even more standards questions. And the 1996 Telecommunication Act's V-chip provision took standards concerns to a whole new height, implanting content-blocking mechanisms in consumer receivers—assuming any consumer takes the trouble to program them.

Where the pendulum will swing from here is difficult to predict. But it is probably a fair assumption that standards or continuity acceptance staff will be an important factor in at least the major electronic media enterprises for the foreseeable future. With even middle-market licenses and franchises worth multiple millions of dollars, a competent and pragmatic standards executive can detect content situations likely to put those investments in jeopardy and then take preventative action. Civil lawsuits arising from allegedly deceptive commercials and listener/viewer protests about tawdry or disparaging programming are both threats to the bottom line. The skilled standards executive is paid to protect the station or network from such eventualities without inhibiting the sometimes edgy expressiveness that can attract the interest of saleable audiences and the advertisers striving to reach them.

Matthew Margo
Vice President, Program Practices, CBS

Although the television networks are not licensed, they act as proxies for their owned and affiliated stations, which are required to act in the public interest. As such, the networks bear the responsibility of administering governmental rules and regulations and self-regulatory guidelines and standards related to program and nonprogram material.

The Program Practices Department is an integral part of fulfilling the foregoing responsibility, as well as supporting CBS's commitment to providing viewers with broadcast material of the highest caliber and credibility.

Over the years, this commitment and responsibility have resulted in the development and application by Program Practices of guidelines and standards that our department applies to program and nonprogram material (the latter including commercials, public service announcements, and on-air promotions).

CBS program standards articulate important principles of responsibility to the audience, but in a manner that respects the creativity of the programs presented. The review process involves an interaction between producers, program executives, and Program Practices executives. These individuals ensure that CBS entertainment programs conform to generally accepted standards of public taste. As those standards evolve over time, CBS strives to be contemporary.

The extent and degree of Program Practices involvement in ongoing series—ranging from general consultation to in-depth episode review and supervision—depend on the nature and subject manner of the series as a whole and of particular episodes.

CBS takes into account the suitability of material for the particular time period in which that material is to be broadcast and the corresponding differences in audience composition and expectation from one daypart to another. Program Practices executives review material for excessive or gratuitous violence, sexuality, nudity, and inappropriate language. They ensure that character portrayals are sensitive to current ethnic, religious, sexual, and other significant social concerns. When controversial issues are addressed in entertainment programming, Program Practices executives review the material for balance and accuracy.

Program Practices reviews approximately 45,000 advertising submissions annually. This review process is based on public interest considerations as well as advertisers' demands that competitors be required to substantiate claims prior to broadcast. In addition to ensuring that advertising is truthful, lawful, and tasteful, Program Practices makes certain that such advertising does not promote competitive programming and is not otherwise inimical to CBS's competitive interests.

Program Practices also evaluates requests for public service time on the network. If the organization is determined to be eligible and the content of the messages ("PSAs") is acceptable, the PSAs will generally be scheduled in the network's regular public service rotation. Program Practices, in close partnership with CBS's Promotions Department, also supervises the development of all PSAs for the network's "CBS Cares" campaign. These messages, which feature CBS entertainment talent, tackle various causes, including HIV/AIDS, mentoring, diversity, cancer prevention, alcohol abuse, and violence prevention.

Standards vary widely from media company to company, of course. At one end of the continuum are the broadcast networks with licenses to protect and affiliates to keep happy. As public airwave vehicles available to anyone with a receiver, they must be especially cognizant of the impact of their programming on children and on "tonnage" advertisers, who are playing to reach large numbers of people without giving offense. "We need to please a broad and heterogeneous audience," NBC's top standards executive Alan Wurtzel points out. "It's a huge responsibility."[16] Yet, even within the broadcast network arena, there are standards differences. The edgy and younger-skewing Fox has always been willing to air material that an older-skewing network like CBS may choose to avoid.

Basic (advertiser-supported) cable networks do not use the open airwaves. Instead, consumers voluntarily pay to have them brought into their homes. Cable programmers therefore have greater latitude in their content choices. Among cable and DBS networks, standards vary even more widely than they do among broadcast chains. MTV and Comedy Central aim to attract much younger audiences than do A&E or the History Channel—so they put in place guidelines that are much more liberal. Finally, pay and VOD (video on demand) options maintain the most permissive guidelines of all, as presumably adult consumers specifically buy such content and there are no advertiser sensitivities with which to contend. (The wide and wild world of the Internet, of course, evidences the widest standards range—up to and including no apparent standards at all.)

Any of these standards are, of course, only guidelines that must be applied on a case-by-case basis. In each such instance, it is the evaluative ability of the standards executive that is most crucial to whether the guidelines work fairly or capriciously. "Ultimately, I know of no way to remove some degree of subjectivity from the process," says CBS executive vice president Martin Franks, in talking about his network's evaluation of commercials. "However, our decisions on any ad are the product of a review by more than one person, and we do our best to be guided by precedent, consistency and a desire to serve the sometimes competing interests of our three constituencies: our audience, our affiliates and our advertisers."[17] Of course, these same constituencies must be served in evaluating program content as well.

It must be made clear that in respected media companies, standards and practices staff do *not* oversee news operations. A hallmark of electronic journalism is that it should be discrete from advertising and entertainment functions—and this includes freedom from content oversight. Thus, news directors must be their own standards officers, calling on the additional evaluative skills of their communications attorneys as circumstances warrant.

As long as they occupy the public airwaves and flow unfettered into every functioning radio and television set, broadcasters will have to contend with content concerns beyond those faced by their print, cable, DBS, and Web colleagues. The content of a book or magazine can be skimmed by a consumer prior to bringing it into the house. And cable or DBS service, whether pay or basic, can be refused simply by not hooking up. But broadcast material cannot be previewed, and offensive stations cannot be clipped from a receiver's tuner. V-chips are set-embedded to allow viewers to block violent or sex-laden shows—but standards officials must still react to both pre- and postcoding situations. In fact, the presence in house of intelligent standards executives is a much more flexible and precise mechanism for addressing content responsibilities than any blanket electronic censoring circuit.

Lobbyists and Public Relations Executives

Ultimately, standards officials are responsible for protecting the image of the outlet that employs them. For most businesses in our profession, that image is critical to marketplace distinctiveness and success. Lobbyists and public relations executives engage in similar labors—evaluating the essence of a company's brand and the potential dangers to it and then taking action as a result of these evaluations.

All of the networks, some major station groups, and several key MSOs employ lobbyists to advance the continual stating of their case before congressional and regulatory decision makers. In addition, every branch of the electronic media has its trade association, for which lobbying is a major if not singular activity. For broadcasters, of course, it is the National Association of Broadcasters (NAB). For cable, it is the National Cable Television Association (NCTA) or the smaller Cable & Telecommunications Association (CTAM). Public broadcasters have the Association of America's Public Television Stations (AAPTS). The interests of satellite communicators are pressed by the Satellite Broadcasting & Communications Association of America (SBCA), and those of low-power television stations are represented by the Community Broadcasters Association (CBA). Dwarfing these forces, however, are the more than 150 lobbyists working for the various telephone companies. These advocates are on the front line of the struggle to win for the telcos a greater role in providing electronic media delivery services.

Lobbyists serve a positive function in informing government about trends, developments, dangers, and possibilities that pertain to their industry and its contribution to the public good. There is no question that such communications are self-serving and are conducted to benefit the respective lobbyist group's cause. But if lobbyists can show that their association's interests parallel the interests of the public in the matter under consideration, they have promoted a mutually valuable service. For example, in successfully pushing for the abolishment of the financial interest and syndication (*fin/syn*) rules that prohibited them from owning and distributing most of the prime-time shows they carried, network advocates demonstrated how these rules were working against the adequate financing of high-quality "free" TV programming. On the other side, the Motion Picture Association of America (MPAA) and some independent producer representatives argued that lifting the rules would again give the broadcast networks dominant power and undermine the market for a wider range of creative product. In this as in other industry issues, the various sides assembled as much researched evidence as they could to support their assertions, thereby giving government officials and their staffs more evaluative data than they would be able to uncover on their own.

Whether lobbying before legislators or regulatory agency personnel, professional advocates as a whole probably spend at least as much time trying to convince government *not to do* something as they do in attempting to initiate some action. Mature industries that have benefited handsomely from the status quo obviously try to maintain that status quo as long as possible. On the other hand, young upstart businesses and newly coalesced citizens' groups are more likely to promote regulatory or legislative *change* that will enhance their prospects or social agendas. Nonetheless, an established entity sometimes may seek a new restrictive rule to protect its competitive position, while a less-developed interest may try to block that rule in order to open up the game. Either way, reputable lobbyists aggressively but openly pursue their employers' objectives in whatever forums are available.

Because of their inherently public nature and their centerpiece role in the lives of the citizenry, the electronic media are lobbied about or lobbied against to a much greater degree than are some much larger but less spotlighted industries. Citizens groups like Action for Children's Television (ACT), the Office of Communications of the United Church of Christ, Accuracy in Media (AIM), Action for Smoking and Health (ASH), and a variety of locally based organizations have, over the years, made a significant impact on electronic media operations and regulations.

Virtually everyone listens to the radio and watches television—and undertakes such activities in their prized leisure moments. So a rate increase by a cable MSO is much more likely to be called by the constituency to their Congress members' attention than is a Federal Reserve Board policy that will depress the interest rates paid on their savings accounts. Very few people understand macroeconomics, despite its importance to their financial status. But everybody thinks they know all about television and how its systems should be operated. Thus, our business could not survive without a vigilant lobbying effort. When one operates in a fishbowl, it is important to keep turbulence to a minimum so nothing gets spilled.

Lobbyists work outside of government, and other important influence wielders work inside. For the electronic media, one such entity is the National Telecommunications and Information Administration (NTIA). A part of the Commerce Department (and thus of the executive branch), the NTIA has the potential to impact an administration's plans and policies as regards the electronic media. Successor to the Nixon era's Office of Telecommunications Policy, the NTIA not only disburses grants to public broadcasting and assigns spectrum space to federal users but also represents the current administration's views at the FCC and on Capitol Hill. For a president with a hands-on interest in telecommunications matters, the NTIA can play a substantial role in advocating the adding or subtracting of regulatory policies and can either accelerate or retard the application of new communications technologies.

In a similar manner, the Department of Justice can, if a president chooses, exercise considerable influence in the telecommunications arena. An activist chief executive may permit or encourage the Justice Department to pursue antitrust investigations designed to retard the formation or to initiate the breakup of consolidations in various aspects of our business. The department may also be energized to expand oversight of Equal Employment Opportunity activities by media businesses and may push the FCC toward similar undertakings. Occasionally, the Justice Department may take the FCC to court if it disagrees with a commission action.

Many times, the department's real power comes not from bringing suit but from behind-the-scenes influencing of involved parties through either the threat of such suits or the assurance that no legal action is contemplated. Although conducted within the fabric of government, such influencing is at least as powerful as what any concerted lobbying effort from outside interest groups can accomplish and may itself have been subtly impacted by such lobbying in other corners of the administration.

A special, more continuous, and more visible lobbying effort is conducted by public relations professionals who work in the field of political communications. Whether they serve as press officers on a government agency's or politician's staff, or work outside for corporate/business interests, the job of these communicators is to interact positively with the media in order to reach that media's public.

Press secretaries and holders of similar titles must be adept at making their boss or governmental agency look good but without engaging in deception when crises arise. As with lobbyists, political communicators are expected to put the best possible face on their employer's actions and positions. Journalists recognize this bias and take it into account when preparing their stories. Relations sour, however, when mutual trust is violated. A press officer who deliberately misleads a reporter or the press corps in general will have no further credibility. A reporter who betrays a confidence by publicizing an off-the-record comment or breaking a story before an agreed-on release time will be frozen out in the future. Each party is in the business of analyzing and evaluating the expected actions of the other in order to present its position or story most effectively to the mass audience. The skill by which each plays the game without breaking the rules determines the career success or failure of press officer and reporter alike.

Political communicators thus engage in a very delicate series of dances with the reporters on whom they depend for access to the public. As professors Kathleen Jamieson and Karlyn Campbell explain:

> There is a constant tension between those assigned to ferret out the news and those assigned to control press access to news sources. Press secretaries, public information officers, and public relations firms are hired to ease press access to favorable information and to minimize or block press access to unfavorable information. When negative information seeps out, it is the function of news managers to prevent or lessen its spread through the media and to cast it in the best possible light.[18]

For a press officer, then, it is critically important to be able to evaluate when an unfavorable story already is substantially out and to proceed to disseminate whatever damage control information is necessary. Waiting too long to say anything not only loses a press officer the chance to be heard but also is perceived by reporters as stonewalling—an action only slightly less harmful than deception to the reputation of a political spokesperson. Waiting too long is especially dangerous in the era of the electronic media's twenty-four-hour news cycle. Electronic news is always "on"—most notably via cable news networks and the Internet. The public relations spokesperson in our industry does not have the luxury of talking only to print journalists who publish but once a day or once a week.

In the case of communicators working for corporations as well as those working for politicians, knowing *when* is often as essential as knowing *how*. As Exxon's oil spill in Alaska demonstrated, a crisis situation involving a company (or an officeholder) demands a strategic communication response within a very short time. Failure to respond expeditiously only deepens the crisis and layers a communication disaster atop the physical or economic one. "Silence is the crime Exxon is paying for," asserted American Association of Advertising Agencies president John O'Toole just after the accident, "and that's a helluva price to pay for not running an ad the very next day."[19] Given the near-instantaneous communication that can be provided by the electronic media, a corporate foul-up or political misstep can suddenly affect the perceptions of millions of people. Thus, public communicators in a crisis situation must evaluate and respond wisely and quickly.

Most professionals in the field agree that the best strategy for dealing with a crisis is to anticipate it through *proactive communication*. "Before a crisis strikes, image advertising, issue advertising and PR can build a reservoir of good will," explains public relations executive Steven Fink. "Then, when the crisis hits, the public will trust you more, forgive you faster."[20] Certainly, if a press secretary or corporate spokesperson has maintained positive proactive links with the media, then the task of crisis management will be easier. The topic must still be addressed directly and quickly, however; and speed must not be confused with impulsiveness. Skilled practitioners know that they must never disseminate a message that tells the truth without taking the time to evaluate and cast that truth in a lucid, understandable nugget. "You can over explain and get yourself in trouble," warns John Burke of Burson-Marsteller. "You can, for example, send a biochemist to talk about a chemical; he will tell the truth, and he'll over tell the truth. He may create unfounded concerns about other issues."[21]

Stephen Serkaian
President, Kolt & Serkaian Communications

Meeting the communication needs of a variety of clients in the public and private sectors requires balancing client interests with media needs, substance with style, and reality with perception. Trying to control the message can be difficult because there are more wrong ways than right ways to do it. And while we can control the delivery of the message, we can't control the media through which the message is communicated. Therein lie some of the many challenges of practicing public relations.

My firm offers a variety of communication counseling services including public and media relations, strategic planning and policy development, advertising, executive communications training, crisis communications training, conference and special-event planning, public opinion and organizational research, and promotional video production.

My prior staff experience of providing public relations services for a U.S. Senator, a state Department of Commerce director, and a state Speaker of the House helped prepare me for the day-to-day rigors of the public relations industry.

As a public relations consultant, I work for many clients for whom I've designed and implemented comprehensive public relations strategies. For example, I serve as a liaison to print and broadcast reporters, trying to balance what's best for my clients with giving timely and accurate information to journalists. I have to anticipate questions that reporters might ask my clients, and I must be prepared to recommend answers to those questions.

It's my job to focus public and private sector executives in their responses to media inquiries. I always advise my clients not to tell reporters everything they know about a particular issue, but to stay on message and communicate only what the reporter needs to know.

I also have to balance being a cheerleader for my clients with being a straight shooter, always remembering that even if I can't tell reporters everything I know about an issue, I must never tell a lie. If journalists ever question my credibility, my service to clients and my relationship with the media instantly become ineffectual.

In addition, I recommend to clients different ways to publicize their stories, products, and events, whether it's through press releases, news conferences, or enhancing an event with camera-pleasing visuals and backdrops. Many issues simply receive better news coverage with the added visual punch.

Besides managing my accounts, I have to manage the work of our firm's staff members and make sure that project work hours are logged properly so that invoices can be sent to clients in a timely manner. I also have to stay on top of important issues every day by reading local, state, and national newspapers, as well as trade publications involving my clients' industries, along with public relations industry publications.

I try to meet the need-it-now challenges that clients and reporters demand, whether it's writing a press release at the last minute or answering phone calls from newspersons on deadline who want more information for their stories.

We celebrate when we attract all the media outlets in town to cover our client's story and it airs as the lead piece on the evening news or is placed on the front page of the local newspaper. It more than makes up for the times when our stories receive less than deserved media coverage. Either way, a career in public relations is an exhilarating and rewarding experience.

Figure 7-1. Herbert Hoover's whistle-stop presidential campaign in Salem, Illinois, November 2, 1928. Today's campaigns run instead on instantanous electronic rails. *Courtesy of the Smithsonian Institution, photo no. 79-88.*

Brevity and precision are the hallmarks of most informative electronic media messages, and political/corporate conveyances are no different. Press secretaries and similar spokespersons soon learn that to serve their employers well they must also serve the legitimate needs of stations and networks for credible and timely information. It thus is no accident that many political and public relations communicators worked in a media outlet at some earlier point in their careers. By acquiring firsthand familiarity with how news organizations function, they can better anticipate and satisfy the requests and questions that will be directed their way. In our journalistic tradition, reporters and spokespersons are usually adversaries, but this does not mean they must be antagonists. True professionals in these endeavors know how to build mutual respect while pursuing different objectives. The successful public relations executive knows how to evaluate a message on the basis of its accuracy, advocacy value to the client, and usability value to the media.

Unlike advertising, public relations efforts are much less likely to surface in easily identifiable chunks of airtime. Instead, public relations usually involves the dissemination of information via a variety of mass and nonmass means in such a way that the attitude it expresses becomes intertwined with the information and attitudes being expressed by journalists, decision makers, and opinion leaders. For example, the hairnet industry's public relations efforts had a lot to do with mandated hairnet use by food service employees—but when was the last time you saw a commercial for a hairnet?

Today, political and public relations operatives have more tools but less time than ever before to craft their messages. On the political side, yesteryear's localized and intermittent—but much more controllable—whistle-stop campaign (see figure 7-1) has been replaced by continual electronic monitoring of everything a candidate says and does. And rather than being restricted to the boundaries of a railway station, today's monitoring is projected on a national, even international platform.

On the corporate side, the growth of new media like the Internet has set up the expectation that a firm will be constantly visible and available to electronic news gatherers and the general public alike. "The Internet is also helping break down the boundaries between public relations and advertising," observes trade reporter Wendy Marx. "Indeed, several commercial World Wide Web sites are products not of advertising agencies but of PR agencies and in-house PR departments."[22] All in all, electronic public relations practitioners can look forward to a greater and more time-sensitive variety of challenges in the years to come.

Critics and Commentators

Just as media lobbyists and public relations professionals try to evaluate the image landscape for politicians and corporations, so do critics and commentators attempt to evaluate the performance of the media themselves. Because our delivery systems are so central to people's lives, electronic media critics and their observations are of fundamental importance in offering guidance to the public about how to use these media more enjoyably and productively. Industry-attuned critics who know how to fashion reasonable media analyses and suggestions for improvement can, in Richard Blackmur's words, "make bridges between the society and the arts."[23] By opening up lines of communication linking creator and consumer (outlet and audience), a critic makes it easier for each to comprehend the needs and limitations of the other.

Listeners and viewers, for example, can learn from the knowledgeable critic why the electronic media must take some actions and cannot take certain others. This commentator might point out that a cable system is forced to black out a sporting event because of a local exclusivity clause enforced by the team or league. Similarly, that critic might discuss economic and affiliation factors that enable one station to offer a complete noon newscast while another must make do with a sixty-second newsbreak.

Meanwhile, from the opposite perspective, a credible critic can let a cable system know that a number of subscribers are upset with its choice to move a particular program channel to an upper service tier or can alert a station that shifting a popular news anchor from the late- to the early-evening newscast may cost that outlet upscale viewers who aren't home to watch the 5:00 'cast. If the commentator passing this information along is respected by media managers as being well versed in our industry, these executives may give that critic's bridge-building observations serious consideration for the benefit of everybody concerned.

Successful electronic media commentators thus must acquire an intimate understanding of their subject if they are to be taken seriously by both consumers and media practitioners. As former *Washington Post* columnist Lawrence Laurent once observed, "This complete critic must be something of an electronics engineer, an expert on our governmental processes, and an electrician. He must have a grasp of advertising and marketing principles. He must be able to evaluate all of the art forms; to comprehend each of the messages conveyed, on every subject under the sun."[24]

Thus, an authoritative electronic media critic takes the responsibility to be broadly conversant with industry systems and issues in addition to program content. This is especially important in media criticism because media consumers seldom have the contacts and the time to acquire and to update a thorough knowledge of the media for themselves. According to former *New York Times* writer Jack Gould, "Critics in a sense are the proxies of the viewers. This does not mean that viewers necessarily will agree with the reviewers. But it does mean that there is the common bond of an independent opinion."[25] Gould's *Chicago Tribune* counterpart, Larry Wolters, adds: "The critic alone can serve as a watchdog for the viewer [or listener] who cannot always speak effectively by himself."²⁶ Finally, we must realize that broadcast/cable reviewers, in analyzing and evaluating an overwhelmingly entertainment-oriented industry, must be entertainers themselves if they are to maximize their exposure. The breadth and diversity of their public (as compared, say, with that of the ballet critic) require that their output be expressed in a manner that is interesting, concise, and even fun to listen to or read. The *Minneapolis Tribune's* Will Jones candidly admitted to this aspect of his work when he wrote that his columns were supposed "to serve as an entertainment feature for the paper."²⁷ Nor is this a newly acquired expectation. Four decades ago, the trade publication *Television Age* editorialized that "the critic's function today seems to be to amuse and entertain viewers prospective and actual with wit, if not malice."²⁸

David Bianculli
TV Critic, New York Daily News

"Television critic" is a job everyone does for free, and many people would like to do for money—at least if the wide-eyed enthusiasm of grade-school kids when I visit classrooms on Career Day is any indication. When you watch TV and write about it for a living, though, you can't just watch what you want. You have to take the bad with the good—and, as with any other art form, the bad outnumbers the good by about nine to one.

One useful attribute for being a TV critic, therefore, is the ability to remain optimistic under duress, and to derive more joy from discovering something wonderful than from ridiculing something awful. Poking fun at lame material is easy; offering true criticism, by putting material in context and offering opinions, insights, and suggestions without revealing too much of the plot is harder. It's also better.

The biggest and best secret about being a TV critic is how much fun it is, even at its most demanding. Sooner or later, everything I care about shows up on television—music, politics, theater, literature—and, sooner or later, I get to write about it.

Every day. That includes everything from a *Nova* documentary on lightning and a six-hour documentary on the Beatles to weekly doses of *The West Wing* and *The Simpsons*. And if I'm encountering an unfamiliar subject, I get to learn on the job.

For those who have opted for a career in criticism but are straddling the fence between film and TV, let me say this: Roger Ebert never gets calls at home telling him to rush to the theaters to see a late-breaking documentary film. On the TV side of the fence, the only constant is unpredictability. The beat's responsibility covers not only preview tapes sent out in advance (during premiere and sweeps months at the rate of about fifteen per day) but late-breaking news like earthquakes and terrorist bombings, as well as live TV events such as the Oscars, the Emmys, and certain episodes of *Saturday Night Live*.

To do all of this viewing most effectively, I work exclusively out of my home, in an office that includes several TV sets, each attached to a separate VCR, pulling in signals from antenna, cable, and satellite TV. That way, when breaking news happens, I can compare instantly—on deadline—to track who's covering what, and how. My workdays, counting all the viewing and writing I do, generally start at 6:00 a.m. and run past midnight.

The "up" side of all this is that, after twenty years on the job, I'm not bored yet. The "down" side is that, to do my job the way I feel it should be done, I have no life.

Finally, a little free advice for anyone still not discouraged from traversing this particular career path. As preparation for this job, don't learn only about TV and journalism. During college, study as much as you can about as much as you can. My

Evaluative Functions

graduate course in statistics, though I hated it at the time, proved invaluable when it came to deciphering Nielsen ratings and shares—and over the years, from science to Shakespeare, I've been able to recycle something valuable from just about every course I ever took, with the possible exception of physical education.

Today's effective critics have learned that wit helps attract an audience, whereas malice alienates the media that critic is paid to objectively assess. Certainly, reviewers should be detached from the media professionals whose programs and operations they are scrutinizing. But a detached relationship still leaves the lines of communication open so that the critic is given access to data that are helpful in arriving at informed judgments. Once a critic assumes the role of media foe, however, these lines are shut down, and neither the public nor the industry can profit from the *uninformed* pronouncements that are certain to result. If reviewers and media decision makers become enemies, the profession that both serve is the first casualty.

Rather than risk this possibility, some overzealous media apologists would prefer a situation in which there is no systematic criticism. The broadcast and wired media come directly into one's home, they argue, so there is no need for commentary like that necessary to propel people to this or that concert hall, theater, gallery, movie house, or bookstore. Further, these anticritics continue, because media content is so readily and even automatically available, everybody can make unaided choices in the privacy and convenience of their own homes. All that is needed, they maintain, is a program schedule to serve as a menu from which to select. For radio, a list of stations and their formats should suffice as should a basic website catalogue for the Internet.

Certainly, there is no arguing that electronic media are popular, or that they are easily usable by virtually every inhabitant of the United States. There is also no debating that listeners and viewers each make thousands of programming and source choices every year and that many of these consumers and most of their choices are not motivated by a trained critic's counsel. Yet it is exactly this availability and popularity that require the attention of astute professional critics who neither disdain nor pander to the media they cover.

Former FCC commissioner Lee Loevinger once observed that "broadcasting is popular and universal because it is elemental, responsive to popular taste, and gives the audience a sense of contact with the world around it which is greater than that provided by any other medium."[29] This sense of contact, however, can be severely warped if listeners, viewers, and online surfers lack the observational training and guidance necessary to (1) make their media choices wisely; and then (2) evaluate the success of those choices.

Perceptive commentators encourage perceptive programming and media operation. They therefore help our industry improve much more than is possible via any law or governmental policy. Regulators can only require or prohibit. Through their evaluative efforts, critics, on the other hand, can reward and stimulate.

Librarians and Teachers

As critics understand, the electronic media collectively constitute a vast information industry and one that consumers preeminently rely on for concise and immediate enlightenment. However, our profession cannot convey information efficiently to other people if we cannot keep track of it ourselves. Working in a variety of electronic media settings, *librarians* arrange and locate the data on which so many other professionals rely to accomplish their own jobs.

Some librarians function within industry trade associations. This is a very efficient deployment because individual or corporate members from around the country thereby have a common resource on which they can draw and a much more comprehensive resource than any single company could construct on its own. Trade association librarians concentrate on compiling and accessing the kinds of materials that people in their particular branch of the industry are more likely to need. Because the scope of their subject is relatively limited, they acquire in-depth expertise in it and can identify and assemble materials of greatest importance to their memberships.

Other librarians work for governmental agencies or communications law firms. In these instances, of course, the focus is on collecting and retrieving legal documents and public filings. Ideally, the information they uncover helps a citizen or client deal more efficiently and responsibly with government so that both parties' interests are served. The administrative or legal librarian may also be asked to provide documents that will help a party defend itself in a government adjudicatory proceeding (like an FCC license-renewal hearing) or better prepare its case in a civil or criminal trial. Even the brightest communications attorney can do a client little good without swift access to the materials and case studies that bear on the filing or proceeding being undertaken. A skilled legal librarian helps ensure such access.

Large news-gathering operations, particularly at the network level, also use librarian expertise, of course, in order to maintain that wealth of background material necessary to round out news coverage and provide the source for interesting historical or sidebar pieces on a breaking story. Much of this material may be audio/visual in nature (photographs, audio recordings, film and video footage, and digitally stored images), which can mirror that sense of actuality audiences have come to expect from electronic journalism.

When library professionals move beyond print matter or database collections into these other realms, they are usually referred to as *media librarians*. In addition to news organizations, media librarians are found at industry archives such as the Museum of Radio and Television (New York) or the Museum of Broadcast Communications (Chicago); within college, university, and school district libraries and resource centers; and at companies that specialize in the rental of large collections of recorded material. A notable aspect of this last enterprise is the *stock photography* business, which accumulates and cross-indexes vast quantities of still photos and motion picture footage. News departments, studio and independent producers, and advertising agencies use the services of a stock house to obtain photos, slides, or rushes (motion picture/video segments) to be used as elements within their various projects. Increasingly, visual content is stored and/or shipped in computer disk format for more efficient and true-to-the-original reproduction. Whatever the format, if a needed item can be located by a knowledgeable stock librarian, obtaining the rights for its use is much more cost efficient than sending a crew or photographer to shoot it fresh. And in the case of historical pictures or rushes, of course, the stock archive becomes irreplaceable.

Finally, large advertising agencies also employ in-house librarians. In most instances, their collections include both print and audio/visual matter that account, media, and creative departments can consult in helping with client research, audience and marketing studies, and campaign development. New-business development executives especially may rely on their agency libraries to pull together the product category and company profiles necessary to targeting the best prospective clients.

If you previously thought of librarians only in conjunction with your municipal or school facilities, the variety of situations we have just cited demonstrate the spectrum of other library-related enterprises. Speaking of schools, they constitute the environment in which our final category of evaluators—educators—perform their important instructive functions.

Broadly speaking, media teachers are divided into two groups: (1) those who teach *via* the elec-

Figure 7-2. Students at the Virginia Ridge one-room school at Philo, Ohio, listen to a lesson from the *Ohio School of the Air. Courtesy of the State of Ohio Department of Education.*

tronic media; and (2) those who teach *about* those media. The former group traces its heritage at least as far back as 1930, with the Ohio School of the Air and similar projects (see figure 7-2). During the subsequent seventy-five years, and with varying degrees of success and public acceptance, instructors from a variety of disciplines have been put on the air in an attempt to share their expertise with larger or more dispersed student bodies. The surge of interest about in-school instructional television (ITV) during the late 1950s and 1960s gradually ebbed as the concept of *educational* broadcasting gave way to that of *public* broadcasting following passage of the Public Broadcasting Act of 1967. Now the focus was on building a noncommercial television network to serve general informational and cultural needs of the citizenry as a whole rather than curriculum-based in-school activities.

However, even though over-the-air instructional broadcasting declined markedly, the availability of low-cost, reliable video equipment provided, by the 1980s, a number of off-air options for teaching by television. National, regional, and local programs can be recorded on cassette or disk and distributed for playback on the machines that are present in virtually every school. Larger school districts are able, with minimal expenditure, to put their own master teachers on tape as supplementary resources for classrooms throughout their communities. Particularly when combined with an educational access channel provided by the local cable system as part of its franchise agreement, these lessons can even be fed to the community at large and to homebound students as well. Increasingly, adjacent school districts are combining forces to provide *interactive* services for sharing specialized classes that no single district could afford to offer. Some of these services provide talkback circuits through which participating students in other school buildings can converse with the presenting teacher in his or her own classroom.

Through the use of now-abundant satellite capacity, video teleconferencing is another teaching tool. Especially as used at the college level, teleconferencing can deliver instructional experiences that are even international in nature with students, faculty, and other experts joining together for global lessons and seminars. Facilitated by such consortia as the pioneering Mind Extension University, entire degree programs can be offered via satellite and tap the cooperative resources of dozens of major universities. More conventionally, and with the support of the Annenberg Foundation and other agencies, college-level coursework has been presented as public broadcasting series programming, with students able to secure credit through a participating campus in their area.

As digital conversion permits public stations to transmit several content streams simultaneously, opportunities for such curricular activities increase.

The essential ingredient in all of these projects is the articulate faculty member. Beyond being masters of their own subject area, these teachers must be able to structure and deliver this material clearly and do so in a way that projects well through the screen. Thus, they must carefully analyze their own field to isolate the topics most essential to the course at hand and then evaluate which particular lessons and materials would be most effective in the tele-viewing or Web-delivered setting. Poor teaching will never make good viewing, and ill-conceived electronic delivery will not authentically convey great teaching. Instructors who appear on the tube must be intimately aware of the interdependency of content and electronic carriage in facilitating distance learning.

Sometimes outside subject matter specialists are assisted and tutored by other instructors whose discipline *is* the electronic media itself. *Media studies faculty* members, our second category of media teachers, strive to increase understanding of the role, content, and techniques of the electronic media profession. Although primarily found at the postsecondary level, media studies teachers are becoming increasingly prevalent in certain high school settings as more districts recognize the importance of the electronic media in the present and future lives of their students.

Unfortunately, such recognition has been slow in coming. Our educational system prides itself on its ability to turn out productive citizens who have learned how to earn a wage, evaluate a purchase, balance a checkbook, and make some sense out of the printed word. Yet the purchase of electronic media content through the prioritized expenditure of one's own time—and the analysis and evaluation of that content via one's own intellect—are subjects that are still all too rare in the standard school curriculum.

How many book reports, for example, were you called on to complete before leaving high school? On the other hand, how many electronic media listening or viewing reports were assigned? That there were probably many times more of the former than the latter directly contradicts the media-use pattern that dominates most people's lives. Sadly, a significant number of graduates seldom read more than a book or two a year. But all absorb the collective equivalent of thousands of book-length narratives through continual consumption of electronic media content.

This immense discrepancy between classroom training and real-world behavior exists in the college as well as the secondary school. "To say that the communications media are central to the functioning of our society is to state the obvious," observes professor Everette Dennis. "However, American undergraduate education almost completely ignores the study of mass communication. Unless students major in communications, journalism, or media studies, they can go through college without acquiring more than fragmentary knowledge about mass communication."[30] Recently, print literacy has grudgingly had to share the spotlight with computer literacy, but *media literacy* remains a largely unmet need.

One important task for electronic media educators, then, is to convince curricular authorities of the importance of their discipline so that they can proceed to educate a generation of discerning listeners and viewers. The second key task for these faculty members is to train skilled and dedicated media *practitioners* who will use our high-tech tools to their full potential. If, as we contend throughout this book, electronic media personnel collectively constitute a *profession,* then these personnel must be thoroughly and authoritatively educated. This is the ultimate analytical and evaluative duty because it will shape the people who themselves will shape the destiny of our entire enterprise.

"Broadcasting," "telecommunications," or "media studies" instructors, as they are variously known, are found in college/university, trade/technical school and, less often, in secondary school settings. The high school teacher's task may involve classes in media appreciation for the student body as a whole as well as preprofessional training for students interested in pursuing media careers. Trade and technical school faculty take the process one step further by offering detailed post-secondary instruction designed to qualify graduates for entry-level jobs somewhere in our industry. The limitation of some trade school curricula, however, is that their narrowness and brevity prepare students only for their *first* job. No real attention is given to the broader educational perspectives that a person needs in order to move to higher and more managerial roles.

For their part, four-year college and university media professors seek to integrate profession-specific insights with the more comprehensive instruction from other academic departments. The goal is to fashion bachelor's and graduate degree programs that can underpin a person's entire career. Most faculty strive, therefore, for a curricular balance between technical training and theoretical education. In this way, both short- and long-term career and life considerations can be raised. As with law, medicine, or any other endeavor, the profession that trains its members only in its own affairs does not prepare them to function successfully in the larger society. "A conventional university-based professional education combines practical and intellectual skills," state media professors George Pollard and Peter Johansen; "[R]ecruits learn not only to do the work, but also think about its implications."[31]

Glenda C. Williams
Media Professor,
University of Alabama

"Why *do* you teach?" The question was posed to me recently by a friend in the industry, a friend very knowledgeable about students and higher education. "Why would you leave the industry to work with college students?"

Truly, the description is deceptive. It's hard to explain exactly what goes into teaching at the college level: several hours of lecturing each week, hours of prep time for those lectures, grading papers, counseling students on career goals and opportunities, staying up-to-date on industry trends and techniques, writing research papers for publication, producing creative pieces. It requires self-discipline and motivation, creativity, patience, and long hours. A sense of humor helps.

The hours are somewhat flexible, though I still sometimes pull "all-nighters" to get projects graded or to finish a paper. There are occasional travel opportunities, budget permitting, to conferences full of friends. But mostly, teaching is about helping students discover what they are capable of doing and being.

I spent seven years working full-time in mid-management at a cable network before I moved into teaching. During that time, most of the station interns spent at least half of their time in my department with me. I taught many of them how to write and supervised their rotation throughout the network. It was this interaction, more than anything else, that gave me a desire to do the same thing in the college classroom.

In our program, I primarily teach management and sales, with occasional classes in the creative area. My focus is to help students discover the many careers available outside news or production. Far too many of our students start college with the goal of either reporting from in front of the camera or shooting from behind the camera. They have no idea how many other jobs and careers are possible in our industry. I spend four years of their lives (and mine) introducing them to the possibilities and helping them discover where they fit into the equation. Many, of course, stay in news or production. But it's always exciting when a student finds a different niche that's absolutely perfect for them.

A former professor and teaching mentor once told me, "Always take your subject seriously, but never take *yourself* seriously." I've always tried to remember that. I think learning can be fun and should be exciting. I feel passionately about my subjects and truly want to share that passion with my undergraduates. I love it when students tell me they're watching commercials with a different eye because of what they learned in my class. I know then that they are really *learning*, not just trying to pass a class.

So why do I teach? Because of the students who send notes telling me I made a difference in their lives. Because of the student who cried and thanked me for believing in her when no one else did. Because of those rare moments when, in the middle of a lecture, I see a light go on in one student's eyes—and know they really got it. Teaching is a calling and a passion—and I'm proud to say it's mine.

Astute electronic media faculty at all levels realize the value of instruction in specific skills. But they also perceive the broader picture. In his book *Technology and the Academics,* Sir Eric Ashby wrote that one of the strongest desires of human beings seems to involve learning to perform at least one thing thoroughly and with a high degree of expertise. In a highly technological age, the close cooperation of applied arts such as media studies with more traditional liberal arts perspectives can construct curricula in which specialist training provides the viable and valuable core for a modern liberal education. As electronic media become even more prominent in interconnecting our life experiences, and as concern about the content and responsibilities of these media becomes more central to the concerns of society as a whole, media studies themselves can constitute the nucleus of instruction for the educated person of the twenty-first century.

Electronic media teachers whose backgrounds reflect a blend of industry and academic experience are maximally qualified to serve the immediate and long-range preparatory needs of their students. Graduates of such a faculty's curriculum learn to manipulate the devices and protocols common to our media. But these students also acquire perspectives on the analytical and evaluative processes needed in *any* field in which the abilities to formulate and to communicate policy action are at a premium. The unemployed liberal arts graduate and the out-of-work computer engineer are poignant illustrations of the human fallout that occurs when the people who design instruction ignore either a person's wish to do one thing well or his or her eventual need to flexibly adapt to new conditions.

In the past seven chapters, we have explored the major career roles that the electronic media encompass. Whether you find yourself in any of these occupations or in one of the myriad of other positions that exist today or will exist tomorrow, try to develop and maintain the broadest possible professional perspective. This perspective will help you to detect the full range of your own potentialities, and it will also optimize your value to our industry. Because of their power, cost, and complexity, the electronic media will always be group endeavors. Individuals who possess the widest view of where the group needs to be headed are most likely to avoid disasters and amplify opportunities.

Chapter Flashback

Evaluators function both inside and outside the electronic media. *Consultants and program researchers* are outside specialists called in by a media company to improve its performance or operation in one or more ways. They include program consultants, sales consultants, and systems consultants. Good consultants base their recommendations on solid research. Specialist program researchers are often called upon to deliver detailed information about consumer behaviors and attitudes. *Media analysts*, in contrast, concentrate on media businesses themselves, compiling financial profiles on the health of these businesses in order to advise potential investors and lenders on which enterprises constitute the best investment choices.

Although they are most often perceived as enforcers, *regulators* also serve a protective role in creating a stable operating environment in which the electronic media can operate with consistency. While the FCC is the most obvious regulator of much of our industry, the Congress, courts, and a variety of other federal agencies such as the FDA and FAA also exercise authority, as do state and local units like attorneys general and cable franchise awarders. *Communications attorneys* are experts at evaluating electronic media law and regulation. They seek to prevent problems before they are required to litigate them. The most successful lawyers practice with a keen appreciation of communication law's private and public interests. Such interests are also protected by *standards and practices officials,* who work within networks and station groups. Despite the fact that they are sometimes negatively referred to as censors, skillful clearance executives serve an important function in guarding the images of their outlet and its advertisers while not inhibiting creative expressiveness. Although permissions standards vary depending on the audience a given network or program is seeking, a common goal is the maintenance of evaluative standards that are consistent within a particular media unit.

Lobbyists evaluate the political climate and keep government informed about trends and needs in their employer's branch of the media. They strive to demonstrate that their media client's policies parallel the best interests of society. While many lobbyists work for industry companies, others represent citizen's groups like the Consumers Union or governmental agencies like the NTIA. *Public relations executives* are more visible lobbyists who strive to represent their company's or political candidate's positions to and through the media as a means of positively influencing private and public opinion. *Critics* and *commentators* independently evaluate the performance of the electronic media and offer guidance to the public on media use and interpretation. The respected critic thus stimulates both industry improvements and increased public enjoyment of the products of that industry.

Librarians preserve, organize, and retrieve the data and program content elements on which other media professionals rely to accomplish their own tasks. When librarian responsibilities move beyond print matter or database collections into pictorial or audio holdings, they are often more precisely titled *media librarians*. Finally, *electronic media teachers* perform one of two roles in the service of education. Some use these media as tools to extend their own academic discipline (English, mathematics, etc.); others teach actual media techniques and issues. These latter educators strive to enhance media literacy among the general student body and/or to prepare future professionals for media employment.

Review Probes

1. What is the relationship between media consultants and program researchers?
2. Why are media analysts sometimes referred to as the most powerful people in our profession?
3. List three examples of regulators performing a media *protective* function.
4. In what ways is the role of communications attorney similar to and different from that of a conventional lawyer?
5. What protective function does each of the following occupations provide for media outlets: communications attorney, standards and practices official, lobbyist, critic?
6. In what ways must media teachers be consummate *evaluators* in order to properly fulfil their roles?

Suggested Background Explorations

Balas, Glenda. *Recovering a Public Vision for Public Television.* Lanham, MD: Rowman & Littlefield, 2003.

Bryant, Jennings, and Dolf Zillmann. *Media Effects: Advances in Theory and Research.* 2nd ed. Mahwah, NJ: Lawrence Erlbaum, 2002.

Buddenbaum, Judith, and Katherine Novak. *Applied Communication Research.* Ames: Iowa State University Press, 2001.

Christ, William, ed. *Media Education Assessment Handbook.* Mahwah, NJ: Lawrence Erlbaum, 1997.

Croteau, David, and William Hoynes. *Industries, Images, and Audiences.* 3rd ed. Thousand Oaks, CA: Sage, 2002.

Day, Louis. *Ethics in Media Communication: Cases and Controversies.* 4th ed. Belmont, CA: Wadsworth, 2003.

de Mooij, Marieke. *Consumer Behavior and Culture: Consequences for Global Marketing and Advertising.* Thousand Oaks, CA: Sage, 2003.

Dickson, Tom. *Mass Media Education in Transition: Preparing for the 21st Century.* Mahwah, NJ: Lawrence Erlbaum, 2000.

Elmer, Greg, ed. *Critical Perspectives on the Internet.* Lanham, MD: Rowman & Littlefield, 2002.

Feintuck, Mike. *Media Regulation, Public Interest, and the Law.* New York: Columbia University Press, 1999.

Giles, David. *Media Psychology.* Mahwah, NJ: Lawrence Erlbaum, 2003.

Gripsrud, Jostein. *Understanding Media Culture.* New York: Oxford University Press, 2003.

Lewis, Justin. *Constructing Public Opinion: How Political Elites Do What They Like and Why We Seem to Go Along with It.* New York: Columbia University Press, 2001.

Lindlof, Thomas, and Bryan Taylor. *Qualitative Communication Research Methods.* 2nd ed. Thousand Oaks, CA: Sage, 2002.

Mickey, Thomas. *Deconstructing Public Relations: Public Relations Criticism.* Mahwah, NJ: Lawrence Erlbaum, 2003.

Middleton, Kent, William Lee, and Bill Chamberlain. *The Law of Public Communication.* 6th ed. Boston: Allyn & Bacon, 2004.

Morrison, Margaret, et al. *Using Qualitative Research in Advertising: Strategies, Techniques, and Applications.* Thousand Oaks, CA: Sage, 2002.

Noll, A. Michael, ed. *Crisis Communications: Lessons from September 11.* Lanham, MD: Rowman & Littlefield, 2003.

Orlik, Peter. *Electronic Media Criticism: Applied Perspectives.* 2nd ed. Mahwah, NJ: Lawrence Erlbaum, 2001.

Overbeck, Wayne. *Major Principles of Media Law.* Belmont, CA: Wadsworth, 2004.

Pare, Daniel. *Internet Governance in Transition: Who Is the Master of This Domain?* Lanham, MD: Rowman & Littlefield, 2003.

Parsons, Patricia. *A Manager's Guide to PR Projects: A Practical Approach.* Mahwah, NJ: Lawrence Erlbaum, 2003.

Russomanno, Joseph. *Speaking Our Minds: Conversations with the People behind Landmark First Amendment Cases.* Mahwah, NJ: Lawrence Erlbaum, 2002.

Stempel, Guido, David Weaver, and G. Cleveland Wilhoit. *Mass Communication Research and Theory.* Boston: Allyn & Bacon, 2003.

Tillinghast, Charles. *American Broadcast Regulation and the First Amendment: Another Look.* Ames: Iowa State University Press, 2000.

Trend, David. *Welcome to Cyberschool: Education at the Crossroads in the Information Age.* Lanham, MD: Rowman & Littlefield, 2001.

Wimmer, Roger, and Joseph Dominick. *Mass Media Research: An Introduction.* 7th ed. Belmont, CA: Wadsworth, 2003.

CHAPTER 8

Cueing Up Your Career

One Hundred Suggestions for Breaking into the Profession

The first seven chapters introduced a variety of specific occupations within the electronic media industry. In this last chapter, we provide you with advice for entering each of ten broad areas of our profession. These tips have been prepared by individuals who know these areas well and who, like you, once sought a way into the industry. In reading these suggestions, you will find several pieces of advice that are common to multiple career areas as well as some that are more area-specific. But separately and collectively, the recommendations that follow are intended to provide a candid view of the challenges you will face in pursuing the wealth of opportunities that the electronic media offer.

We present these career areas in a progression that roughly approximates the order in which we previously introduced the key occupations of which each is comprised:

1. On-Air Talent
2. Electronic Journalism
3. Advertising
4. Online Media
5. Engineering
6. Production
7. Corporate Media
8. Sales
9. Programming
10. Management

Start by reading the tips in the area most of interest to you. Reflect on how these tips relate to activities in which you are now engaged or have the opportunity to become engaged. Then explore the suggestions from the other career areas in order to pick up common themes and interrelationships. Finally, construct an *action plan* for yourself based on the insights that these ten highly regarded professionals have provided for your consideration and guidance.

Use this opportunity to help fashion a path to realize your own professional vision—because no vision can be achieved without careful self-assessment and development. Gene Jankowski, former president of the CBS Broadcast Group, put it this way:

> The story is told about the day Robin Hood was dying. He was ninety-nine years old, on his death bed, when he called Marian to open the window. She opened the window. He said, "Bring me my bow and arrow." She brought the bow and arrow. Robin said, "I'll shoot the arrow in the air. Where it lands, I know not where, so bury me there!" Which goes to explain why Robin Hood is buried in his closet!
>
> The moral of the story is this: if you are going to have a vision, be sure you have the ability to carry it out.[1]

The following recommendations are intended to help you isolate your key abilities and uncover methods by which they can be honed to professional standards.

Ten Career Tips for On-Air Talent

Jennifer Cotter
Senior Vice President of Development, Oxygen Media

1. *It's not easy!* Just because you are funny, spontaneous, attractive or know a lot about a particular subject *does not* mean you'll be able to convey that subject when the light on the camera or mic switch goes on. Be prepared. You will have to work at it to be yourself on air.
2. *Be yourself.* Everybody wants to be the next Oprah! Unfortunately, you can't be. You have to reflect the world through *your* lens if you are ever going to make a mark. Mimicking your idol won't make you a star.
3. *Know your strengths.* Whether it's acting, interviews, pop culture, or news, focus on and continue to hone what you are good at while continuing to strengthen your weak areas.
4. *Learn the mechanics of the job.* Don't make the mistake of thinking being "on air" is just being personable. This is a *job*—and like all jobs there are some skills you need to develop. A host who can't hit the mark, throw to commercial, or speak at the proper pace is not going to last very long. Practice the skills needed in the field in which you want to succeed.
5. *Don't just read.* The key to being a good on-air personality is *connecting* with your audience. While it's great to be able to memorize or read from a teleprompter, the best on-air personalities are the ones that are passionate about what they do *and* what they are talking about.
6. *Develop calling cards.* Even when you are just starting out, have a headshot, resume, and a tape if at all possible. People need something to remember you by. These calling cards don't have to be perfect and don't have to cost a lot to prepare.

7. *Perfect a few on-air "personalities."* If you are going for a news anchor position, you want to be a "news" person in your audition. But there are not too many other places on TV that you can get away with that type of performance. Imagine if Katie Couric spoke like a local news anchor. You need to adapt your style to the job. Work on a few different versions of how you will perform on air, so when you are asked to audition you can match the personality to the job.
8. *Network and follow up.* It's a bad business to get started in, but contrary to popular belief, there are a lot of nice people who don't mind helping someone who is trying to get started. Get out there and meet people in your field (and keep in touch with the people you meet). Casting agents, managers, and executives meet many people in a day. You want them to remember you. But let's be clear. Being obnoxious *never* works. Sending crazy gifts, a hundred headshots, or calling once a week is not showing respect for the person's time. Keep in touch, but be careful not to overdo it!
9. *Keep working at it!* Even when you get a performance job and you feel like you've hit your stride, you can *always* get better! Look at the people who are the best at what they do. They are always maturing, learning, and staying passionate about being on the air.
10. *Don't give up!* This process may take only one audition or it may take three *years* of auditioning to land one job. *If you love it and want to do it, don't give up!*

Ten Career Tips for Electronic Journalism

Lauren Stanton
News Anchor, WZZM 13, Grand Rapids, Michigan

1. Be prepared to make squat in the beginning. Even though broadcast journalism on the surface seems like a glamorous profession, with the likes of Katie Couric pulling in millions, no one-man-band green reporter should expect to make a whole lot of green. Typically, you'll earn less than $20,000 a year. Personally, I made $5.50 an hour as an on-air reporter in my first gig in market 106 in Lansing, Michigan. The salary jumps can be quick, though. Your next market jump could easily double your wages, or more.
2. Be prepared to know all jobs in the newsroom. If you think you'd someday like to become a reporter or anchor, you should also think about being a producer, a videographer, and an editor. When you start off in smaller markets, you'll be expected to wear many hats. You'll likely have to be your own photographer and edit your own stories. In my first job, I was a "one-woman band," often setting up my camera on a tripod to shoot a stand-up, hoping it'd be in focus when I got back to edit at the station. I also filled in as the "weather specialist," was taught to run a live truck, and came in at midnight to produce a morning show I anchored six hours later.

3. Be prepared to work crummy hours and holidays. This can change as you climb your way up, but expect to work a few holidays regardless. When you're a swinging single, that means the 11:00 news on New Year's Eve, and when you have little ones, you may have to work on Christmas. But coworkers usually work together on this one to come up with compromises. Also, hours can vary. From working weekends to early hours to late night, news is a twenty-four-hour business, and you can expect to work every shift at one point or another in your career.
4. Be prepared to roll with the punches when stories change, deadlines get moved up, or when all hell breaks loose. I can't tell you how many times I worked all day on a story, and a half hour before the show, there's breaking news and my story had to be dropped—so I could run out and cover the breaking story. It's frustrating, but you have to roll with the punches. Especially in smaller markets, shows can go to hell in a handbag. Tapes may not roll, scripts may not print, the Teleprompter will go down, and you have to be able to think fast and still communicate effectively to your crew and especially to your viewers. It's high stress, but you have to be able to remain calm.
5. Be prepared to be all-knowing, or at least know a little about a lot. You may be assigned to a specific beat as a reporter, but you'll inevitably be thrown out on a story you know nothing about. It's good to read the paper, read the wires, watch other news channels, and be as up-to-date on major news stories as possible. Also know how to use the Internet to research stories. But remember, the Internet can also get facts wrong. So check your facts.
6. Be prepared to be in a very competitive, sometimes cutthroat business. It's a given that you'll be competing against other stations in your market, going after stories and interviews, all in an effort to win in the ratings or nudge out another reporter for a major exclusive. Whether it's for personal satisfaction, an award, or for a bigger career move, being competitive is the name of the game in broadcast journalism. But be aware, even though working as a team is key in TV news, it's just as competitive amongst coworkers; anchor against anchor and reporter against reporter. It's up to you how you handle it—and hopefully, thrive on it.
7. Be prepared to be picked apart hair by hair. You're under a microscope. People will e-mail and call to say they don't like your clothes, your hair, your makeup, your voice, your laugh—you get the picture. You have to have thick skin in this business or it'll bring you down. It gets easier the longer you're in the business. But beware, every aspect of your life is up for scrutiny.
8. Be prepared to be in a good mood even when you're *really* not. As a TV news reporter or anchor, people will recognize you. Many will feel like they know you. So when you're at the local Wal-Mart in your sweatpants arguing with your spouse, be prepared for someone to say, "Aren't you that lady on TV?" In all seriousness, you represent the station you work for, and therefore, you must always be on your best behavior, even if you think no one will recognize you.
9. Know that in the beginning you'll always be wishing you were in a bigger market or had a better job. But once you get there, you look back at those early days as some of the best times. This is cliché, but the grass is always greener on the other side. Once you get to the other side or that bigger market, you'll realize the positive aspects of starting in a small market. Know you'll have some great times, you'll learn *a lot*, and you'll make your biggest mistakes with the fewest people watching!
10. Be prepared for one of the most exciting careers imaginable. With all of these reality checks I've given, you may be rethinking a career in broadcast journalism. But know that there's no

business like it. Every day is different. You won't sit at a cubical all day. You are "in the know" on every big story in town. You get to interview the "who's who" in your particular city, state, and someday even national news makers and celebrities. The environment is fast paced and sometimes stressful, but it's also fun and energetic, where you're surrounded by sociable and creative people. You get to meet amazing people in your community and feel good about telling compelling stories that touch your audience. There are days that seem routine. But then there are the days that you can make a difference by doing reports that can change laws, uncover wrongdoings, expose criminals, and even save lives.

Ten Career Tips for Advertising

Karl Bastian
President, Greenlight Marketing

1. Learn as much of the industry as possible. There are many career opportunities within advertising: broadcast producer, copywriter, art director, account executive. The more you know about each and how they work together to create commercials, the more versatile, and therefore valuable, you will be as an employee.
2. Get an internship. Class work is only part of the education equation. You need practical, real-world experience. A semester or two working at a TV or radio station or advertising agency will give you valuable experience you simply can't get from the classroom.
3. Watch TV. Listen to the radio. Sometimes the best research comes from sitting in a La-Z-Boy with a remote control in your hand. As you watch or listen to commercials, analyze the video and audio techniques. How does a commercial grab your attention? What does it tell you about the product it's selling? Watching and listening can teach you a great deal about what and what not to do in advertising.
4. Prepare a portfolio. If you are going to work as an art director or copywriter on the creative side of the industry, you will need a portfolio. This will include samples of print ads, radio spots, and TV storyboards you developed either in class or independently. Whether or not the work is produced is not important. The quality of the work and how it is presented, however, is critical. A professionally prepared portfolio full of thoughtful, compelling work is your biggest key to success as an advertising creative.
5. Be curious. The most successful people in advertising never settle for easy answers. They listen to experts. They read industry publications. They explore and embrace new developments in the industry. They are passionate about advertising. A successful career in advertising requires constant learning.

6. Be a brand. In advertising, it's all about selling brands. Especially when it comes to getting a job. Through appearance and presentation, you need to position yourself as an employee no reasonable company could do without. This means looking and acting professional and being prepared to meet the demands of the job.
7. Be prepared to start at the bottom. No one in advertising ever started out with a corner office. Any position that offers you the opportunity to learn and advance is worth exploring.
8. Stay up on industry developments. Like many professions, advertising is being greatly impacted by the Internet and interactive technologies. The more you know about these, the better positioned you will be for the future.
9. Get involved. Getting ahead means getting involved, even while you're still in school. Attend workshops. Join an ad club. Volunteer to help charity organizations with their advertising needs. You'll meet people and learn things you never could on your own.
10. Stay positive. Advertising is a very competitive field. Good jobs can be hard to find. Stay prepared, positive, and persistent.

Ten Career Tips for Online Media

David Antil
Director, Postmerger Integration, ETAS, Inc.

1. Maintain a broad perspective. The online media industry is new, large, and growing. It is important to know the short history of the online sector you are pursuing and keep your eyes on its future.
2. Understand that there are no traditional ways to get into this business. Unlike most other media industries, this one is only about a decade old. People get here many different ways. The industry has drawn people, methods, and business models from nearly every other media arena. Your background in any of these can be applied somewhere in online media. Find out where by learning the business and who does what. Locate someone who has a job you think may be fun and find out how they got there. You may eventually get there, too. But be assured, it will not be via a straight path.
3. Toss out the five-year career plan. This business will change in a way that will be unrecognizable to you in five years. Instead, plan on learning the various business models of the companies on your radar screen. Be open to seemingly odd career paths. In my own fifteen-year career, I can point to four complete industry and career changes, from network TV program development to CD-ROM production to Internet/e-business consulting to postmerger integration and strategic marketing. I'm sure there will be more.

4. If you are a creative—learn the technology. If you are a technologist—learn to appreciate the creative process. You don't have to be a jack-of-all-trades, but you have to know what the other guy or gal sitting next to you is doing because you have to work together.
5. Learn how to translate. I can't tell you how many times I've had to literally stand between a software engineer/programmer (who craves logic and details) and a creative person (who lives in abstractions and shades of gray) and explain to them what the other one wants or is thinking. It takes some time to learn the different languages and appreciate the nuances of each situation. But if you can do this you will be indispensable.
6. Know how your company (or how the company you want to work for) makes money. Understand the "business." Cool technology and beautiful pictures alone don't cut it anymore. Learn how your company can benefit through the creative application of technology.
7. Be pleasant and fun to work with. Long ago, someone gave me that great piece of advice and I believe it applies to every lawful career. The working world is hard enough without the jerks. Who wants to work with them? Be competent, be nice, and people will want to work with and for you. If you are both competent and a pleasure to be around, you will have a big advantage over the competent jerks.
8. Be an expert in something. Specialize. Be the person who knows more than anyone else about something that is—or will be—important to your company. People will seek you out. In an industry that is so new and changing daily this is not as hard as it sounds.
9. Be flexible and adapt to changes. If you don't love change, then the online world is not the place for you.
10. Think globally. The world is getting smaller. Online media have an international audience, whether intended or not. It is likely the company you will work for will be part of a larger global conglomerate or will have a distinct international audience segment. Global experience and sensitivities will be a requisite for success in most careers—especially one with as far-reaching an audience as the online industry.

Ten Career Tips for Engineering

Jeff Kimble
*Principal Sales Engineer–North America,
NEWSkies Satellites, Inc.*

1. Take as many business classes such as accounting, human relations, and sales as you can. To further your career you'll eventually enter management and be faced with balancing budgets, human resources issues, etc. The immediate benefit is that you will have an idea of how management and the business side works and what is expected of you from the beginning.

2. Be prepared for working as an apprentice when you first start out in the field. The length of time you'll spend as an apprentice varies based on your previous experience and your ability to show that you can handle the assignments given to you and do the work that is expected. Also be prepared to be on call 24/7/365 and work all the weird and graveyard shifts—including holidays—for the first part of your career. Take that time to learn as much about your craft as possible, including how certain products/services work. Also accept as many hands-on projects as you can get to develop your knowledge base.
3. Work on your interpersonal skills and positive interactions with other people. Engineers were once considered strange and in their own little world (some still are). Now, however, they are a vital part of a media team along with the operations, accounting, and sales departments. Everyone needs to help generate more revenue from existing company assets through a better program design—or more productivity through automation and a safer working environment. Nevertheless, engineers still must speak a totally different language and maintain a different mindset than the rest of the world. You will be the human interface to that world, and you will find yourself explaining something very technical to customers, to coworkers, and/or to people in management. Remember that you know this stuff—they don't—so explain kindly and professionally.
4. Practice time- and stress-management skills: how to prioritize projects in real time and under pressure situations. Also learn how to relieve stress when situations get too crazy. Above all, do not bring your work stress home with you. In the engineering world the biggest stresses are time and money for any project you work on. You will never have enough of one and/or the other. This is especially true of live remotes. Problems and requests do not come at you one at a time; they always come in packs like wolves. You must be able to think on the fly to provide the best solutions as they are presented to you and communicate your strategy/solutions back as professionally as possible. It is an unwritten rule that engineers will help fellow engineers regardless of who they work for (competitors or not). Because the next time it may be you in desperate need of help. Besides, you may just be helping your future boss (who may be impressed with how you handled the situation and later hires you).
5. Take courses in industrial safety and how to work safely and correctly in the workplace. Have a very *healthy respect* for electricity and RF energy. Learn everything you can about the equipment you'll be working with. What you don't know, or even assume, will kill you! When working around this type of high-energy environment, *always* have a spotter standing in the same area. The spotter's only job is to watch you work and be prepared to get you out of danger should something go wrong. This is true no matter how much experience you may have with the equipment. All it takes is one mistake—familiarity breeds contempt—always be on guard. All clichés. But in this case, all very true.
6. The engineering field can be very big or very small, depending on the areas of expertise you choose. In either case, your reputation will be built on your work ethic, on the way you interact with others in the media community, and on the mistakes you make along the way. We all make mistakes. It's how we deal with them—or not—that makes the difference. Your reputation, and your attitude, can mean the difference between quick advancement and more responsibilities and not being able to find a job. Work hard at all times on your reputation and your professional attitude. Protect them no matter what. Don't let anybody convince you to do or say anything that will harm either one of these qualities even remotely. Your reputation and attitude—good or bad—are the only things you will carry with you throughout your career.

Cueing Up Your Career

7. The media engineering field and the applications used in it change rapidly—especially in the digital world. Get the best electronics education you can and be prepared to update your knowledge consistently and constantly for the rest of your career. Read as many periodicals as you can in your chosen field. Also look into joining the various professional organizations such as Society of Broadcast Engineers (SBE) and Society of Motion Picture and Television Engineers (SMPTE). They will keep you updated. Joining these types of formal organizations and meeting their strict standards also tells your current and prospective employers that you are fully qualified and take your engineering career seriously.
8. Any project you work on will have an effect on other people: producers, directors, account executives, regional operations managers, etc. Interact early and frequently with your project team and anyone else affected by your project. Update them *as things change*. They in turn will provide valuable feedback about their expectations/interaction with your project. With that new information, you will be in a position to catch any problems and errors/omissions more quickly. This will make changes easier and less expensive than waiting until the end of the project to solicit comments.
9. You will need to know about logistics detail and backup planning. As a broadcast engineer, for example, it will be your responsibility to technically plan out events such as live remotes and also provide the backup plan should some natural/man-made disaster happen during the event. When working on "live" productions, *don't ever* take the "little things" for granted, and *don't ever assume anything*. Do you have enough satellite/transmission time at the beginning and end of the event? In case of failure or disaster, what is the minimum amount of equipment you need to still be on the air? Always ask questions and test everything until it meets your satisfaction. The most common avoidable problem that happens during a live remote is loss of power to equipment. Pay attention to primary electrical current draw. Make sure that you have enough electricity on your power circuits to not blow circuit breakers. I've been there—done that—and it's not fun.

    As time and money permit, try to schedule setup and dress rehearsal the day before. *Use all of the equipment as you intend to use it on* the live broadcast: cameras operating, "recorders recording," satellite truck/equipment transmitting, etc. Give everything a chance to warm up and thermally settle in. Then measure current draw on all primary-feed AC circuits—ideally at the same time of day as your live remote. Make sure you have at least 25% more power than actually needed on each circuit. If you cannot schedule a dress rehearsal the day before, schedule the setup as early as possible, and start the dress rehearsal power-up as soon as you can.
10. You may not get rich as an engineer. However, the demand for engineers is constant and the pay is good, depending on experience. In the beginning of your career, you will not make the big bucks. Therefore, make a dynamic life/career plan—what do you want to do ten, twenty, thirty, and forty years from now? Revisit the plan each time your life and/or career changes and update accordingly. This will help you avoid making bad career choices such as taking a job for more money with few or no career advancement possibilities vs. one with a little lower pay in the immediate term but with great promotion potential.

Ten Career Tips for Production

Michael Franks
*Director of Photography, Michael Franks Enterprises,
Los Angeles*

1. *Have a goal.* Make a goal for yourself, both short and long term. This goal is based on your dream or vision for yourself in production. Don't be afraid to aim high. Then, once you've achieved it, make another.
2. *Be patient.* Once you have a goal, don't be discouraged that you haven't achieved it as quickly as you first thought. Some stages of your success will take time, particularly in a team-oriented endeavor like production.
3. *Be persistent and committed.* Along with patience comes being persistent. No matter how discouraging the business can get, if you really want to succeed you must be *unrelenting*. (Sometimes this is more important than having a great technical talent.)
4. *Enjoy your work.* Create an environment around you that has others on the set or in the studio enjoy working with and for you.
5. *Make no enemies, burn no bridges.* The production industry is a very small world. Create lasting friendships and make choices that don't ruin relationships. Projects come and go, but the friendships you make can last a lifetime.
6. *Be open to learning.* No matter where you are in your career, there is always more to learn: new technologies, ideas, ways of working. Be open to what's new. A different approach will keep your work and workday from getting stale.
7. *Seek the advice and support from those you most admire.* Production is not a solo business. You would be surprised at how willing someone you hold in great esteem is to help you with your struggle.
8. *Do your work and be prepared.* You will be more flexible, more creative, and spontaneous on the set if you've done your work and prepared yourself. Sometimes, for one reason or another, what you've planned gets thrown out. If you have done your preparation, you have something to depart from and can find a new way to your destination.
9. *Always do your best work.* Put 100% of yourself into what you do. No production job is too small to do well, and a job well done will most likely lead to another.
10. *Be confident and believe in yourself and your vision.* Although this may sound like a cliché, confidence is the key to accomplishing. No one can give this to you, it comes from within. If you feel you can't accomplish a project—you usually will be right. The opposite is also true.

Ten Career Tips for Corporate Media

Scott Wallace
Production Manager, Summit Training Source, Inc.

1. Develop a wide knowledge base and skill set. Due to downsizing, cost cutting, competition, and lower profit margins, corporate media departments no longer consist of a larger number of specialists, but rather a few multitalented individuals.
2. Select a good media employer. Select companies to work for where what you do/produce is a major contributing factor to the success of the firm. This will help preserve your employment if downsizing occurs.
3. Stay out of corporate politics. Don't play "the game" at the expense of other people. Be yourself and do your job well.
4. Stay current. Keep yourself up-to-date on all new technology and developments in your field.
5. Perform all work and make all decisions as if you owned the company you are working for. If you would not want an employee of yours to do it, then your company probably would not want you to do it, either.
6. Track your results. Always be ready and able to communicate your worth to the company. This may include cost savings, number of products produced, comments from customers, sales figures, etc.
7. Treat everyone the same. Always treat everyone with respect and in a professional manner. Do not treat people based solely on their position within the company. You never know where your friends and enemies may be hiding.
8. Experience is very valuable. Learn something from every project and person you work with.
9. Be flexible. Be willing to take on new challenges and explore new opportunities that could contribute to the success of the company for which you work.
10. Always give it your best. Do the best job you possibly can on every project you undertake. In a perfect world, your last project would always be your best project.

Ten Career Tips for Sales

Tim Hygh
Director of Regional Advertising, WJR/WPLT/WDRQ, Detroit, Michigan

1. *Extracurricular activities.* Participate in everything you can at school stations, including on-air, traffic, promotions, and management. For example, news gathering skills are similar to sales skills. Take as many news shifts as possible to learn how to investigate, write, create, and produce—all within tight deadlines.
2. *Social.* Get involved in social organizations. Be a people watcher, help produce events. This is your opportunity to discover how people outside the electronic media business work. It's also a chance for you to sharpen organizational and event management skills.
3. *Service.* Donate your time to a service organization to cultivate a "service" mentality. Salespeople are service providers. They provide solutions to problems that are not always advertising related. This is also an effective way to meet people and network.
4. *Self-sufficiency.* Start your own business. You need to learn the ins and outs of being an entrepreneur. This also provides additional income and a cushion to use between jobs and during slow months.
5. *Speak the language.* You have to be able to create a personal bond with your clients. You have to know more than just sales jargon. In short, you have to be able to speak the vernacular of the *Wall Street Journal*, *Advertising Age*, *People Magazine,* and *Sports Illustrated*. The best sales position interview in which I participated didn't center on spots and dots. The interviewer wanted to know if I knew the prime interest rate, who my local government representatives were, and who won last year's Super Bowl. He wanted to know if I knew the things that were important to clients.
6. *Other education skills.* Take as many business courses as possible, including accounting. Media sales revolve around budgets. You have to know the basics. Don't take your general education classes for granted. It was nice to remember some basic biology while working out a long-term advising plan with the head of Pfizer in Michigan. He didn't care about cost per thousand. You must also be able to use your computer for producing presentations, letters, and e-mails. In addition, speech and theater classes are effective skill builders that you will use the rest of your life.
7. *Expectation levels.* Put expectation levels in perspective. Sales will be harder than you think. You will want to quit every first quarter of the year when business is slow. You cannot quit, because longer tenures equate to escalating success for your clients and for yourself.
8. *Things you can't learn in school.* Learn some "classic" sales techniques from nonmedia sales gurus like Tom Hopkins and Zig Ziegler. Yes, they can be corny, but they are very helpful in understanding the sales process. These are available in cassette tape form at your local library and are well worth your time.

9. *Surround yourself with other successful people.* Stay in touch with your classmates. Network with like-minded professionals. Find people who can be trusted to share ideas, ideals, and goals.
10. *Get a life.* Don't get too wrapped up in your electronic media sales career. Be passionate and be extraordinary, but don't forget to find a life outside of the sales arena. It will keep you fresh and lessen the anxiety that is inherent in this profession.

Ten Career Tips for Programming

Mike Donovan
Director of Marketing and Educational Services,
National Association of Television Program Executives

1. There are many doorways to the programming field (which is, by the way, the reason you shouldn't make the objective on your first résumé too specific). Don't just think local station. In fact, other than some news and public affairs, many local TV stations don't produce any of their own programming. So if news and public affairs are not what you're interested in, a TV station may not be the place to go.
2. Something I would consider if I were starting out today would be to find a way to get to Los Angeles or New York City and apply for and accept any job I could get at one of the major studios (Warner Bros., Paramount, Fox, etc.) or TV networks (ABC, CBS, NBC, etc.). No matter what direction you hope to go in the industry, these companies are directly or indirectly involved in those directions, and you're putting yourself in the environment filled with people from whom you can learn a lot and with whom you can share ideas.
3. Another great place to start is in research with a station rep firm. A year at a rep firm is like an advanced degree in TV programming. (If you don't know what a station rep firm is, review chapter 4.)
4. In an internship or job interview, one of the ways to stand out from the crowd is by sounding smart. Learn the vocabulary and concepts of the industry. For example, be sure to know the differences between broadcast TV and cable TV; between a local broadcast station and a broadcast network; a local cable system and a cable network; and the distinctions between a network program, a syndicated program, and a local program.
5. As soon as possible, start to act and think like an industry professional. Professionals subscribe to industry trade publication like *Broadcasting & Cable* and *TV Weekly*. Professionals are aware of trends in the industry such as the current practice of running a program on a cable network a week after running it on a broadcast network. Professionals are aware of regulatory changes that could affect the way business is done in the programming industry. Professionals stay "connected" through memberships in industry organizations and associations.
6. Often professional associations offer very useful publications. For example, the National Association of Television Program Executives (NATPE) publishes the *Media Content*

Directory and the *Guide to North American Media*. Both are full of potential job-search information. Professional associations also often maintain informative websites and sometimes offer student memberships. Check them out.

7. This may seem like an obvious suggestion, but watch a lot of television. Be familiar with as many types of TV programming as possible. Any discussion of new or future programming is almost always based on previous programming ("It's like *Friends* but with dogs."). The more you know about what's on and what's been on, the less likely you'll have that vague look on your face when someone references a program. If you're interested in educational programming for children, there are FCC rules and regulations regarding a TV station's use of this kind of programming. Look them up and be aware of them.
8. When you watch, pay attention to the credits at the end of the programs, especially the final credits (produced by Air Pirates Productions in association with Paramount). These are the companies that produce programming. Maybe they're the companies you want to work for.
9. Networking with as many industry professionals as possible (through attendance at conferences, seminars, informational interviews) is one of the best moves you can make. But if you do get the chance to meet and talk with some industry people and they give you their business card and tell you to stay in touch, the next time they hear from you should *not* be you asking for a job. In fact, try to avoid ever directly asking someone for a job and putting them in the uncomfortable position of having to say no. Instead, follow up your meeting with a "thanks for taking the time to talk to me and would you be willing to look over and critique my résumé when it's ready" note. If they are willing to look over your résumé, make sure to send it to them. And when you do, ask them if they have any suggestions as to where or to whom you should send it.
10. Finally, here's the most important piece of information to remember. The television programming industry is a business. The purpose of a business is to make money. Therefore, the purpose of the television programming industry is to make money. It's great if a program informs, educates, and/or entertains as long as it makes money. Like it or not, the bottom line is: the bottom line.

Ten Career Tips for Management

Bill Parris
General Manager, Multicultural Media Broadcasting, Inc.

1. *Know your profession.* The strongest credential for media management is superior knowledge of the media to be managed. Study the concept behind each industry trend and master the management mechanics of your company. Read industry trades. Media knowledge is a continuing education.
2. *Think conceptually.* Leaders know *why* a phenomenon occurred, followers often only know *how*. Learn what stimulates your personal creative thinking and treat yourself to a repetitive and regular regimen of that stimulus.

3. *Love the language.* Master the liberal art of the spoken word. Communications efficiency, including precise phrasing and word use, is a powerful professional asset.
4. *Be a model employee.* Always display a positive attitude and superior work ethic. Most negative personnel actions in media industries are caused by basic personnel problems present in any workplace. Conversely, management promotions are often based and built on a foundation of trust, dependability, and predictability.
5. *Manage emotion.* The self-discipline to manage instant impulse is often an advantage in the job environment. Those in "performance and creative" positions must develop a thick skin and learn to minimize the emotional damage often inflicted by criticism of their creative product.
6. *Resist office politics.* Focus your energy on the job. Time invested in trying to decipher interpersonal influences and motives is better invested in enhancing your performance. Concentrate on that performance and avoid politics. A negative comment shared with a coworker empowers that political player to quote you at any time, to their advantage. Never give others that power.
7. *Pursue perfection.* Consistently high standards attract positive attention, enhancing professional reputation. Develop the self-discipline to produce the best possible product.
8. *Study leadership.* Energy and leadership fuel promotions and propel upward mobility. Be perceived as a center of positive, creative energy. Management positions can be assigned, but respect and leadership qualities must be earned. At the same time, leadership *techniques* can be learned.
9. *Have high moral standards.* Trust is the foundation upon which ownership selects management. A reputation for integrity is an informal nomination for promotion.
10. *Love the art in yourself, not yourself in the art.* Creativity is a gift to be cherished, encouraged, and protected, while egos are often nourished by fleeting flattery. Never assume glory will last. Life is subject to change without notice. Always appreciate the employment opportunity that allows you to practice your creativity while leading others.

A Final Flashback

The authors of these career suggestions, as well as those who wrote the occupational profiles found in chapters 1–7, all exhibit an unquestionable enthusiasm—indeed, a love—for the electronic media profession. It is our hope that, through the material presented in this book, you have begun to cultivate that same enthusiasm and love. For it is these deep-seated feelings that will enable you to prosper in your career. They will fuel the drive necessary to skillfully and honestly serve the publics who expect so much both of you and of the media tools with which you will be entrusted.

Despite its obvious opportunities, this is not an easy field in which to succeed. Much depends on your own good judgment and perseverance. As respected media educator Louis Day observes:

> With today's electronic information explosion, the diversity of job opportunities in the future should be impressive. However, in the short term my enthusiasm is tempered by the knowledge that my students are joining an industry in which professionalism is often elusive and the financial rewards are still woefully inadequate for most entry-level degree holders. This is the most frustrating part of my job. Students come to me full of idealism and energy. The quandary confronting me, as a media teacher, is whether to quell this enthusiasm with brutal candor early in the student's program or to temper my criticisms and to allow students to make their own judgments once they have joined the ranks of media practitioners.[2]

Hopefully, this chapter and this book as a whole have nurtured your passion for a media career while also leavening that passion with a realistic perspective on our field and its challenges.

Legendary advertising creator and agency founder George Lois provides a frank assessment of a media professional's most important personal attributes with this summary counsel:

> My advice to anyone who goes into the communicating business is this: Be true to your talent, develop it, push it to its limits. Believe in yourself and the work you do. If you're talented, you can make it. Every job is an icon, the most important job in the world. Put in the hours; you have to be competitive to do great work, and energy begets energy. . . . Produce the best you can and remember that this drive has to continue every second of your life. Make manly or womanly decisions, not cowardly ones.[3]

All good wishes for success in your own decision making now and in the future. Thanks for reading!

Review Probes

1. What are some key differences between public perceptions about electronic media careers and the reality of these jobs?
2. What does the story of Robin Hood's death teach you about your own specific career aspirations?
3. What is the difference between professional *vision* and professional *passion*?
4. What does Louis Day believe are the greatest drawbacks to an electronic media career, and what are some specific examples of each of these drawbacks?
5. What are the two or three tips that appear most often in the chapter's "ten tips" lists?
6. List five ways in which your perception of the electronic media profession has changed as a result of reading this text.

Suggested Background Explorations

Berger, Arthur. *Media and Society: A Critical Perspective.* Lanham, MD: Rowman & Littlefield, 2003.
Bivins, Thomas. *Mixed Media: Moral Distinctions in Advertising, Public Relations, and Journalism.* Mahwah, NJ: Lawrence Erlbaum, 2003.
Block, Bruce. *The Visual Story: Seeing the Structure of Film, TV, and New Media.* Woburn, MA: Focal Press, 2001.
Bognar, Desi. *International Dictionary of Broadcasting and Film.* 2nd ed. Woburn, MA: Focal Press, 1999.
Corwin, Norman, and Douglas Bell. *Years of the Electric Ear: Norman Corwin.* Lanham, MD: Scarecrow Press, 1994.
Friedman, Barbara. *Untangling the Web: How to Find Anyone or Anything on Deadline.* Mahwah, NJ: Lawrence Erlbaum, 2004.
Gunter, Barrie. *News and the Net.* Mahwah, NJ: Lawrence Erlbaum, 2003.
Hart, Colin. *Television Program Making.* Woburn, MA: Focal Press, 1999.
Kubey, Robert. *Creating Television: Conversations with the People behind 50 Years of American TV.* Mahwah, NJ: Lawrence Erlbaum, 2003.
Lois, George. *What's the Big Idea?* New York: Penguin, 1991.
Macdonald, Myra. *Exploring Media Discourse.* New York: Oxford University Press, 2003.
Mindich, David. *Tuned Out: Why Young People Don't Follow the News.* New York: Oxford University Press, 2004.
Murray, Michael. *Indelible Images: Women of Local Television.* Ames: Iowa State University Press, 2001.

Orlebar, Jeremy. *The Practical Media Dictionary.* New York: Oxford University Press, 2003.

Orlik, Peter. *Electronic Media Criticism: Applied Perspectives.* 2nd ed. Mahwah, NJ: Lawrence Erlbaum, 2001.

O'Sullivan, Tim, Brian Dutton, and Philip Rayner. *Studying the Media.* 3rd ed. New York: Oxford University Press, 2003.

Potter, W. James. *Media Literacy.* 2nd ed. Thousand Oaks, CA: Sage, 2001.

Randazzo, Sal. *Mythmaking on Madison Avenue.* Chicago: Probus Publishing, 1993.

Silvia, Tony, and Nancy Kaplan. *Student Television in America: Channels of Change.* Ames: Iowa State University Press, 1998.

Sturken, Marita, and Lisa Cartwright. *Practices of Looking: An Introduction to Visual Culture.* New York: Oxford University Press, 2001.

Wasko, Janet. *How Hollywood Works.* Thousand Oaks, CA: Sage, 2003.

Appendixes

Key Electronic Media Professional Associations

Academy of Television Arts and Sciences (ATAS). Supports the Los Angeles television production community and provides student internships and faculty fellowships. 5220 Lankersham Boulevard, North Hollywood, CA 91601.

American Advertising Federation (AAF). Promotes the study of advertising through national college competitions and college chapters. 1101 Vermont Avenue N.W., Suite 500, Washington, DC 20005.

American Association of Advertising Agencies (AAAA). Trade and lobbying organization for agencies and their top executives. 405 Lexington Avenue, New York, NY 10174.

American Cinema Editors, Inc. (ACE). Honorary professional organization founded in 1950 to give recognition to outstanding film editors and to advance their art. 100 Universal City Plaza, Bldg. 2282, Rm. 234, Universal City, CA 91608.

American Meteorological Society (AMS). Promotes the meteorological profession, including on-air practitioners. 45 Beacon Street, Boston, MA 02108.

American Society of TV Cameramen, Inc. Professional organization for those making their living shooting video for television. 2520 Lotus Hill Drive, Las Vegas, NV 89134.

American Sportscasters Association. Set up to further the interests of talent who cover sporting events for radio and television. 225 Broadway, Suite 2030, New York, NY 10007.

American Women in Radio and Television, Inc. (AWRT). Professional society for persons (of both sexes) throughout the electronic media; offers special services, seminars, and scholarships for students. 8405 Greensboro Drive, Suite 800, McLean, VA 22102.

Association for Education in Journalism and Mass Communication (AEJMC). Academic organization of journalism, advertising, and public relations faculty. 234 Outlet Point Boulevard, Suite A, Columbia, SC 29210.

Association of America's Public Television Stations (AAPTS). Spokes-organization for noncommercial television outlets and their executives. 666 11th Street N.W., Washington, DC 20001.

Association of National Advertisers, Inc. (ANA). Trade group for the large companies that buy the bulk of national advertising time and space. 708 Third Avenue, New York, NY 10017.

Audio Engineering Society, Inc. (AES). Main group for audio-recording technologists and developers. 60 E. 42nd Street, Rm. 2520, New York, NY 10165.

Broadcast Cable Financial Management Association (BCFM). Organization for electronic media business officers and comptrollers; aids in improving and standardizing financial procedures and reporting mechanisms. 932 Lee Street, Suite 204, Des Plaines, IL 60016.

Broadcast Education Association (BEA). Principal academic organization for college and university electronic media faculty; offers several annual scholarships to students. 1771 N Street N.W., Washington, DC 20036.

Broadcasters' Foundation Inc. Group of veteran electronic media professionals who promote a number of industry-benefiting causes, including library services and student scholarships. 7 Lincoln Avenue, Greenwich, CT 06830.

Cable & Telecommunications Association (CTAM). Trade association enhancing the management and promotion of the cable medium. 201 N. Union Street, Suite 440, Alexandria, VA 22314.

Cabletelevision Advertising Bureau, Inc. (CAB). Trade group set up to promote the use of its medium by advertisers. 830 Third Avenue, New York, NY 10022.

Center for Communication, Inc. Nonprofit organization designed to further the study and understanding of mass communications through seminars and other educational enterprises. 271 Madison Avenue, Suite 700, New York, NY 10016.

Community Broadcasters Association (CBA). Trade organization for low-power television (LPTV) outlets and executives. Box 4300, Hopkinsville, KY 42241.

Country Radio Broadcasters Inc. (CRB). Nonprofit organization to serve country-format radio stations. 819 18th Avenue S., Nashville, TN 37203

Direct Marketing Association (DMA). Group that advances the cause and study of direct-to-the-consumer selling communications. Offers student seminars and scholarships. 1120 Avenue of the Americas, New York, NY 10036.

Electronic Media Rating Council (EMRC). Independent organization that monitors the accuracy and research design of audience measurement services. 200 W. 57th Street, Suite 204, New York, NY 10019.

Federal Communications Bar Association (FCBA). Professional organization for communications attorneys, particularly those eligible to represent clients before the FCC. 1020 19th Street N.W., Suite 325, Washington, DC 20036.

Interactive Advertising Bureau (IAB). Trade association to promote online advertising. 200 Park Avenue S., Suite 501, New York, NY 10003.

Intercollegiate Broadcasting System, Inc. (IBS). Cooperative exchange for student-operated broadcast facilities. 367 Windsor Highway, New Windsor, NY 12553.

International Radio and Television Society Foundation, Inc. (IRTS). Electronic media forum and development organization supported by top companies and executives in the industry. Hosts key conferences, faculty and student seminars, and internships. 420 Lexington Avenue, Suite 1601, New York, NY 10170.

Media Communications Association International (MCA/I). Organization of corporate/industrial electronic media professionals. 401 N. Michigan Avenue, Chicago, IL 60611.

Motion Picture Association of America (MPAA). Trade and lobbying group for the major film production studios. 15503 Ventura Boulevard, Encino, CA 91436.

Museum of Broadcast Communications. Nonprofit organization for the preservation and recognition of excellence in radio and television. 78 E. Washington Street, Chicago, IL 60602.

Museum of Television and Radio. Similar objectives to the above, but this older resource is located to serve the East Coast and has a heavier emphasis on network history. 25 W. 52nd Street, New York, NY 10019.

National Academy of Television Arts and Sciences (NATAS). New York–based society to support and recognize creative achievement in television program content and technique. 111 W. 57th Street, Suite 600, New York, NY 10019.

Appendixes

National Academy of Television Journalists (NATJ). Organization to promote professionalism among video reporters. Box 31, Salisbury, MD 21803.

National Association of Broadcasters (NAB). Chief trade and lobbying group for the commercial broadcasting industry. 1771 N Street N.W., Washington, DC 20036.

National Association of College Broadcasters (NACB). Cooperative organization of student-operated media facilities and their staffers designed to improve student outlets and to advance career preparation. 71 George Street, Providence, RI 02912.

National Association of Farm Broadcasters (NAFB). Organization for stations serving rural America. Box 500, Platte City, MO 64079.

National Association of TV Program Executives International (NATPE). Trade organization bringing together program creators, syndicators, and station program executives. 2425 Olympic Boulevard, Suite 550E, Santa Monica, CA 90404.

National Broadcasting Society/Alpha Epsilon Rho (NBS/AERho). Cooperative organization of electronic media professionals and students preparing to enter the industry. Box 4206, Chesterfield, MO 63006.

National Cable and Telecommunications Association, Inc. (NCTA). Chief trade and lobbying association for the cable industry. 1724 Massachusetts Avenue N.W., Washington, DC 20036.

National Religious Broadcasters (NRB). Lobbying and trade association for operators of religious program outlets; administers a code of ethics pertaining to fund-raising and related monetary activities. 9510 Technology Drive, Manassas, VA 20110.

PROMAX & BDA. Key association for radio/television promotion and creative services personnel as well as electronic media designers. 2029 Century Park E., Suite 555, Los Angeles, CA 90067.

Public Relations Society of America (PRSA). Main organization to advance the standards and practice of public relations as a profession; also supports student chapters. 33 Irving Place, 3rd floor, New York, NY 10003.

Radio Advertising Bureau (RAB). Trade group designed to promote radio as an advertising vehicle; hosts frequent seminars to improve time-selling and copy techniques. 261 Madison Avenue, New York, NY 10016.

Radio-Television News Directors Association (RTNDA). Main organization and sounding board for electronic journalism executives; also maintains student chapters. 1600 K Street N.W., Suite 700, Washington, DC 20006.

Recording Industry Association of America, Inc. (RIAA). Chief trade group for the producers and distributors of music and other recorded product for the consumer market. 1330 Connecticut Avenue N.W., Washington, DC 20036.

Satellite Broadcasting and Communications Association of America (SBCA). Main trade and lobbying organization for the home satellite dish industry. 225 Reinekers Lane, Suite 600, Alexandria, VA 22314.

Society of Broadcast Engineers, Inc. (SBE). Organization representing the interests and certification of technical staff working in the industry. 9247 N. Meridian Street, Suite 305, Indianapolis, IN 46260.

Society of Cable Telecommunications Engineers, Inc. (SCTE). Professional association offering information, developmental resources, and standards to cable engineers and associated professionals. 140 Philips Road, Exton, PA 19341.

Society of Motion Picture and Television Engineers (SMPTE). Influential forum for leading film and television equipment designers and developers; aids in refinement and specification setting for new devices and procedures. 595 W. Hartsdale Avenue, White Plains, NY 10607.

Society of Professional Journalists/Sigma Delta Chi (SPJ/SDX). Organization for the advancement and protection of the practice of journalism; facilitates involvement by students as well as working professionals. 3909 N. Meridian Street, Indianapolis, IN 46208.

Station Representatives Association (SRA). Trade organization of rep firms and their executives. 16 W. 77th Street, Suite 9E, New York, NY 10024.

Syndicated Network Television Association (SNTA). Main trade association for sellers of advertiser-supported programming. 630 Fifth Avenue, Suite 2320, New York, NY 10111.

Television Bureau of Advertising (TVB). Industry office for the promotion and expansion of broadcast television time sales and monitoring of advertising trends. 3 E. 54th Street, 10th floor, New York, NY 10022.

Television Critics Association (TCA). Alliance of writers who regularly prepare television critiques for media outlets. c/o *Wichita Eagle,* 825 E. Douglas Avenue, Wichita, KS 67202

Wireless Communications Association International, Inc. (WCA). Trade group for MMDS and related television delivery systems. 1140 Connecticut Avenue N.W., Suite 810, Washington, DC 20036.

Women in Cable & Telecommunications Communications, Inc. (WCTC). Professional society designed to enhance the industry and career advancement for women in it; sponsors student seminars and scholarships. 14555 Avion Parkway, Suite 250, Chantilly, VA 20151.

Major Unions Active in the Electronic Media

American Federation of Musicians, United States and Canada (AFM). Musical performers working both in media and live venues. 1501 Broadway, Suite 600, New York, NY 10036.

American Federation of Television and Radio Artists (AFTRA). On-air talent including entertainers, commercial actors, announcers, and newscasters. 260 Madison Avenue, New York, NY 10016.

Directors Guild of America, Inc. (DGA). Directors working in all phases of film and the electronic media. 7920 Sunset Boulevard, Los Angeles, CA 90046.

International Alliance of Theatrical Stage Employees and Moving Picture Machine Operators of the United States and Canada (IATSE). Stagehands, lighting personnel, and other technical trades in film, video, and stage. 1515 Broadway, Suite 601, New York, NY 10036.

International Brotherhood of Electrical Workers (IBEW). Engineers and technicians. 1125 15th Street N.W., Washington, DC 20005.

National Association of Broadcast Employees and Technicians (NABET). Engineers and technicians plus other support personnel. 501 Third Street, 8th Floor, Washington, DC 20001.

Screen Actors Guild (SAG). Performers appearing in motion picture and television entertainment programming and commercials. 5757 Wilshire Boulevard, Los Angeles, CA 90036.

Writers Guild of America (WGA). Writers of radio/television entertainment programming and motion pictures; divided into East (New York) and West (Los Angeles) groups. 555 W. 57th Street, New York, NY 10019; 7000 W. Third Street, Los Angeles, CA 90048.

Notes

Preface

1. George Pollard and Peter Johansen, "Professionalism among Canadian Radio Announcers: The Impact of Organizational Control and Social Attributes," *Journal of Broadcasting & Electronic Media* (Summer 1998): 357.
2. Price Hicks, "Changing More Than Channels," *Debut* 12 (1999–2000): 8.
3. Dale Myers, remarks to the Great Lakes Broadcasting Convention, Lansing, MI, March 8, 2004.

Chapter 1

1. Tim Moore, "Producing the Image" (remarks to the Great Lakes Radio Conference, Mount Pleasant, MI, April 25, 1981).
2. Tyree Ford, "Monday Memo," *Broadcasting,* May 11, 1987, 18.
3. Bruce Holberg, remarks to the Great Lakes Radio Conference, Mount Pleasant, MI, April 4, 1992.
4. Ford, "Monday Memo," 18.
5. Ibid.
6. Fred Jacobs, "Role of the Air Personality" (speech to the Great Lakes Radio Conference, Mount Pleasant, MI, April 3, 1993).
7. Steven Shields, "Creativity and Creative Control in the Work of American Music Radio Announcers" (paper presented to the Association for Education in Journalism and Mass Communication Convention, Minneapolis, August 1990).
8. Steven Shields, "So You Want to Be in Radio," *Feedback* (Summer 1990): 4.
9. Ed Shane, "Challenges Facing the Radio Industry," *Journal of Radio Studies* (December 2002): iv.
10. Tom Carpenter, "Pitfalls of Voice Tracking," *Broadcasting & Cable,* September 9, 2002, 42.
11. Bill McConnell, "Mays Makes Himself Clear," *Broadcasting & Cable,* September 2, 2002, 22.
12. Susan Eastman, Sydney Head, and Lewis Klein, *Broadcast/Cable Programming,* 3rd ed. (Belmont, CA: Wadsworth, 1989), 446.
13. Eric Bogosian, "I Hear America Screaming," *Adweek,* May 3, 1993, 30.
14. Chriss Scherer, "You Want to Hear Something?" *B E Radio* (July 2000): 30–31.
15. Donna Petrozzello, "Michael Jackson: King of Talk," *Broadcasting & Cable,* June 6, 1997, 34.
16. Wayne Munson, *All Talk* (Philadelphia: Temple University Press, 1993), 152.
17. Walter Lippmann, *Public Opinion* (New York: Free Press, 1967), 25.
18. Dan Trigoboff, "The Gender Trap," *Broadcasting & Cable,* April 5, 1999, 23.
19. Sharon O'Malley, "Star Qualities: How to Spot a Potential Anchor in Your Newsroom," *Communicator* (April 1999): 44.
20. Melissa Holmberg and John Roach, "Top TV Critics: Gulf War Is Made-for-TV Drama," *TV Guide,* February 2, 1991, 25.

21. Mark Harmon, "Television News Anchor Longevity, Name Recall, Parasocial Interaction, and Paracommunity Orientation," *Feedback* (Spring 1997): 7.
22. Peter Orlik, "Systemic Limitations to Irish Broadcast Journalism," *Journal of Broadcasting* (Fall 1976): 473.
23. Shirley Biagi, ed., *Newstalk II* (Belmont, CA: Wadsworth, 1987), 116.
24. Ibid., 9–10.
25. Marc Fisher, "Blackout on the Dial," *AJR News Link,* June 23–29, 1998, 2.
26. Ibid., 8.
27. Deborah Potter, "The Body Count," *Feedback* (November 2002): 1.
28. Dan Trigoboff, "Building a Better Weathercast," *Broadcasting & Cable,* December 21, 1998, 43.
29. Glen Dickson, "Weather Systems Get Automated," *Broadcasting & Cable,* October 27, 1997, 74.
30. Elayne Rapping, *The Looking Glass World of Nonfiction TV* (Boston: South End Press, 1987), 58–59.
31. Andrew Bowser, "Weather Fronts Local News," *Broadcasting & Cable,* October 27, 1997, 66.
32. Trigoboff, "Better Weathercast," 44.
33. Michael Seidel, "Field and Screen: Baseball and Television," in Kathleen Henderson and Joseph Mazzeo, eds., *Meanings of the Medium* (New York: Praeger, 1990), 53.
34. Raymond Carroll, Laurie Lattimore, and Bill Erwin, "Proximity, Prominence, and Diversity in the Local TV News Sports Coverage" (paper presented to the Broadcast Education Association Convention, Las Vegas, April 4, 1997).
35. Sarah Ruth Kozloff, "Narrative Theory and Television," in Robert Allen, ed., *Channels of Discourse* (Chapel Hill: University of North Carolina Press, 1987), 63.
36. Marc Gunther, "The Wide World of Roone Arledge," *TV Guide,* December 28, 2002, 32.
37. John Houseman, "Is TV-Acting Inferior?" *TV Guide,* September 1, 1979, 25.
38. Horace Newcomb, *TV: The Most Popular Art* (Garden City, NY: Anchor Books, 1974), 249.
39. George Schaefer, comments in *Television Makers,* Newton E. Meltzer–produced televised PBS documentary, 1986.
40. Ken Howard, "Why I Left Prime-Time TV for Harvard," *TV Guide,* February 21, 1987, 6.
41. Steven Levitt, letter to Peter Orlik, November 6, 1995.
42. Brandon Tartikoff, "Tartikoff's Tricks of the Trade," *TV Guide,* October 17, 1992, 15.
43. Merrill Panitt, "Wheel of Fortune," *TV Guide,* October 5, 1985, 1.
44. Elaine Warren, "Why Are These Women Smiling?" *TV Guide,* April 9, 1998, 51–52.
45. Carl Levine, "Casting Talent," *Video Systems* (November 1991): 109.
46. Fred Cohn, "Actors Strut Their Stuff on the Corporate Stage," *Corporate Video Decisions* (November 1989): 32.
47. Ibid., 29.
48. Levine, "Casting Talent," 109.
49. Ned Soseman, "There Goes the Neighborhood," *Video Systems* (July 1993): 4.
50. Robin Pride, "The Game of Corporate Television," *Video Systems* (June 1991): 14.
51. Alec Foege, "A Separate Peace," *Adweek,* January 1, 2001, 13.
52. Ibid.
53. Scott Jones, "Secret Agent Man," *Digital TV* (March 2003): 78.
54. Sam Haskell, "Making a Difference," *The Caucus Quarterly* (Fall 1998): 24.

Chapter 2

1. "The Marriage of Radio Advertising and New Product Lines," *Broadcasting,* June 22, 1987, 43.
2. "Stakelin Accentuates the Positive of Radio," *Broadcasting,* May 20, 1985, 85.
3. "Testifying to Radio's Powers," *Broadcasting,* June 24, 1985, 51.
4. Jim Dale, "Budget TV," *ADS Magazine* (June 1985): 96.

Notes

5. Charles Kuralt, remarks to the Writing for Television Seminar, Center for Communications, New York, November 12, 1985.
6. "Reliving the Revolution," *one* (Spring 2003): 6.
7. Sally Hogshead, "Work Ethic, Ugh," *Adweek,* September 16, 2002, 10.
8. "Storyboards That Really Control Production and Save Money Too," *ASAP* (May–June 1988): 12.
9. J. Roberto Whitaker-Penteado, "Making It Too Easy," *Adweek,* April 25, 1994, 14.
10. Lynne Grasz, in Peter Orlik, *The Electronic Media: An Introduction to the Profession*, 2nd ed. (Ames: Iowa State University Press, 1997), 424.
11. Larry Gerber, "Face Value," *emmy* (October 1999): 24.
12. Damon Rarey, "A Brief History of the TV Graphics Revolution," *Feedback* (Spring 1990): 9–10.
13. Paul Young, "Hollywood Squeezing Scripters," *Variety,* April 23–27, 1995, 1, 50.
14. Daniel Cerone, "The Comedy Crisis," *TV Guide,* October 4, 1997, 34.
15. "The Write Stuff," *Broadcasting,* December 9, 1985, 119.
16. "Hugh Wilson: On Top of the World," *Broadcasting,* November 27, 1989, 95.
17. Betsy Sharkey, "The Devil You Know," *Adweek,* August 3, 1992, 12.
18. "Creators Break Trend with Hit Like *Northern Exposure*," *(Mt. Pleasant, MI) Morning Sun,* October 12, 1992, 10.
19. Peter Roth, remarks to the International Radio and Television Society Faculty/Industry Seminar, New York, February 8, 1991.
20. Richard Schickel, "Show Them the Money," *Brill's Content* (November 2000): 132.
21. Dave McNary, "Scribe Tribe: Les Misérables," *Variety,* December 23, 2002–January 5, 2003, 42.
22. Danielle Claman, in Peter Orlik, *The Electronic Media: An Introduction to the Profession*, 2nd ed. (Ames: Iowa State University Press, 1997), 328.
23. Eileen Meehan, "Conceptualizing Culture as Commodity: The Problem of Television," *Critical Studies in Mass Communication* (December 1986): 448–449.
24. "Hall of Fame: The Class of 1994," *Broadcasting & Cable,* November 14, 1994, 19.
25. Aaron Sorkin, remarks to the NATPE Convention, Las Vegas, January 23, 2001.
26. Chris Jones, "Riding the Waves of Change," *Video Systems* (September 1997): 76.
27. John Morley, "Working with Scriptwriters," *Audio-Visual Communications* (December 1986): 45.
28. Bill Miller, "Working with Scripts and Scriptwriters," *Video Systems* (January 2003): 26.
29. Carl DeSantis and Phyllis Camesano, "The Professional Draft," *Audio-Visual Communications* (March 1979): 24.
30. Bill Miller, "Shooting with Style," *Video Systems* (December 1998): 78.
31. Robin Pride, "The Game of Corporate Television," *Video Systems* (June 1991): 14.
32. Phillip Stella, "Surviving the Business of Business Television," *Video Systems* (September 1997): 50, 52.
33. Robert Emmett Dolan, *Music in Modern Media* (New York: G. Schirmer, 1967), xi.
34. Robert MacKenzie, "Highway to Heaven," *TV Guide,* December 29, 1984, 40.
35. Robert Jenkins, "Music, Music, Music," *BPME Image* (October 1991): 4.
36. Matthew Mirapaul, "In a Nod to Lush Film Scores, Game Music Gains Texture," *New York Times Online,* October 9, 2003, 1.
37. Ibid., 3.
38. Hairong Li and Janice Bukovac, "Cognitive Impact of Banner Ad Characteristics: An Experimental Study," *Journalism & Mass Communication Quarterly* (Summer 1999): 341.
39. Jon Mandel, remarks to the Broadcast Education Association Convention, Las Vegas, April 21, 2001.
40. Richard Tedesco, "Web Ads Get Glitzy, Savvy," *Broadcasting & Cable,* November 8, 1999, 48.
41. Antone Silvia, "Preparing Students for the Multimedia Newsroom: MSNBC and Interactive Storytelling," *Feedback* (Summer 1997): 24.
42. Carl Lindemann, "Eight Tips for Web Producers," *Broadcasting & Cable,* March 19, 2001, 46.
43. Stan Ferguson, remarks to the Broadcast Education Association Convention, Las Vegas, April 6, 2000.
44. Chad Rea, "Talent Is King," *Adweek,* September 30, 2002, 10.

Chapter 3

1. "Robotic Cameras: Cutting out the Middle-Men," *Broadcasting,* December 7, 1987, 96.
2. Dan Trigoboff, "John Henry versus the Supercomputer," *Broadcasting & Cable,* December 14, 1998, 28.
3. "Endangered Species," *Broadcasting,* January 15, 1990, 146.
4. Ken Kerschbaumer, "Ira Goldstone: Tribune's Golden Boy," *Broadcasting & Cable,* April 25, 2001, SR8.
5. Barry Thomas, "Tomorrow's Technical Staff," *B E Radio* (November–December 1999): 12.
6. National TeleConsultants, "Reviewing Staff for Skill Sets and Tool Kits," *Broadcasting & Cable,* November 1, 1993, D16.
7. David Scheirman, "Education in the Sound Industry," *Recording Engineer/Producer* (March 1992): 45.
8. Julian McBrowne, "Studio Engineering Etiquette," *EQ* (May 1994): 40.
9. William Moylan, "Employment Trends," *Recording Engineer/Producer* (December 1988): 38.
10. Erin Caslavka, "Sound Hound," *emmy* (April 1999): 18.
11. Moylan, "Employment Trends," 38.
12. Herbert Zettl, *Sight, Sound, Motion: Applied Media Aesthetics* (Belmont, CA: Wadsworth, 1973), 44.
13. Ibid., 38.
14. Joseph Tawail, "Lighting with Pencil and Paper—the Light Plot," *Educational and Industrial Television* (April 1975): 22.
15. Bill Holshevnikoff, "The Lighting Plan," *Video Systems* (January 1996): 71.
16. Mary Coffman, "Going Solo," *Communicator* (April 2000): 34.
17. Ibid., 34–35.
18. Jac Holzman, "Monday Memo," *Broadcasting,* July 19, 1989, 24.
19. "The Role of Film in TV Programming," *Broadcast Engineering* (May 1979): 77.
20. Holzman, "Monday Memo," 24.
21. Bob Fisher, "A Conversation with John Toll, ASC," *International Photographer* (July 1999): 31.
22. Bob Fisher, "Through the Looking Glass," *emmy* (June 1999): 34.
23. Bill Hines, "#101 Factors Affecting the DP and Operator Relationship," *International Photographer* (July 1999): 24.
24. Bill Miller, "To Hire or Not," *Video Systems* (August 2003): 21.
25. Carl Levine, "Bob Tur: A Veteran News Hound Looks Ahead at HDTV," *Video Systems* (May 1998): 129.
26. Larry Gerber, "Invisible Art, Visible Artists," *emmy* (April 2002): 62.
27. Terry Kattleman, "Bob Fisk of Phoenix Editorial," *Creativity* (June 2002): 82.
28. Gerber, "Invisible Art," 63.
29. Ibid.
30. Ibid.
31. James Gartner, in Peter Orlik, *The Electronic Media: An Introduction to the Profession,* 2nd ed. (Ames: Iowa State University Press, 1997), 474.
32. Sarah Kozloff, "Narrative Theory and Television," in Robert Allen, ed., *Channels of Discourse* (Chapel Hill: University of North Carolina Press, 1987), 63.
33. Timothy Dwyer, "The Guy Who Really Calls the Shots," *TV Guide,* January 29, 1994, 22.
34. "John Finger Talks about the Challenges of Filming *Cheers,*" *NATPE Programmer* (January 1986): 152.
35. Neal Koch, "TV's New Ruling Class," *Channels* (May 1989): 31.
36. Mill Roseman, "Joy Radio," *Communication Arts* (January–February 1989): 66.
37. A. William Bluem and Roger Manvell, eds., *Television: The Creative Experience* (New York: Hastings House, 1967), 48.
38. Ibid., 42.
39. Stephen Barr, "Directing: Auteurs Need Not Apply," *Corporate Video Decisions* (April 1990): 40.
40. Gartner, 474.

Notes

Chapter 4

1. Paul Boscarino, letter to Peter Orlik, October 23, 1995.
2. Joe Mandese, "The Buying and Selling," *Advertising Age: 50 Years of TV Advertising* (Spring 1995): 20.
3. "Networks and TV's Significant Others Line Up for Fall," *Broadcasting,* September 4, 1989, 27.
4. Walt Wurfel, "How Some of America's Successful Salespeople Operate," *NAB Radio Week,* January 15, 1990, 4.
5. Bruce Stoller, remarks to the Great Lakes Radio Conference, Mount Pleasant, MI, April 3, 1993.
6. Sondra Michaelson, "Empathy vs. Sympathy as a Radio Sales Approach," *Broadcasting,* February 8, 1988, 42.
7. James Greenwald, remarks to the International Radio and Television Society Faculty/Industry Seminar, New York, February 19, 1993.
8. Steve McClellan, "Katz's Beloyianis: The Never-resting Rep," *Broadcasting & Cable,* July 17, 1995, 65.
9. Catherine Taylor, "Web Disconnect," *Adweek,* September 9, 2002, I.Q. 26.
10. David Kaplan, "Survey: TV Ad Model Has Promise for Web," *Adweek,* September 29, 2003, 12.
11. Chuck Ross, "CBS Chief Talks about NFL, Future," *Advertising Age,* February 2, 1998, 11.
12. Don Acree, "Radio Marketing and Promotions: An Emerging Career Field," *College Broadcaster* (April–May 1990): 44.
13. Joe Flint, "Local TV Sales: Making Money on Main Street," *Broadcasting,* December 2, 1991, 59.
14. Nancy Smith, "Monday Memo," *Broadcasting & Cable,* June 21, 1993, 70.
15. Vince Manzi, "Servicing What You Sell" (remarks to the National Association of Television Program Executives Convention, New Orleans, January 17, 1990).

Chapter 5

1. Keith Gould, "Leaders of the Pack," *Adweek,* March 3, 1997, 22.
2. Susan Eastman, "What Programming Paradigm?" (paper presented to the Broadcast Education Convention, Las Vegas, April 5, 2002).
3. Bill Hennes, "Fighting Copycat Syndromes among Radio Stations," *Broadcasting,* April 14, 1986, 22.
4. John Silliman Dodge, letter to Peter Orlik, July 23, 2003.
5. Ed Shane, "The State of the Industry: Radio's Shifting Paradigm," *Journal of Radio Studies* (Summer 1998): 2.
6. Stuart Elliott, "Ad Buyers Cold to TV Season but Still Plan to Spend Big," *New York Times Online,* September 15, 2003.
7. Clark Smidt, "Monday Memo," *Broadcasting,* May 8, 1989, 22.
8. Steve McClellan, "WNYW-TV PD Post Axed," *Broadcasting,* April 6, 1992, 26.
9. Eastman, "Paradigm."
10. Deborah McAdams, "The Gatekeepers," *Broadcasting & Cable,* July 19, 1999, 24.
11. Robert Pittman, remarks to the International Radio and Television Society Faculty/Industry Seminar, New York, February 7, 1991.
12. Steven Goldstein, "A Report from the Programming Department," *Feedback* (Winter 2001): 23.
13. Martin Antonelli, "Monday Memo," *Broadcasting,* January 6, 1989, 38.
14. Ed Shane and Ed Cohen, "JRS Forum," *Journal of Radio Studies* (June 2002): 5.
15. Ibid.
16. Steve McClellan, "NBC, Affils Forecast DTV Weather Channel," *Broadcasting & Cable,* November 17, 2003, 38.
17. William McCavitt and Peter Pringle, *Electronic Media Management* (Newton, MA: Focal Press, 1986), 58.
18. "What's in a Cable System?" *Broadcasting & Cable,* February 18, 2002, 4T.

19. William Foster, "Monday Memo," *Broadcasting,* December 19, 1988, 34.
20. Larry Patrick, remarks to Central Michigan University broadcast management students, Mt. Pleasant, MI, December 3, 2003.
21. John Rodman, "Learning from Our Mistakes," *Communicator* (April 1991): 28.
22. Chris Tuohey, "Producer Survey," *American Journalism Review* (November 2001): 57.
23. Mark Effron, "The Third Way," *Broadcasting & Cable,* November 17, 1997, 54.
24. Tuohey, "Survey," 61.
25. Tom Kirby, remarks as part of "Number One in News" panel, National Association of Broadcasters Convention, Las Vegas, April 12, 1988.
26. Rolla Cleaver, remarks as part of "Number One in News" panel, National Association of Broadcasters Convention, Las Vegas, April 12, 1988.
27. John Abel, "Reinventing Broadcasting" (speech to the Broadcast Education Association Convention, Las Vegas, April 7, 1995).
28. Eric Taub, "Bleed, Speed, Greed," *emmy* (April 2003): 50.
29. John von Soosten, remarks to a faculty study group in conjunction with the International Radio and Television Society Faculty/Industry Seminar, New York, February 18, 1993.
30. Steve McClellan, "The Venerable AP is Mulling New Curley Cues," *Broadcasting & Cable,* October 27, 2003, 35.
31. Dan Trigoboff, "NDs Thrive on GM Track," *Broadcasting & Cable,* September 27, 1999, 54.
32. Neal Koch, "TV's New Ruling Class," *Channels* (May 1989): 30–31.
33. Kathy Brown, "The King and Queen of Comedy," *Adweek,* June 19, 1989, B.T. 52–53.
34. Horace Newcomb and Robert Alley, *The Producer's Medium* (New York: Oxford University Press, 1983), 239.
35. David Antil, remarks to Central Michigan University broadcasting students, Mt. Pleasant, MI, October 21, 1994.
36. David Leathers, "The Line Producer's Pre-production Checklist," *Video Systems* (August 1992): 46.
37. Newcomb and Alley, *The Producer's Medium,* 183–184.
38. Chris Carter, remarks to the National Association of Television Program Executives Convention, New Orleans, January 15, 1997.
39. Barry Garron, "Wolf Says NBC Tightening Programming Budget," *Broadcasting & Cable,* January 26, 1998, 30, 32.
40. "Wednesday: Suzanne DePasse," *NATPE Access* (November 1995): 4.
41. Mary Connors, "Looking for 'Tom McElligott,'" *Adweek,* June 10, 1991, 42.
42. Ann Cooper, "Bernbach's Children Come of Age," *Adweek,* March 25, 1996, 34.
43. Paul Goldsmith, "Monday Memo," *Broadcasting,* September 22, 1980, 12.
44. Donald David, *I Wish Somebody Had Told Me That the Day I Started* (Detroit: Campbell-Ewald Advertising internal publication, 1965), 8.
45. Roger Bodo, in Peter Orlik, *The Electronic Media: An Introduction to the Profession* (Needham, MA: Allyn & Bacon, 1992), 383–384.
46. Cooper, "Bernbach's Children," 37.
47. Rick Boyko, "Boy, oh Boy, oh Boyko," *Adweek,* January 22, 2001, 6.
48. Richard Morgan, "Peter Principle: Should a CD's Desire to Make Ads Be a Fireable Offense?" *Adweek,* June 19, 1989, 2.
49. Ibid.
50. Richard Morgan, "Commerce, Creativity Collide in Minneapolis," *Adweek,* December 19, 1988, 52.
51. Tim Leon, letter to the editor, *Adweek,* December 3, 1990, 8.
52. Cooper, "Bernbach's Children," 33.
53. Richard Morgan, "As Ad Biz Is Overrun with Free Agents, the Personality of the Players Is Lost," *Adweek,* July 24, 1989, 2.

Notes

Chapter 6

1. Noreen O'Leary, "In Search of a New Definition," *Adweek,* April 22, 1996, 28.
2. Jennifer Comiteau, "Taking One for the Team," *Adweek,* May 5, 1997, 27.
3. Noreen O'Leary, "The Incredible Shrinking Account Exec," *Adweek,* May 26, 2003, 23.
4. Jack Taylor, letter to the editor, *Adweek,* December 18, 1995, 13.
5. J. J. Jordan, "Want the Best Service? Look for Great Work," *Adweek,* April 22, 1996, 23.
6. Craig Reiss, "The Two Faces of Media," *Adweek,* January 22, 1996, 30.
7. Michael Drexler, "Unraveling the Media Myth," *Advertising Age,* November 11, 2002, 22.
8. Cristina Merrill, "What's in a Name? Lots!" *Adweek,* August 24, 1998, 21.
9. Nancy Bishop, "A New Channel for Dallas' Media Mover," *Adweek,* July 24, 1989, 25.
10. Chuck Ross, "TN Media Tool Adds to Optimizers," *Advertising Age,* January 26, 1998, 6.
11. Erwin Ephron, "Media Audit's Time Has Come," *Advertising Age,* September 2, 2002, 16.
12. Lawrence Teherani-Ami, "Diamonds in the Rough," *Adweek,* June 23, 2003, 22.
13. Kemba Johnson, "Motivating Seller to Go Extra Mile," *Advertising Age,* September 11, 1995, S-10.
14. Owen Charlebois, "Is a Radio Research Renaissance Coming?" *Journal of Radio Studies* (June 2002): v.
15. Linda Moss, "A Constant Companion," *Broadcasting & Cable,* February 11, 2002, 17.
16. Herbert Howard and Michael Kievman, *Radio and TV Programming* (Columbus, OH: Grid Publishing, 1983), 157.
17. Steve McClellan, "Nielsen Rates New Box Ready for 2004 Rollout," *Broadcasting & Cable,* December 9, 2002, 38.
18. Hasson Fattah, "The Metrics System," *Adweek,* November 13, 2000, 98.
19. Ibid., 102.
20. Gary Stevens, in Peter Orlik, *The Electronic Media: An Introduction to the Profession,* 2nd ed. (Ames: Iowa State University Press, 1997), 586.
21. Geoffrey Foisie, "TV Stations Can't Bank on Wall Street," *Broadcasting,* June 1, 1992, 36.
22. "The Business of Brokering/89," *Broadcasting,* August 7, 1989, 37.
23. James Gammon, in Peter Orlik, *The Electronic Media: An Introduction to the Profession* (Needham, MA: Allyn & Bacon, 1992), 418.
24. Peter Witkin, "Live Images Transmitted across Ocean for First Time," *New York Times,* July 11, 1962, 16.
25. Ann Mack, "Grabbing Market Share," *Adweek,* November 24, 2003, I.Q. 4.
26. Alice Flaherty, "Writing Like Crazy: A Word on the Brain," *Chronicle of Higher Education,* November 21, 2003, B6.
27. "Creating an Award-Winning Website," *Video Systems* (August 2003): 62.
28. Russell Shaw, "Plain Vanilla, Please," *Broadcasting & Cable,* February 12, 2001, 40.
29. Richard Morgan, "Agency Rainmakers Suffer a Drought While Prospects Get Savvy to Tricks," *Adweek,* June 25, 1990, 2.
30. Don Peppers, in Peter Orlik, *The Electronic Media: An Introduction to the Profession* (Needham, MA: Allyn & Bacon, 1992), 424.

Chapter 7

1. Karrie Jacobs, "Watching a Radio Consultant Consult," *Adweek's Radio 1985 Issue* (July 1985): 8.
2. Russell Mouritsen, in Peter Orlik, *The Electronic Media: An Introduction to the Profession* (Needham, MA: Allyn & Bacon, 1992), 427.
3. Michael Keith, *Radio Programming: Consultancy and Formatics* (Newton, MA: Focal Press, 1987), 8–9.
4. Kira Green, "TV's Test Pilots," *Broadcasting & Cable,* July 17, 2000, 54.
5. Debra Goldman, "The Shape of Things to Come," *Adweek Media Quarterly,* September 18, 1995, 6.
6. John Higgins and Price Colman, "The Powers That Buy," *Broadcasting & Cable,* August 17, 1998, 28.

7. Harry Jessell, "Trygve Myhren: On Cable's Shifting Balance of Power," *Broadcasting & Cable,* November 14, 1994, 40.
8. Erwin Krasnow and Lawrence Longley, *The Politics of Broadcast Regulation,* 2nd ed. (New York: St. Martin's Press, 1978), 38–39.
9. Arthur Goodkind, "Broadcast De-regulation and Self Defense," *Broadcasting,* February 13, 1984, 36.
10. Rachelle Chong, in Peter Orlik, *The Electronic Media: An Introduction to the Profession*, 2nd ed. (Ames: Iowa State University Press, 1997), 563.
11. Glen Robinson, *Communications for Tomorrow: Policy Perspectives for the 1980s* (New York: Praeger, 1978), 384–385.
12. Krasnow and Longley, *The Politics of Broadcast Regulation*, 39–40.
13. "The Washington Lawyer: Power Behind the Powers That Be," *Broadcasting,* June 16, 1980, 57.
14. Ibid., 32.
15. Judy Smith, remarks to the International Radio and Television Society Faculty/Industry Seminar, New York, February 8, 1996.
16. Joan Voigt and Wendy Melillo, "Rough Cut," *Adweek,* March 11, 2002, 29.
17. Ibid., 28–29.
18. Kathleen Jamieson and Karlyn Campbell, *The Interplay of Influence* (Belmont, CA: Wadsworth, 1983), 81.
19. Richard Morgan, "Exxon Learns the Hard Way," *Adweek,* April 24, 1989, 3.
20. Teresa Tritch, "Crisis Ads: When All Hell Breaks Loose," *Adweek,* October 17, 1988, 48.
21. John Burke, "Crisis Public Relations: Tylenol and Other Headaches" (remarks to the Center for Communications, New York, November 13, 1986).
22. Wendy Marx, "PR Joins the Interactive Parade," *Advertising Age,* April 17, 1995, 15.
23. Richard Blackmur, "A Burden for Critics," in Richard Blackmur, ed., *The Lion and the Honeycomb* (New York: Harcourt Brace, 1955), 206.
24. Lawrence Laurent, "Wanted: The Complete Television Critic," in *The Eighth Art* (New York: Holt, Rinehart & Winston, 1962), 156.
25. Jack Gould, "A Critical Reply," *New York Times,* May 26, 1957.
26. George Brandenburg, "TV Critic's Role Is Middleman—Wolters," *Editor and Publisher,* December 23, 1961, 39.
27. George Condon, "Critic's Choice," *Television Quarterly* (November 1962): 27.
28. "But Who Listens?" *Television Age,* September 27, 1965, 19.
29. Lee Loevinger, "The Ambiguous Mirror: The Reflective-Projective Theory of Broadcasting and Mass Communications," *Journal of Broadcasting* (Spring 1968): 110–111.
30. Everette Dennis, "Undergraduate Education Should Stop Ignoring the Importance of Media," *Chronicle of Higher Education,* February 4, 1987, 36.
31. George Pollard and Peter Johansen, "Professionalism among Canadian Radio Announcers: The Impact of Organizational Control and Social Attributes," *Journal of Broadcasting & Electronic Media* (Summer 1998): 367.

Chapter 8

1. Gene Jankowski, in Peter Orlik, *The Electronic Media: An Introduction to the Profession*, 2nd ed. (Ames: Iowa State University Press, 1997), 647.
2. Louis Day, in Orlik, *The Electronic Media*, 611.
3. George Lois, in Orlik, *The Electronic Media*, 656–657.